FRP-strengthened RC Structures

FRP-strengthened RC Structures

J. G. Teng
The Hong Kong Polytechnic University, China

J. F. Chen
The University of Nottingham, UK

S. T. Smith
The Hong Kong Polytechnic University, China

L. Lam
The Hong Kong Polytechnic University, China

JOHN WILEY & SONS, LTD

Baffins Lane, Chichester,
West Sussex, PO19 1UD, England

National 01243 779777
International (+44) 1243 779777
e-mail (for orders and customer service enquiries): cs-books@wiley.co.uk
Visit our Home Page on http://www.wiley.co.uk
or http://www.wiley.com

Other Wiley Editorial Offices

John Wiley & Sons, Inc., 605 Third Avenue,
New York, NY 10158-0012, USA

WILEY-VCH Verlag GmbH
Pappelallee 3, D-69469 Weinheim, Germany

John Wiley & Sons Australia, Ltd
33 Park Road, Milton, Queensland 4064, Australia

John Wiley & Sons (Canada) Ltd, 22 Worcester Road
Rexdale, Ontario, M9W 1L1, Canada

John Wiley & Sons (Asia) Pte Ltd, 2 Clementi Loop #02-01,
Jin Xing Distripark, Singapore 129809

Library in Congress Cataloging-in-Publication Data

FRP-strengthened RC structures / J. G. Teng ... et al.
p. cm.
Includes bibliographical references and index.
ISBN 0-471-48706-6 (alk. paper)
1. Composite materials. 2. Reinforced concrete, Fiber. 3. Civil engineering. I. Teng, J. G.

TH223 .F77 2001
624.1′8341 – dc21

2001026893

British Library Cataloguing in Publication Data

A catalogue record for this book is available from the British Library

ISBN 0-471-48706-6

Typeset in 10/12pt Times Roman by Laser Words Pvt. Ltd., Chennai, India.
Printed and bound by Antony Rowe Ltd, Eastbourne
This book is printed on acid-free paper responsibly manufactured from sustainable forestry, in which at least two trees are planted for each one used for paper production.

Contents

Preface xi

Notation xiii

1 FRP Composites for Strengthening RC Structures 1
1.1 Introduction 1
1.2 Methods of Forming FRP Composites 2
1.3 Mechanical Properties of FRP Composites 3
1.4 Resins 5
1.5 Comparison of Performance for Strengthening of Structures 5
1.6 Partial Safety Factors 6
1.7 Concluding Remarks 7
References 8

2 Bond Strength of FRP-to-concrete Joints 11
2.1 Introduction 11
2.2 Test Methods, Behaviour and Failure Mode 13
2.3 Existing Bond Strength Models 14
2.3.1 Empirical Models 14
2.3.2 Fracture-mechanics-based Models 15
2.3.3 Design Proposals 18
2.3.4 Performance of Bond Strength Models 19
2.4 Chen and Teng's Model 20
2.5 Design Recommendation 26
2.6 Conclusions 26
References 27

3 Flexural Strengthening of Beams 31
3.1 Introduction 31
3.2 Methods of Strengthening 32
3.2.1 General 32

3.2.2 Unstressed Soffit Plates 33
3.2.3 End Anchorage for Unstressed Soffit Plates 33
3.2.4 Prestressed Soffit Plates 34
3.2.5 Additional Remarks 35
3.3 Failure Modes and Typical Behaviour 36
3.3.1 Classification of Failure Modes 36
3.3.2 Flexural Failure 36
3.3.3 Shear Failure 39
3.3.4 Plate-end Debonding Failures 39
3.3.5 Intermediate Crack-induced Interfacial Debonding 42
3.3.6 Other Debonding Failure Modes 44
3.3.7 Other Aspects of Debonding 44
3.3.8 U Strip End Anchorage 45
3.4 Flexural Strength 46
3.4.1 General 46
3.4.2 Design Equations 47
3.4.3 Presence of Preloading 51
3.5 Interfacial Stresses 53
3.5.1 General 53
3.5.2 Approximate Analytical Solutions 53
3.5.3 Finite-element Results 54
3.5.4 Higher Order Analytical Solutions 56
3.5.5 Remarks 56
3.6 Plate-end Debonding Strength Models 57
3.6.1 General 57
3.6.2 Review of Debonding Strength Models 58
3.6.3 Experimental Data 61
3.6.4 Performance of Debonding Strength Models 63
3.6.5 Smith and Teng's Model 71
3.7 Strength Model for Intermediate Crack-induced Interfacial Debonding 73
3.7.1 General 73
3.7.2 Intermediate Flexural Crack-induced Debonding 74
3.7.3 Intermediate Flexural Shear Crack-induced Debonding 77
3.8 Design Recommendation 78
3.8.1 Critical Sections and Plate-end Anchorage 78
3.8.2 Strength Check 79
3.8.3 Partial Safety Factors 80
3.8.4 Application of the Design Procedure to Indeterminate Beams 80
3.9 Conclusions 80
3.10 Appendix A 81
3.11 Appendix B 83

3.11.1 Shear-capacity-based Models 83
3.11.2 Concrete Tooth Models 86
3.11.3 Interfacial Stress-based Models 90
References 95

4 Shear Strengthening of Beams 103
4.1 Introduction 103
4.2 Methods of Strengthening 104
4.2.1 Strengthening Schemes 104
4.2.2 Selection of Strengthening Schemes 107
4.3 Failure Modes 109
4.3.1 General 109
4.3.2 Shear Failure with FRP Rupture 109
4.3.3 Shear Failure without FRP Rupture 110
4.3.4 Shear Failure due to FRP Debonding 110
4.3.5 Failures Near Mechanical Anchors 112
4.4 Existing Research on Shear Strength Models 112
4.5 Chen and Teng's Shear Strength Model 115
4.5.1 General Expressions 115
4.5.2 Shear Failure Controlled by FRP Rupture 117
4.5.3 Shear Failure Controlled by FRP Debonding 119
4.6 Design Recommendation 123
4.6.1 General Equations 123
4.6.2 Shear Failure Controlled by FRP Rupture 124
4.6.3 Shear Failure Controlled by FRP Debonding 124
4.6.4 Spacing Requirement for FRP Strips 126
4.6.5 Strengthening Design for Indeterminate Beams 127
4.7 Comparison with Experimental Observations 127
4.8 Conclusions 130
References 130

5 Flexural Strengthening of Slabs 135
5.1 Introduction 135
5.2 Methods of Strengthening 135
5.2.1 General 135
5.2.2 Simply Supported One-way or Two-way Slabs 136
5.2.3 Cantilever Slabs 138
5.3 Failure Modes and Typical Behaviour 140
5.3.1 Cantilever Slabs and One-way Slabs 140
5.3.2 Two-way Slabs 142
5.4 Design Recommendation for One-way Slabs 143
5.5 Design Recommendation for Two-way Slabs 143
5.5.1 Flexural Strength by the Moment Coefficient Method 143

5.5.2 Flexural Strength by Yield Line Analysis 144
5.5.3 Debonding Strengths 144
5.5.4 Anchorage of FRP Strips/Sheets 145
5.6 Conclusions 145
References 145

6 Strengthening of Axially and Eccentrically Loaded Columns 147
6.1 Introduction 147
6.2 Methods of Strengthening 148
6.2.1 General 148
6.2.2 Wrapping 148
6.2.3 Filament Winding 149
6.2.4 Prefabricated Shell Jacketing 149
6.2.5 Comparison of Strengthening Methods 150
6.2.6 Constructional Aspects 150
6.2.7 Shape Modification 151
6.3 Failure Modes and Typical Behaviour of FRP-confined Concrete Columns 153
6.3.1 Circular Concrete Columns 153
6.3.2 Rectangular Concrete Columns 155
6.3.3 Elliptical Concrete Columns 156
6.4 Compressive Strength of FRP-confined Concrete 158
6.4.1 General 158
6.4.2 Review of Strength Models 159
6.4.3 Experimental Data 161
6.4.4 Trends of Experimental Data 162
6.4.5 Performance of Strength Models 164
6.4.6 Lam and Teng's Model 167
6.5 Stress–Strain Behaviour of FRP-confined Concrete 168
6.5.1 General 168
6.5.2 Review of Stress–Strain Models 169
6.5.3 Experimental Data 175
6.5.4 Trends of Experimental Data 176
6.5.5 Performance of Existing Stress–Strain Models 182
6.5.6 Lam and Teng's Stress–Strain Model 185
6.6 FRP-confined Rectangular Concrete Columns 190
6.6.1 General 190
6.6.2 Existing Models 190
6.6.3 Experimental Data 193
6.6.4 Lam and Teng's Model 195
6.7 FRP-confined Elliptical Columns 199
6.8 Design Recommendation for Axially Loaded Columns 200

6.8.1 General Expression for the Axial Strength of FRP-confined Concrete 200
6.8.2 Ultimate Strength of FRP-confined RC Columns under Axial Load 200
6.9 Design Recommendation for Eccentrically Loaded Columns 201
6.9.1 General 201
6.9.2 Existing Studies on FRP-confined RC Columns under Eccentric Loads 201
6.9.3 Ultimate Strength of FRP-confined RC Columns under Eccentric Loads 202
6.9.4 Moment–Curvature Analysis of FRP-confined RC Columns 204
6.9.5 Typical Interaction Curves 205
6.10 Conclusions 206
References 208

7 Seismic Retrofit of Columns 213
7.1 Introduction 213
7.2 Typical Failure Modes of RC Columns under Seismic Attack 215
7.2.1 General 215
7.2.2 Shear Failure 215
7.2.3 Flexural Plastic Hinge Failure 216
7.2.4 Lap Splice Failure 216
7.2.5 Flexural Shear Failure of Columns with Longitudinal Reinforcement Cut-offs 217
7.2.6 Footing Failure 219
7.3 Methods of Retrofit 219
7.3.1 FRP Jackets with Horizontally Oriented Fibres 219
7.3.2 Longitudinal Bonding of FRP 220
7.4 Flexural Ductility of Retrofitted Columns 220
7.4.1 Ductility Factors 220
7.4.2 Ductility of Retrofitted RC Columns 222
7.5 Ultimate Conditions of Retrofitted Columns 225
7.5.1 General 225
7.5.2 Shear Failure 226
7.5.3 Flexural Plastic Hinge Failure 226
7.5.4 Lap Splice Failure 227
7.5.5 Flexural Failure of Columns with Longitudinal FRP 227
7.6 Design Recommendations 227
7.6.1 General 227
7.6.2 Shear Strength 228
7.6.3 Flexural Plastic Hinge Confinement 232

7.6.4 Lap Splice Clamping 233
7.6.5 Detailing Considerations 234
7.7 Conclusions 235
References 236

Index

Preface

Strengthening of reinforced concrete (RC) and other structures using advanced fibre-reinforced polymer/plastic (FRP) composites has become very popular in the last few years. This has been due both to the needs for maintaining and upgrading essential infrastructure in all parts of the world, and the well-known advantages of FRP composites including both good corrosion resistance and ease for site handling due to their light weight. The continuous reduction in the material cost of FRP composites has also contributed to their popularity.

A decade ago, the technology of strengthening structures using FRP composites was still in its infancy, with very few publications on the technology available. Today, one can hardly keep pace with all the papers appearing in both journals and conference proceedings. This exponential growth of research is testimony to the wide recognition of the importance of this new technology. Of particular significance is the fact that many practical applications have preceded the development of design methods. This is probably unique in the application of high-tech solutions in the construction industry and clearly illustrates the need for extensive research to advance the technology to a new level. Such research is expected to lead to the development of innovative strengthening methods using FRP composites and rational design methods for FRP-strengthened structures. Rational design methods will ensure the safe and economic use of this new technology.

While a great amount of research now exists in the published literature, predominantly on the strengthening of RC structures using FRP composites, it is scattered in various journals and conference proceedings. This makes it difficult for engineers to exploit and for new researchers to assimilate. The aim of this book is to provide a state-of-the-art summary of the existing research on FRP strengthening of RC structures, with the emphasis on structural behaviour and strength models. The main topics covered include bond behaviour (Chapter 2), flexural and shear strengthening of beams (Chapters 3 and 4), flexural strengthening of slabs (Chapter 5), strengthening of columns subject to axial or eccentric loads (Chapter 6) and seismic retrofit of columns (Chapter 7). For each topic treated, the methods of strengthening are first discussed, followed by a description of behaviour and failure modes and then the presentation of rational design recommendations which are suitable for future inclusion in design codes. The authors hope that the book will be of interest to a wide range of readers, such as new and more experienced researchers, writers of design codes, engineers, and graduate and senior undergraduate students, as a consolidated source of up-to-date information on this important topic.

This book has grown out of research carried out by the authors in several projects led by the first author and funded by The Hong Kong Polytechnic University over the last few years. The authors are grateful to the university for this generous support. The authors also wish to thank Professor J. M. Ko, former Head of the Department of Civil and Structural Engineering and current Dean of the Faculty of Construction and Land Use for nurturing this research area and constant encouragement, and to Professor M. Anson, former Dean of the Faculty for his appreciation and support of the work. Thanks are also due to The University of Wolverhampton and The University of Nottingham in the UK for supporting the second author's collaborative work with the other authors at The Hong Kong Polytechnic University.

This research has also received generous support from L & M Specialist Construction Ltd, Fosroc Hong Kong Ltd and Sika Hong Kong Ltd who donated FRP materials for experimental work. Thanks are extended to these firms and their representatives Dr. J. S. Wang, Mr. A. Au and Mr. S. Fan for this support.

The authors are indebted to Dr. D. J. Oehlers of the University of Adelaide in Australia for his valuable comments and discussions, particularly on a draft of Chapter 3, and Dr. Y. Xiao of the University of Southern California, USA, for his valuable comments on the final draft of Chapter 7, which led to a number of improvements to these chapters. The authors are also indebted to Dr. Xiao for willingly providing them with some of his test data for inclusion in the book.

The authors wish to thank Dr D. Hughes, Mrs W. Hunter, Ms S. Lambert and Ms S. Corney at John Wiley & Sons, Ltd and their copyeditor Mr N. Hankins for their professionalism, promptness and understanding in dealing with all requests and queries from the authors.

Last but not least, the authors wish to express their profound gratitude to their families for their understanding and support during the preparation of this book.

J. G. Teng
J. F. Chen
S. T. Smith
L. Lam

Notation

The symbols used in this book, except those which are used in appendices only, are defined below. Where a symbol has more than one definition, Chapter numbers are specified for those definitions which are used only in a single chapter but not for those definitions used in more than one chapter.

A_{bl} = cross-sectional area of a single longitudinal steel reinforcing bar in column
A_c = cross-sectional area of concrete in column
A_e = cross-sectional area of effectively confined concrete (Chapter 6)
= effective shear area (Chapter 7)
A_{frp} = cross-sectional area of FRP (Chapter 3)
= cross-sectional area of a pair of FRP strips (Chapter 4)
A_g = gross cross-sectional area of column
A_s = cross-sectional area of steel tension reinforcement in beam
A_{sc} = cross-sectional area of longitudinal steel reinforcement in column
A_{sh} = cross-sectional area of transverse steel reinforcement in column
A_{si} = cross-sectional area of ith layer of steel reinforcement in beam (Chapter 3)
= cross-sectional area of ith layer of longitudinal steel reinforcement in column (Chapter 6)
a = length of semi-major axis of elliptical section
a_0 = parameter defined by Equation 2.7f
B = distance from support to nearest concentrated load (shear span)
b = length of semi-minor axis of elliptical section
= shorter side length of rectangular column (Chapter 6)
b_a = width of adhesive
b_c = width of concrete
= width of beam (Chapter 3)
b_f = width of flange of T-beam

b_p = width of plate
b_w = width of web of T-beam
C = total concrete compressive force
c = thickness of concrete cover
c_f = coefficient in Equation 2.4c
D = distance from beam compression face to line of action of concrete compressive force (Chapter 3)
= diameter of equivalent circular column (Chapter 6)
D' = centre-to-centre distance of peripheral hoop or spiral reinforcement (Figure 7.11)
D_{frp} = stress distribution factor for FRP along a shear crack
d = distance from beam compression face to centroid of steel tension reinforcement
= diameter of circular concrete column
= section depth for rectangular column in lateral load direction (Chapter 7)
d_{bl} = diameter of longitudinal steel reinforcing bar in column
d_{frp} = distance from beam compression face to centroid of FRP (Chapter 3)
= distance from beam compression face to lower edge of FRP on sides (Chapter 4)
$d_{frp,t}$ = distance from beam compression face to top edge of FRP on sides
d_h = diameter of transverse steel reinforcing bar in column
d_{si} = distance from beam compression face to centroid of ith layer of steel reinforcement (Chapter 3)
= distance from column compression face to centroid of ith layer of longitudinal steel reinforcement (Chapter 6)
E_a = modulus of elasticity of adhesive
E_c = modulus of elasticity of concrete
E_{frp} = modulus of elasticity of FRP
E_p = modulus of elasticity of plate
E_s = modulus of elasticity of steel
E_{sec} = f'_{cc}/ε_{cc}
E_2 = slope of linear second portion of stress–strain curve
e = eccentricity of an ellipse
f'_c = concrete cylinder compressive strength
f'_{cc} = compressive strength of confined concrete
f'_{co} = compressive strength of unconfined concrete
f_{ct} = concrete-splitting tensile strength
f_{ctm} = surface tensile strength of concrete
f_{cu} = concrete cube compressive strength
f_{frp} = tensile strength of FRP

$f_{frp,e}$ = average/effective stress of FRP at failure
$f_{frp,r}$ = reduced tensile strength of FRP allowing for intermediate crack-induced debonding
f_h = confining pressure provided by transverse steel reinforcement
f_l = lateral confining pressure
f_l' = effective lateral confining pressure
f_o = intercept of stress axis by linear second portion of stress–strain curve (Figure 6.18)
f_p = tensile strength of plate
f_y = yield strength of steel
f_{yh} = yield strength of transverse steel reinforcement in column
f_{yl} = yield strength of longitudinal steel reinforcement in column
G_f = fracture energy for simple shear test
h = depth of beam
= longer side length of rectangular column (Chapter 6)
$h_{frp,e}$ = effective height of FRP bonded on beam sides
I_p = second moment of area of plate
K_n = normal stiffness of adhesive layer
k = coefficient in Equation 7.8
k_g = gap factor
k_p = geometrical factor for simple shear test defined by Equation 2.4d
k_s = shape factor
k_1 = parameter defined by Equation 2.11b
= mean stress factor (Chapter 3)
= confinement effectiveness coefficient (Chapter 6)
k_2 = concrete compressive force centroid factor (Chapter 3)
= strain enhancement coefficient (Chapter 6)
L = bond length (Chapter 2)
= column height (Chapter 6)
= column height for single bending or half column height for double bending (shear span) (Chapter 7)
L_a = anchorage length
L_{c1} = length of primary plastic hinge confinement region
L_{c2} = length of secondary plastic hinge confinement region
L_e = effective bond length
L_{eff} = length of effectively strengthened region
L_{max} = maximum bond length
L_p = total length of bonded soffit plate (Chapter 3)
= length of plastic hinge region (Chapter 7)
L_s = lap splice length
L_v^i = length of shear strengthening region inside plastic hinge
L_v^o = length of shear strengthening region outside plastic hinge
l_{min} = minimum stabilised crack spacing

M = bending moment
M_{db} = predicted debonding moment
$M_{db,end}$ = predicted debonding moment at plate end
$M_{db,f}$ = predicted moment at flexural debonding at plate end
M_{exp} = experimental debonding moment
M_i = ideal flexural strength of column
M_u = ultimate bending moment capacity of plated beam (Chapter 3)
= ultimate bending moment capacity of column (Chapter 6)
M_{uo} = ultimate bending moment capacity of unconfined column
m = modified confinement ratio defined by Equation 6.33
N = axial load
N_u = ultimate axial load capacity of column
N_{uo} = ultimate axial load capacity of unconfined column
n = number of layers of steel reinforcement (Chapter 3)
= number of longitudinal steel reinforcing bars in column
P_u = ultimate bond strength
p = length of internal cracking line surrounded by longitudinal steel bars
R = effective stress (or strain) ratio
R_c = column corner radius
r = parameter in Equation 6.13
s = centre-to-centre spacing of transverse steel reinforcement in column
s_{frp} = centre-to-centre spacing of FRP strips measured along longitudinal axis (Chapter 4)
= clear vertical spacing between FRP straps (Chapter 6)
$s_{frp,max}$ = maximum centre-to-centre spacing of FRP strips measured along longitudinal axis
T_f = thickness of flange of T-beam
t_a = thickness of adhesive
t_c = thickness of concrete
t_{frp} = thickness of FRP
= thickness of FRP or equivalent thickness in the case of wrapped FRP strands/straps with small gaps (Chapter 6)
$t_{frp,g}$ = thickness of sheet plus applied adhesive in wet lay-up
t_p = thickness of plate
t_{sheet} = fibre sheet thickness in wet lay-up
U_{cc} = ultimate strain energy per unit volume of confined concrete
U_{co} = ultimate strain energy per unit volume of unconfined concrete
U_{sc} = ultimate strain energy per unit volume of longitudinal steel reinforcement
U_{sh} = ultimate strain energy per unit volume of confining reinforcement

V = shear force
V_c = contribution of concrete to shear capacity
V_{db} = predicted debonding shear force
$V_{db,end}$ = predicted debonding shear force at plate end
$V_{db,s}$ = predicted shear force at shear debonding at plate end
V_{frp} = contribution of FRP to shear capacity
V_N = contribution of axial load to shear capacity
V_n = shear capacity of shear strengthened beam
V_o = column shear demand based on full flexural overstrength in potential plastic hinge
V_s = contribution of steel shear reinforcement to shear capacity (Chapter 4)
= contribution of column transverse steel reinforcement to shear capacity (Chapter 7)
w_{frp} = width of FRP strip (perpendicular to fibre orientation)
x = distance from compression face to neutral axis of cross-section
= parameter in Equation 6.13
x_{cr} = distance from compression face to neutral axis of cross-section for balanced beam failure
z = downward coordinate from upper tip of shear crack in beam
z_b = coordinate of lower end of effective FRP on sides
z_t = coordinate of upper end of effective FRP on sides
α = multiplier in Equation 3.28
= angle defined by Figure 7.10
α_T = plate-to-concrete tensile stiffness ratio
α_Y = parameter defined by Equation 2.6
β = angle of first fibre orientation measured clockwise from horizontal direction for left side of shear strengthened beam
β_L = bond length coefficient defined by Equation 2.17b
β_p = width coefficient defined by Equation 2.15
β_w = FRP strip width coefficient defined by Equation 4.17b
γ_b = partial safety factor for FRP-to-concrete bond strength
γ_c = partial safety factor for concrete in flexure or axial compression
γ_{frp} = partial safety factor for FRP tensile strength
γ_s = partial safety factor for steel
Δ_p = plastic displacement
Δ_u = displacement at ultimate limit state
Δ_y = displacement at first yield
δ_1 = bond slip at maximum bond shear stress
δ_f = ultimate bond slip
ε_c = compressive strain of concrete
ε_{cc} = compressive strain of confined concrete at peak stress

ε_{cf} = strain at extreme compression fibre of concrete
ε_{co} = compressive strain of unconfined concrete at peak stress
ε_{cu} = ultimate compressive strain of concrete
ε_{frp} = strain of FRP
$\varepsilon_{frp,e}$ = effective (average) strain of FRP at failure
$\varepsilon_{frp,rup}$ = tensile rupture strain of FRP
$\varepsilon_{h,rup}$ = FRP hoop rupture strain from tests on confined concrete specimens
ε_{ini} = strain at tension face of unplated beam from preloading
ε_{max} = maximum usable strain of FRP for shear strengthening
ε_{p} = ultimate strain of plate
ε_{si} = strain in ith layer of steel reinforcement
ε_{su} = strain of steel reinforcement at maximum tensile stress
ε_{t} = strain at transition point in stress–strain curve of FRP-confined concrete
ζ = parameter defined by Equation 4.14
η = multiplier in Equation 3.27
θ = angle of critical shear crack to longitudinal axis
θ_{p} = plastic rotation
λ = parameter defined by Equation 2.14c
= normalised maximum bond length (Chapter 4)
λ_{1} = parameter defined by Equation 2.7c
λ_{2} = parameter defined by Equation 2.7d
μ_{Δ} = displacement ductility factor
μ_{Φ} = curvature ductility factor
ν_{c} = Poisson's ratio of concrete
ρ_{frp} = FRP area or volumetric ratio
$\rho_{frp,cr}$ = critical FRP ratio for balanced beam failure
ρ_{s} = ratio of steel tension reinforcement
ρ_{sc} = ratio of longitudinal steel reinforcement in column
ρ_{sh} = volumetric ratio of transverse steel reinforcement
ρ_{si} = steel reinforcement ratio of ith layer
σ_{c} = compressive stress of concrete
σ_{db} = plate stress at FRP-to-concrete joint failure
σ_{dbic} = predicted plate stress at intermediate crack-induced debonding
σ_{exp} = experimental plate stress at intermediate crack-induced debonding
σ_{frp} = stress of FRP
$\sigma_{frp,max}$ = maximum stress of FRP along a shear crack
σ_{si} = stress in ith layer of steel reinforcement
σ_{x} = longitudinal stress at base of RC beam
σ_{y} = interfacial normal (peeling) stress
τ = interfacial shear stress

τ_f = maximum bond shear stress
$\tau_{max}^{debonding}$ = maximum bond shear stress at failure
τ_u = average ultimate bond shear stress
Φ = curvature
Φ_p = plastic curvature
Φ_u = curvature at ultimate limit state
Φ_y = curvature at first yield
ϕ = angle of second fibre orientation measured clockwise from horizontal direction for left side of shear strengthened beam
ϕ_F = flexural strength reduction factor
ϕ_{frp} = FRP material reduction factor
ϕ_V = shear strength reduction factor

1 FRP Composites for Strengthening RC Structures

1.1 Introduction

The bonding of steel plates to reinforced concrete (RC) structures has been a popular method for strengthening deficient RC structures. In recent years, there has been extensive research on the use of fibre-reinforced polymer (also referred to as fibre-reinforced plastic) (FRP) plates/sheets to replace steel plates in plate bonding. FRPs have also been used widely for column strengthening by external wrapping. At present, there are numerous research teams all over the world undertaking research in this area. Detailed surveys of field applications have recently been undertaken by Shaw and Drewett (1999), Bonacci and Maalej (2000) and Hamilton and Dolan (2000), while general overviews on FRPs for strengthening applications have been presented by Karbhari and Seible (2000) and Neale (2000).

The FRP plate bonding technology was first investigated at the Swiss Federal Laboratory for Materials Testing and Research (EMPA) (Meier *et al.* 1993) where tests on RC beams strengthened with CFRP plates started in 1984. The main advantages of FRP plates are their high strength-to-weight ratio and high corrosion resistance. The former property leads to great ease in site handling, reducing labour cost and interruptions to existing services, while the latter ensures durable performance. FRP plates are normally at least twice but can be over 10 times as strong as steel plates while their weight is only 20 % of that of steel (Darby 1999).

FRP composites have been used in other areas such as the aerospace industry for many years and their superior properties are well known. The limited use of these materials in civil engineering applications has been due to their high cost. Their prices have, however, been coming down rapidly, enabling their wider applications in civil engineering. For application in the strengthening of structures, where the material cost is but one consideration and may be only a small portion of the total cost including labour cost and loss due to interruptions to services, FRP composites often provide the most cost-effective solution overall (Hollaway and Leeming 1999a).

1.2 Methods of Forming FRP Composites

FRP composites are formed by embedding continuous fibres in a resin matrix which binds the fibres together. The common fibres are carbon fibres, glass fibres

(a) (b) (c)

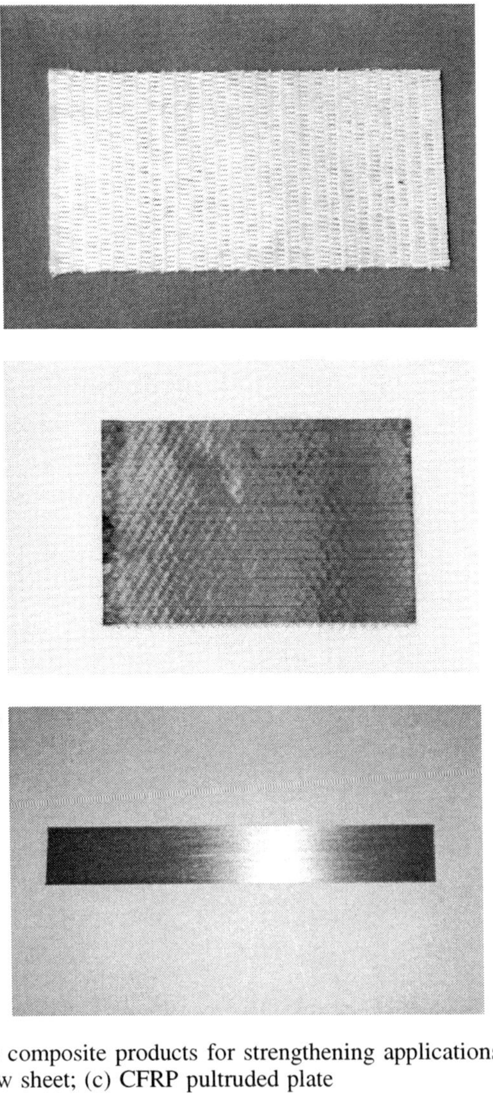

Figure 1.1 Various composite products for strengthening applications: (a) woven glass fabric; (b) carbon tow sheet; (c) CFRP pultruded plate

and aramid fibres, whilst epoxy resins, polyester resins and vinylester resins are the common resins. Depending on the fibres used, FRP composites are classified into three types: glass-fibre-reinforced polymer (GFRP) composites; carbon-fibre-reinforced polymer (CFRP) composites; and aramid-fibre-reinforced polymer (AFRP) composites. A useful general background on the composition of these materials and their mechanical properties can be found in ACI 440R-96 (1996).

Two common methods of forming FRP composites have been used in strengthening RC structures. The first one, generally referred to as the wet lay-up method, which is the most commonly used, involves the *in situ* application of resin to either a woven fabric (Figure 1.1a) or a unidirectional tow sheet (Figure 1.1b). The second method is prefabrication of FRP composites in various forms, including pultrusion for plates (Figure 1.1c) for flexural strengthening and filament winding for shells for column strengthening. The wet lay-up method is more versatile in site application in terms of bonding to curved surfaces and wrapping around corners, while prefabrication allows better quality control. Further details of these forming methods are provided in the following chapters with reference to the specific application under discussion. A third method, namely resin infusion (Karbhari and Zhao 1998), has similar characteristics to those of wet lay-up but is not commonly used.

FRP materials are generally supplied as a package and come with instructions for the end user. Such instructions should be closely followed, particularly in a wet lay-up process.

1.3 Mechanical Properties of FRP Composites

All three types of FRP composites, namely GFRP, CFRP and AFRP composites, have been used for strengthening RC structures in both practical application and research. Table 1.1 taken from Head (1996) for FRP composites with unidirectional fibres illustrates the wide variety of strength and stiffness FRP composites

Table 1.1 Typical mechanical properties of GFRP, CFRP and AFRP composites (Head 1996)

Unidirectional advanced composite materials	Fibre content (% by weight)	Density (kg/m^3)	Longitudinal tensile modulus (GPa)	Tensile strength (MPa)
Glass fibre/polyester GFRP laminate	50–80	1600–2000	20–55	400–1800
Carbon/epoxy CFRP laminate	65–75	1600–1900	120–250	1200–2250
Aramid/epoxy AFRP laminate	60–70	1050–1250	40–125	1000–1800

may possess. It should be noted that the ranges given in this table are indicative, and a particular product may well have properties outside the ranges given here, particularly when the fibre content is different from the ranges considered here. It should further be noted that, in discussing the elastic modulus and tensile strength for an FRP composite formed in a wet lay-up process, the thickness of the FRP composite is generally difficult to control or define precisely. This has led to the use of the fibre sheet thickness (e.g. Ahmed and van Gemert 1999) or a nominal thickness as recommended by the manufacturer (e.g. Teng *et al.* 2000b). Consequently, the elastic modulus and the tensile strength reported are dependent on the definition of thickness and can be outside the values indicated in Table 1.1 by a large amount.

Regardless of the type of fibres or forming method employed, all these three FRP materials have the same stress–strain behaviour: linear elastic up to final brittle rupture when subject to tension. This is a very important property in terms of structural use of FRP composites. Figure 1.2 shows typical stress–strain curves for GFRP, CFRP and mild steel. Apart from illustrating typical strength differences between these materials, these curves give a clear contrast between the brittle behaviour of FRP composites and the ductile behaviour of steel. This has two major structural consequences. First, these materials do not possess the ductility that steels have, and their brittleness may limit the ductile behaviour of RC members strengthened with FRP composites. Nevertheless, when used to provide confinement for concrete, these materials can greatly enhance the strength and ductility of columns. The second implication of the brittle behaviour of FRP

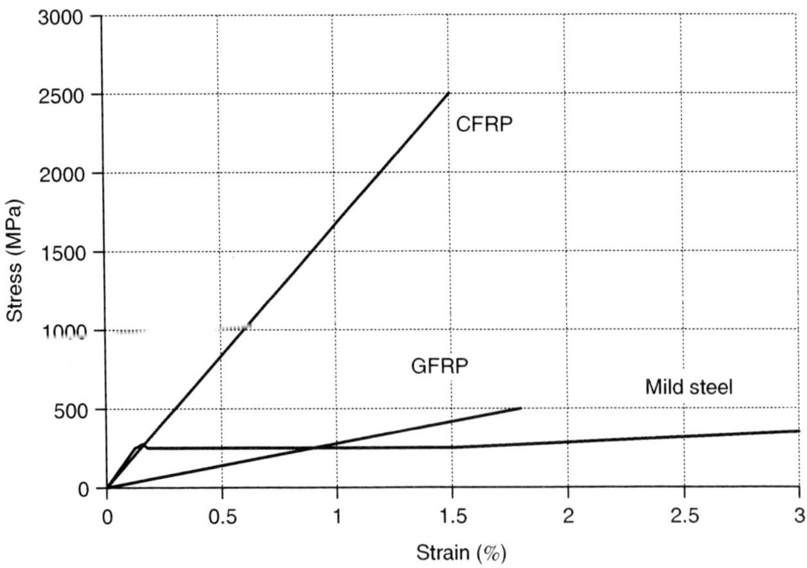

Figure 1.2 Typical FRP and mild-steel stress–strain curves

composites is that redistribution of stresses is restricted owing to this lack of ductility. Consequently, the design of structures bonded with FRP composites cannot follow existing methods for RC structures with FRPs simply treated as equivalent steel reinforcement. Instead, existing design methods for RC structures need to be modified to take this brittleness into account based on extensive research. In addition, completely new problems arise which also require extensive research.

The tensile properties of FRP composites should normally be determined using flat coupons following a suitable test standard (e.g. ASTM D3039/D3039M-95a 1995), but for prefabricated FRP tubes, splitting tensile tests using ring specimens (e.g. ASTM D2290-92 1992) have also been common. Ring-splitting tests are believed to provide closer estimates for the tensile strength of FRP tubes in the hoop direction (e.g. Shahawy *et al.* 2000). In tensile tests, care should be exercised in both the preparation and the testing of the specimen. Tests which are not properly conducted tend to show a large scatter in the test results and produce tensile strengths lower than those from careful tests.

1.4 Resins

Polymeric resins are used both as the matrix for the FRP and as the bonding adhesive between the FRP and the concrete. The latter function is of particular concern here, as weak adhesives can cause interfacial failures. Epoxy resins are generally used when the bonding function of resins is of crucial importance, as in the flexural and shear strengthening of beams. Present-day epoxy resins are so strong that interfacial failures generally occur in the concrete, particularly as the concrete in the structure to be strengthened is in general comparatively weak. All design recommendations in this book concerning debonding failures are based on the assumption that the interfacial strength is dependent on the concrete strength rather than the adhesive strength, with failure either well into the concrete or in the concrete adjacent to the adhesive-to-concrete interface. Nevertheless, when the concrete strength is high, the risk of interfacial failure in the adhesive cannot be ruled out, and should be investigated in detail by a suitable test such as a shear lap test as described in Chapter 2.

The tensile properties of polymeric resins can be determined according to a suitable test standard such as ASTM D638M-93 (1993).

1.5 Comparison of Performance for Strengthening of Structures

CFRP composites have superior properties to GFRP composites, but the latter are cheaper. Table 1.2, taken from Meier and Winistorfer (1995), gives a qualitative comparison of different FRP composites for applications in the strengthening of RC structures.

Table 1.2 Qualitative comparison between E-glass, HT-carbon and aramid fibres. From Meier and Winistorfer (1995). Reproduced by permission of Routledge Spon

Criterion	Fibre composite sheets made of:		
	E-glass fibres	Carbon fibres	Aramid fibres
Tensile strength	Very good	Very good	Very good
Compressive strength	Good	Very good	Inadequate
Young's modulus	Adequate	Very good	Good
Long-term behaviour	Adequate	Very good	Good
Fatigue behaviour	Adequate	Excellent	Good
Bulk density	Adequate	Good	Excellent
Alkaline resistance	Inadequate	Very good	Good
Price	Very good	Adequate	Adequate

1.6 Partial Safety Factors

In this book, design recommendations are made for various failure modes in the limit-state design format. These recommendations are built as far as possible within the framework of the British code for the structural use of concrete (BS 8110 1997). This should not present any significant difficulty for adapting these design recommendations to the format of another design code; often minimal changes are required to achieve this.

The design recommendations include partial safety factors for materials, which are applied either to the best-fit strength model (e.g. columns with FRP confinement and sectional flexural capacity of FRP-strengthened beams) when the mechanism of failure is well understood and the scatter of the test data is limited, or to a lower bound strength model when either or both of these two conditions is not met (e.g. debonding and shear failures). The 95 percentile or similar lower bound of the test data is recommended in the latter case. It should be noted that for statistical analysis of test data or ratios of test-to-predicted results, for simplicity, a normal distribution has been assumed in all studies by the present authors referred to in this book.

To maintain the generality of the design recommendations presented in this book, specific values of partial safety factors will generally not be discussed in the following chapters. This allows different appropriate values of partial safety factors to be used. Nevertheless, as the design recommendations are made within the framework of BS 8110 (1997), the partial safety factors for concrete, γ_c, and steel, γ_s, from BS 8110 (1997) can obviously be used. These partial safety factors are given in Table 1.3.

Two additional partial safety factors, not covered by BS 8110 (1997) or any other RC design code, are also given in Table 1.3: one for the tensile strength of FRP composites, γ_{frp}, and another for the strength of FRP-to-concrete bonded joints subject to tensile forces, γ_b. Both are set at 1.25, based on the best judgement of the authors to give safe and consistent designs. While partial safety

Table 1.3 Partial safety factors for concrete, steel and FRP

Strength	Partial safety factor
Concrete in flexure or in axial compression, γ_c	1.50
Steel reinforcement, γ_s	1.05[a]
FRP tensile strength, γ_{frp}	1.25
FRP-to-concrete bond strength, γ_b	1.25

[a] $\gamma_s = 1.15$ in BS 8110 (1985).

factors based on a rigorous probabilistic basis are desirable, there is at present insufficient information for such factors to be derived. Although the partial safety factors suggested here may be significantly lower than those suggested elsewhere (e.g. Concrete Society 2000), the authors believe that the values suggested here are appropriate for use with the rigorously based design recommendations presented in this book.

It should be mentioned that the partial safety factors recommended here do not account for long term degradation of FRP composites and the FRP-to-concrete bond. The effects of such degradation can be easily incorporated into the design recommendations presented in this book by applying appropriate reduction factors to the relevant FRP and bond properties. The determination of such factors is however beyond the scope of the present book.

1.7 Concluding Remarks

Based on the above discussions on various basic aspects of the use of FRP composites for the strengthening of RC structures, it is clear that the potential application of FRP composites in the strengthening of RC structures is tremendous because of their superior properties. Rapid growth in recent years in both field application of and research on FRP composites for strengthening RC structures is testimony to the importance and popularity of this new technology.

While a great deal of research now exists in the published literature, it is scattered in various journals and conference proceedings. This makes it difficult for engineers to exploit it and for new researchers to assimilate it. The aim of this book is to provide a state-of-the-art summary of the existing research, with the emphasis being on an in-depth treatment of the fundamental behaviour and strength of FRP-strengthened RC structures and the presentation of a comprehensive set of rigorously based design recommendations which are believed to be the most advanced available. The book has grown out of research by the authors over the past few years, and its predecessor appeared in the form of a monograph for limited local distribution (Teng *et al.* 2000a). Most of the design recommendations have been developed by the authors themselves. The main topics covered by the book include bond behaviour (Chapter 2), flexural and shear strengthening of beams (Chapters 3 and 4), flexural strengthening of slabs (Chapter 5), strengthening of columns subject to axial or eccentric loads (Chapter 6) and seismic retrofit of columns (Chapter 7). For each topic treated, the methods of strengthening are first discussed, followed by a description of behaviour and failure modes and then the presentation of design recommendations.

The authors hope that the book will be of interest to a wide range of readers, such as new and more experienced researchers, engineers, and graduate and senior undergraduate students, as a consolidated source of up-to-date information on this important topic. With its in-depth treatment of fundamental behaviour and advanced design recommendations, the book is also believed to be of particular assistance to both users and writers of design codes and guidelines. Code development in this area is still at a very early stage, although much has been done around the world. An example of these activities is the report prepared by the Concrete Society (2000) in the UK providing design guidance for the strengthening of RC structures using FRP composites, which appeared in the final stage of preparing this book. It may be worth noting that the present book provides an extensive treatment of the original research basis for the structural design issues dealt with in the Concrete Society (2000) report as well as many other structural design issues, despite most of the material for the book being developed (Teng *et al.* 2000a) before the report was published.

A number of issues, including constructional considerations, durability, fatigue and fire resistance, are not addressed in this book. Both the recent book edited by Hollaway and Leeming (1999b) and the Concrete Society (2000) report are very useful sources of information on various constructional aspects. For the latter three issues, the limited existing research does not allow an in-depth treatment at present. The recent paper by Hamilton and Dolan (2000) is a very useful source of information for recent research on durability issues, an important area which is attracting increasingly greater research attention.

References

ACI 440R-96 (1996) *State-of-the Art Report on Fiber Reinforced Plastic (FRP) Reinforcement for Concrete Structures*, American Concrete Institute (ACI) Committee 440, Farmington Hills, Michigan, USA.

Ahmed, O. and van Gemert, D. (1999) 'Effect of longitudinal carbon fiber reinforced plastic laminates on shear capacity of reinforced concrete beams', *Proceedings of the Fourth International Symposium on Fiber Reinforced Polymer Reinforcement for Reinforced Concrete Structures*, Baltimore, Maryland, USA, edited by C. W. Dolan, S. H. Rizkalla and A. Nanni, pp. 933–943, American Concrete Institute, Farmington Hills, Michigan, USA.

ASTM D2290-92 (1992) *Standard Test Method for Apparent Tensile Strength of Ring or Tubular Plastics and Reinforced Plastics by Split Disk Method*, American Society for Testing and Materials (ASTM), Philadelphia, USA.

ASTM D638M-93 (1993) *Standard Test Method for Tensile Properties of Plastics (Metric)*, American Society for Testing and Materials (ASTM), Philadelphia, USA.

ASTM D3039/D3039M-95a (1995) *Standard Test Method for Tensile Properties of Polymer Matrix Composite Materials*, American Society for Testing and Materials (ASTM), Philadelphia, USA.

Bonacci, J. F. and Maalej, M. (2000) 'Externally bonded FRP for service-life extension of RC infrastructure', *Journal of Infrastructure Systems, ASCE*, Vol. 6, No. 1, pp. 41–51.

BS 8110 (1985) *Structural Use of Concrete, Part 1. Code of Practice for Design and Construction*, British Standards Institution, London, UK.

BS 8110 (1997) *Structural Use of Concrete, Part 1. Code of Practice for Design and Construction*, British Standards Institution, London, UK.

Concrete Society (2000) *Design Guidance for Strengthening Concrete Structures using Fibre Composite Materials*, Concrete Society Technical Report No. 55, Crowthorne, Berkshire, UK.

Darby, J. J. (1999) 'Role of bonded fibre-reinforced composites in strengthening of structures', *Strengthening of Reinforced Concrete Structures Using Externally-Bonded FRP Composites in Structural and Civil Engineering*, edited by L. C. Hollaway and M. B. Leeming, Woodhead Publishing, Cambridge, UK.

Hamilton, H. R., III and Dolan, C. W. (2000) 'Durability of FRP reinforcements for concrete', *Progress in Structural Engineering and Materials*, Vol. 2, pp. 139–145.

Head, P. R. (1996) 'Advanced composites in civil engineering – a critical overview at the high interest, low usage stage of development', *Advanced Composite Materials in Bridges and Structures, Proceedings of the Second International Conference*, Quebec, Canada, edited by M. M. El-Badry, pp. 3–15, Canadian Society for Civil Engineering, Montreal, Canada.

Hollaway, L. C. and Leeming, M. B. (1999a) 'Review of materials and techniques for plate bonding', *Strengthening of Reinforced Concrete Structures Using Externally-Bonded FRP Composites in Structural and Civil Engineering*, edited by L. C. Hollaway and M. B. Leeming, Woodhead Publishing, Cambridge, UK.

Hollaway, L. C. and Leeming, M. B. (1999b) *Strengthening of Reinforced Concrete Structures Using Externally-Bonded FRP Composites in Structural and Civil Engineering*, Woodhead Publishing, Cambridge, UK.

Karbhari, V. M. and Seible, F. (2000) 'Fiber reinforced composites – advanced materials for the renewal of civil infrastructure', *Applied Composite Materials*, Vol. 7, pp. 95–124.

Karbhari, V. M. and Zhao, L. (1998) 'Issues related to composite plating and environmental exposure effects on composite-concrete interface in external strengthening', *Composite Structures*, Vol. 40, Nos 3–4, pp. 293–304.

Meier, U. and Winistorfer, A. (1995) 'Retrofitting of structures through external bonding of CFRP sheets', *Non-Metallic (FRP) Reinforcement for Concrete Structures, Proceedings of the Second International RILEM Symposium*, Ghent, Belgium, edited by L. Taerwe, pp. 509–516, E & FN Spon, London, UK.

Meier, U., Deuring, M., Meier, H. and Schwegler, G. (1993) 'CFRP bonded sheets', *Fibre-Reinforced-Plastic (FRP) Reinforcement for Concrete Structures: Properties and Applications*, edited by A. Nanni, Elsevier Science, Amsterdam, The Netherlands.

Neale, K. W. (2000) 'FRPs for structural rehabilitation: a survey of recent progress', *Progress in Structural Engineering and Materials*, Vol. 2, pp. 133–138.

Shahawy, M., Mirmiran, A. and Beitelman, T. (2000) 'Tests and modeling of carbon-wrapped concrete columns', *Composites: Part B*, Vol. 31, pp. 471–480.

Shaw, M. A. and Drewett, J. F. (1999) 'Case studies of carbon fibre bonding worldwide', *Strengthening of Reinforced Concrete Structures Using Externally-Bonded FRP Composites in Structural and Civil Engineering*, edited by L. C. Hollaway and M. B. Leeming, Woodhead Publishing, Cambridge, UK.

Teng, J. G., Chen, J. F., Smith, S. T. and Lam, L. (2000a) *RC Structures Strengthened with FRP Composites*, Research Centre for Advanced Technology in Structural Engineering, The Hong Kong Polytechnic University, Hong Kong, China, December.
Teng, J.G., Lam, L., Chan, W. and Wang, J. (2000b) 'Retrofitting of deficient RC cantilever slabs using GFRP strips', *Journal of Composites for Construction, ASCE*, Vol. 4, No. 2, pp. 75–84.

2 Bond Strength of FRP-to-concrete Joints

2.1 Introduction

In the strengthening of RC structures, FRP plates or sheets are externally bonded to RC structures using epoxy resins. For the external FRP to be effective in enhancing the load-carrying capacity of the structure, effective stress transfers between the FRP and the concrete are required.

There are two distinct situations of stress transfer which have the potential to cause interfacial failure or local concrete failure in flexurally strengthened RC beams. This is best illustrated by a beam strengthened with a bonded FRP soffit plate (Figure 2.1). This beam is subject to a single point load at mid-span at which a major flexural crack has formed. For clarity, additional smaller cracks are omitted in Figure 2.1. The FRP plate is highly stressed under the crack, and these stresses have to be transferred to the concrete. Interfacial shear stresses dominate the behaviour of such a situation. Failure is by crack propagation in the concrete parallel to the bonded plate and adjacent to the adhesive-to-concrete interface, starting from the critically stressed position towards one of the ends of the plate. This failure mode will be referred to as intermediate flexural crack-induced interfacial debonding in Chapter 3. On the other hand, high interfacial stresses also develop near the ends of the FRP plate, where stresses in the FRP plate have also to be transferred to the concrete. Here, interfacial peeling stresses (stresses normal to the plate surface) are believed to be as important as interfacial shear stresses. These high interfacial stresses are known to cause premature failures in beams. A detailed discussion of plate-end premature failure modes is beyond the scope of this chapter but will be given in Chapter 3.

In addition to being an important failure mode in flexurally strengthened beams, intermediate crack-induced interfacial debonding is the most common failure

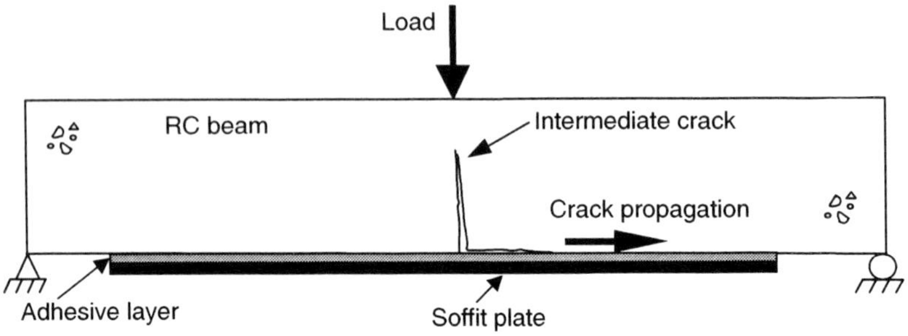

Figure 2.1 Intermediate crack-induced interfacial debonding

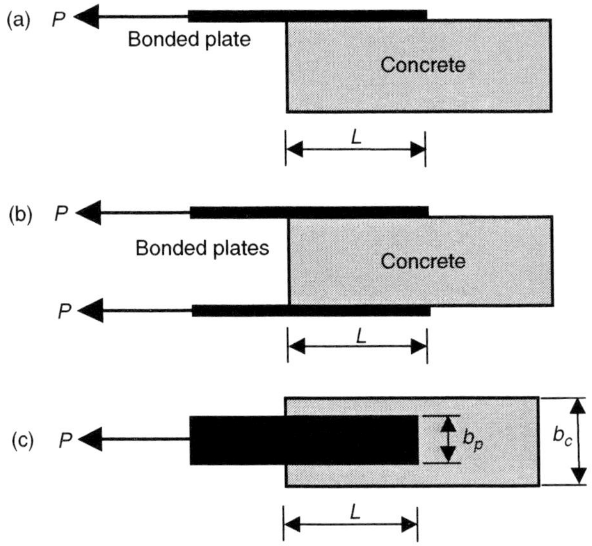

Figure 2.2 Single (a) and double (b) shear tests; (c) plan

mode for plates/strips bonded to the sides of beams for shear strengthening (Chen and Teng 2001a). It can also occur for plates/strips bonded to RC tension members and other RC members such as slabs. This problem is thus of wide importance in the strengthening of RC structures using FRP composites, and in terms of significance may be compared with the end anchorage issue for reinforcing bars. The problem may be idealised and studied as an FRP-to-concrete joint with the FRP subjected to tension (Figure 2.2). In this chapter, the bond behaviour and strength of such idealised joints are discussed, and a design proposal is presented.

Although this book is concerned with the strengthening of RC structures with externally bonded FRPs, the bond strength of steel plate to concrete is also covered in this chapter because their behaviour is similar. It should also be noted that the term 'bond strength' as used here differs in meaning from that used in

RC design books. Here, bond strengths are defined as the force in the bonded plate at joint failure, while in books on RC structures (e.g. Kong and Evans 1987), bond strengths are given in terms of the average shear stress parallel to the reinforcement bar. The conventional definition of bond strength can be misleading for externally bonded plates owing to the existence of an effective bond length, beyond which any increase in the bond length cannot increase the load the bonded plate is able to carry.

2.2 Test Methods, Behaviour and Failure Mode

Substantial experimental and theoretical work exists on the bond strength of FRP or steel plate to concrete joints. Experiments have been carried out using several set-ups, including single shear tests (Figure 2.2a) (e.g. Täljsten 1994, 1997, Chajes *et al.* 1995, 1996, Bizindavyi and Neale 1999), double shear tests (Figure 2.2b) (e.g. van Gemert 1980, Swamy *et al.* 1986, Kobatake *et al.* 1993, Autocon 1994, Brosens and van Gemert 1997, Fukuzawa *et al.* 1997, Hiroyuki and Wu 1997, Maeda *et al.* 1997, Neubauer and Rostásy 1997), and modified beam tests (e.g. van Gemert 1980, Ziraba *et al.* 1995). The discussion in this chapter is chiefly based on simple shear tests, with test data of both steel plate and FRP plate to concrete joints included. A numerical investigation into the differences between these test set-ups can be found in Chen *et al.* (2001).

Theoretical work on the bond strength has included both fracture mechanics analysis (e.g. Triantafillou and Plevris 1992, Holzenkämpfer 1994, Täljsten 1994, Yuan and Wu 1999, Yuan *et al.* 2001) and the development of empirical models based on either regression of experimental data and/or simplistic assumptions (van Gemert 1980, Chaallal *et al.* 1998, Khalifa *et al.* 1998). These works have naturally led to the revelation of some of the characteristics of the behaviour of plate to concrete bonded joints.

An important feature of plate to concrete bonded joints is that the bond strength cannot always increase with an increase in the bond length L (Figure 2.2), and that the ultimate tensile strength of an FRP plate may never be reached however long the bond length is. This leads to the concept of effective bond length, beyond which any increase in the bond length cannot increase the bond strength, as confirmed by many experimental studies (e.g. Täljsten 1994, Chajes *et al.* 1996, Maeda *et al.* 1997) and fracture mechanics analyses (e.g. Holzenkämpfer 1994, Yuan and Wu 1999, Yuan *et al.* 2001). However, a longer bond length can improve the ductility of the failure process. This phenomenon is substantially different from the bond behaviour of internal reinforcement, for which a bond length can always be designed for its full tensile strength if there is sufficient concrete cover. Therefore, this key aspect must be accounted for in the development of bond strength models.

An extensive literature survey of experiments on plate to concrete joints by Chen and Teng (2001b), covering both steel plates and FRP plates, showed that most joints failed by crack propagation in the concrete adjacent to the adhesive-to-concrete interface, starting from the critically stressed position. This chapter is thus primarily concerned with this failure mode. The fact that failure occurs in the concrete allows the same bond strength model to be developed for both steel plate and FRP plate to concrete joints. Bond failure in the adhesive layer is rare and should be avoided by the use of strong adhesives.

2.3 Existing Bond Strength Models

Several bond strength models have been advanced in the last few years. They may be roughly classified into three categories: empirical models based directly on regression of test data (Tanaka 1996, Hiroyuki and Wu 1997, Maeda *et al.* 1997), fracture mechanics models (Täljsten 1994, Holzenkämpfer 1994, Neubauer and Rostásy 1997, Yuan and Wu 1999, Yuan *et al.* 2001), and design proposals which generally make use of some simple assumptions (van Gemert 1980, Chaallal *et al.* 1998, Khalifa *et al.* 1998). This section presents a review of these models and an assessment of their performance, based on the work of Chen and Teng (2001b).

2.3.1 Empirical Models

Hiroyuki and Wu (1997) conducted a set of double shear experiments on RC members strengthened with carbon fibre sheets (CFS), from which they derived the following empirical relationship between the bond length L (cm) and the average bond shear stress at failure τ_u:

$$\tau_u = 5.88L^{-0.669}, \quad \text{MPa} \tag{2.1}$$

Tanaka (1996) presented another simple expression (see Sato *et al.* 1996):

$$\tau_u = 6.13 - \ln L, \quad \text{MPa} \tag{2.2}$$

where L is in mm. The ultimate bond strength of the joint P_u is given by multiplying τ_u by the width b_p and length L of the bond area in the above two models.

Maeda *et al.* (1997) developed a more robust model which considers the effective bond length,

$$\tau_u = 110.2 \times 10^{-6} E_p t_p, \quad \text{MPa} \tag{2.3a}$$

in which E_p (MPa) and t_p (mm) are the modulus of elasticity and thickness of the bonded plate respectively. The ultimate bond strength P_u is obtained by multiplying τ_u by the effective bond area $L_e b_p$. Here the effective bond length L_e is given by

$$L_e = e^{6.13-0.580 \ln E_p t_p}, \quad \text{mm} \tag{2.3b}$$

Note that $E_p t_p$ is in GPa-mm in Equation 2.3b. This model is obviously invalid if $L < L_e$.

2.3.2 Fracture-mechanics-based Models

Holzenkämpfer (1994) investigated the bond strength between steel and concrete using nonlinear fracture mechanics (NLFM). The modified form by Niedermeier (1996) (also see Blaschko *et al.* 1998) calculates the bond strength using

$$P_u = \begin{cases} 0.78 b_p \sqrt{2 G_f E_p t_p} & \text{if } L \geqslant L_e \\ 0.78 b_p \sqrt{2 G_f E_p t_p} \dfrac{L}{L_e}\left(2 - \dfrac{L}{L_e}\right) & \text{if } L < L_e \end{cases} \tag{2.4a}$$

where $E_p t_p$ are in MPa-mm and P_u is in N. The effective bond length L_e and the fracture energy G_f are given by

$$L_e = \sqrt{\frac{E_p t_p}{4 f_{ctm}}}, \quad \text{mm} \tag{2.4b}$$

$$G_f = c_f k_p^2 f_{ctm}, \quad \text{N mm/mm}^2 \tag{2.4c}$$

in which f_{ctm} (MPa) is the average surface tensile strength of the concrete determined in a pull-off test according to DIN 1048 (1991), c_f is a constant determined in a linear regression analysis using the results of double shear or similar tests, k_p is a geometrical factor related to the width of the bonded plate b_p (mm) and the width of the concrete member b_c (mm),

$$k_p = \sqrt{1.125 \frac{2 - b_p/b_c}{1 + b_p/400}} \tag{2.4d}$$

Also by NLFM analysis, Täljsten (1994) developed a similar model

$$P_u = \sqrt{\frac{2 E_p t_p G_f}{1 + \alpha_T}} b_p \tag{2.5a}$$

where

$$\alpha_T = \frac{E_p t_p}{E_c t_c} \tag{2.5b}$$

in which E_c and t_c are the elastic modulus and thickness of the concrete member. Owing to difficulties in accessing Täljsten's (1994) thesis, it is not clear how G_f was defined.

Yuan and Wu (1999) and Yuan *et al.* (2001) studied the bond strength between FRP and concrete using linear elastic fracture mechanics (LEFM) and NLFM. Their LEFM study (Yuan and Wu 1999) resulted in the same equation as Equation 2.5a but included the effect of widths of the plate and the concrete member. That is, α_T in Equation 2.5a is replaced by α_Y given below:

$$\alpha_Y = \frac{b_p E_p t_p}{b_c E_c t_c} \tag{2.6}$$

They also solved the NLFM equation for five different shear stress–slip relationships (Figure 2.3). Among them, the linearly ascending and then descending relationship (Figure 2.3d) may be closest to reality. The maximum load-carrying capacity for this case is (Yuan *et al.* 2001)

$$P_u = \frac{\tau_f b_p}{\lambda_2} \frac{\delta_f}{\delta_f - \delta_1} \sin(\lambda_2 a) \tag{2.7a}$$

where a is obtained by solving

$$\tanh[\lambda_1(L - a)] = \frac{\lambda_2}{\lambda_1} \tan(\lambda_2 a) \tag{2.7b}$$

in which τ_f is the maximum stress on the shear–slip curve of the bond, δ_1 is its corresponding slip, δ_f is the maximum slip, and λ_1 and λ_2 are defined by

$$\lambda_1^2 = \frac{\tau_f}{\delta_1 E_p t_p} (1 + \alpha_Y) \tag{2.7c}$$

$$\lambda_2^2 = \frac{\tau_f}{(\delta_f - \delta_1) E_p t_p} (1 + \alpha_Y) \tag{2.7d}$$

In this solution, the relationship between P_u and L is an implicit function and is dependent on the shear–slip coefficients which are not easy to obtain. Yuan *et al.* (2001) defined the effective bond length as the value corresponding to 97 % of the load-carrying capacity if L is infinite. This gives

$$L_e = a_0 + \frac{1}{2\lambda_1} \ln \frac{\lambda_1 + \lambda_2 \tan(\lambda_2 a_0)}{\lambda_1 - \lambda_2 \tan(\lambda_2 a_0)} \tag{2.7e}$$

where

$$a_0 = \frac{1}{\lambda_2} \sin^{-1}\left(0.97\sqrt{\frac{\delta_f - \delta_1}{\delta_f}}\right) \tag{2.7f}$$

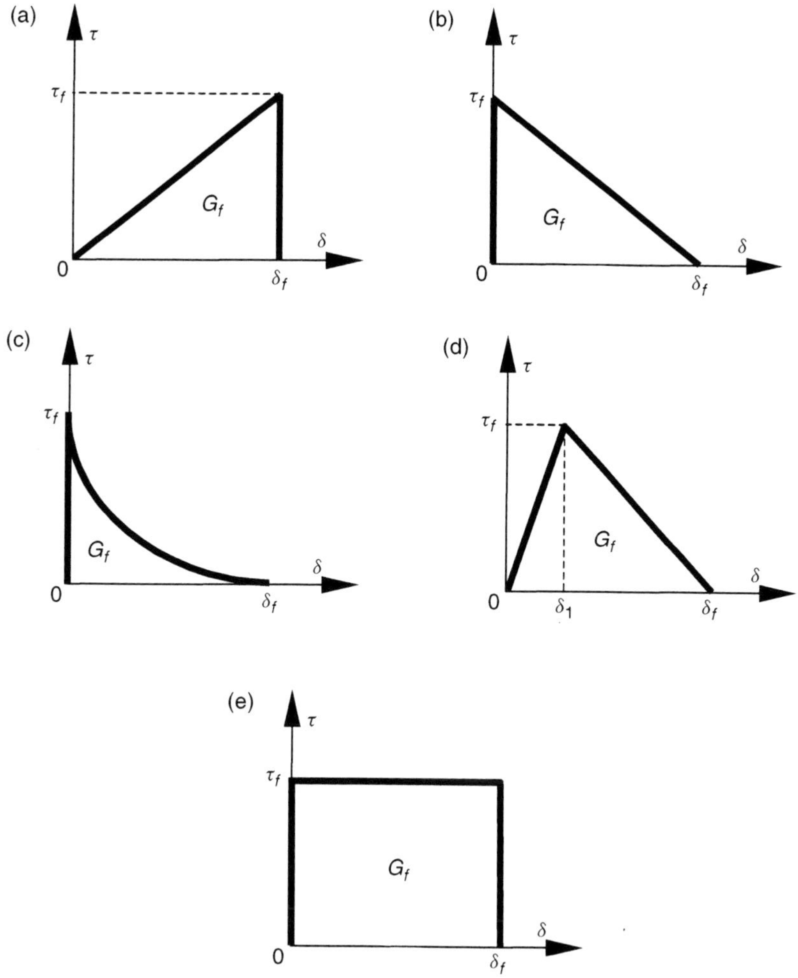

Figure 2.3 Shear–slip models for plate to concrete bonded joints (Yuan and Wu 1999)

Neubauer and Rostásy (1997) conducted a series of double shear tests on CFRP-to-concrete bonded joints. They concluded that, for both concrete fracture and FRP delamination failures, the shear–slip relationship may be represented by the triangular model as shown in Figure 2.3d and the fracture energy can be calculated using

$$G_f = c_f f_{ctm} \tag{2.8}$$

They reported an average value of 0.204 mm with a standard deviation of 0.053 for c_f for 51 specimens. They presented a modified model to Holzenkämpfer

(1994) so that it applies to both CFRP and steel plates:

$$P_u = \begin{cases} 0.64k_p b_p \sqrt{E_p t_p f_{ctm}} & \text{if } L \geqslant L_e \\ 0.64k_p b_p \sqrt{E_p t_p f_{ctm}} \dfrac{L}{L_e}\left(2 - \dfrac{L}{L_e}\right) & \text{if } L < L_e \end{cases} \tag{2.9a}$$

$$L_e = \sqrt{\frac{E_p t_p}{2 f_{ctm}}} \tag{2.9b}$$

The same units as those for Equation 2.4 should be used for Equation 2.9.

2.3.3 Design Proposals

By assuming a triangular shear stress distribution (linearly decreasing from the stressed end to zero at the free end) in the full bond length, van Gemert (1980) proposed a design rule that (see also Brosens and van Gemert 1997)

$$P_u = 0.5 b_p L f_{ctm} \tag{2.10}$$

The only material parameter in Equation 2.10 is the concrete surface tensile strength f_{ctm}. According to Brosens and van Gemert (1997), Equation 2.10 gives the load at the cracking of concrete in the initial force transfer zone but does not take into account the strength reserve after first cracking. While this might be construed to mean that Equation 2.10 always provides a conservative prediction, such an interpretation is invalid as the limit of the effective bond length is not included in this equation. Equation 2.10 implies that any given load P, up to the full tensile strength of the bonded plate, can be carried by a sufficiently long bonded joint. This is conceptually misleading as it contradicts the well-established fact that any additional bond length beyond the effective bond length cannot increase the bond strength.

It may be noted that Volnyy and Pantelides (1999) attempted to derive the effective bond length using the above rule of van Gemert and fracture mechanics. Unfortunately their concept of the 'effective bond length' was incorrect and a number of errors crept into their deduction process, so the result is only an alternative expression of van Gemert's equation.

In the design model of Chaallal *et al.* (1998), it was proposed, based on studies by Roberts (1989) and Varastehpour and Hamelin (1996), that the bond behaves as a Mohr–Coulomb material. For shear strengthening, they assumed that the maximum shear stress $\tau_{max}^{debonding}$ is twice the average stress τ_u and does not exceed the Mohr–Coulomb strength equation given by Varastehpour and Hamelin (1996). This gives

$$\tau_u = \frac{\tau_{max}^{debonding}}{2} = \frac{2.7}{1 + k_1 \tan 33^\circ}, \quad \text{MPa} \tag{2.11a}$$

where the dimensionless parameter k_1 is given by

$$k_1 = t_p \sqrt[4]{\frac{K_n}{4E_p I_p}} \tag{2.11b}$$

$$K_n = E_a \frac{b_a}{t_a} \tag{2.11c}$$

in which E_a, b_a and t_a are the modulus of elasticity, width and thickness of the adhesive respectively, and I_p is the second moment of area of the FRP plate. Because Equation 2.11b was based on limited experimental data and does not relate to the strength of concrete, its applicability is seriously limited. Another drawback of this proposal is that the effective bond length is not considered.

For design use, Neubauer and Rostásy (1997) proposed to reduce the factor 0.64 in Equation 2.9a to 0.5 (a 25 % reduction) for conservative predictions.

Khalifa *et al.* (1998) proposed a modification to Maeda *et al.*'s (1997) model (Equation 2.3) to include the effect of concrete strength, so that it could be used for design. They used the relationship that the bond strength between FRP and concrete is a function of $(f_c')^{2/3}$ (Horiguchi and Saeki 1997), where f_c' (MPa) is the cylinder compressive strength of concrete. Because the cylinder strength of concrete was 42 MPa in the experiments carried out by Maeda *et al.* (1997), the modified equation is thus

$$\tau_u = \frac{110.2}{10^6}\left(\frac{f_c'}{42}\right)^{2/3} E_p t_p, \quad \text{MPa} \tag{2.12}$$

in which $E_p t_p$ is in MPa-mm. The effective bond length is calculated using Equation 2.3b.

2.3.4 Performance of Bond Strength Models

Chen and Teng (2001b) assessed the performance of some of the models reviewed above in predicting the experimental bond strengths for 27 FRP and 23 steel bond test specimens collected by them from the literature (Table 2.1). Their findings are summarised in this section. Purely fracture mechanics based models were not included in their assessment because the fracture energy and the shear-slip parameters were not available. Concrete surface tensile strength f_{ctm} was used in the models proposed by Holzenkämpfer (1994) and Neubauer and Rostásy (1997) but was not reported any of the tests examined by Chen and Teng (2001b) (Table 2.1). The concrete-splitting tensile strength f_{ct} was thus used by Chen and Teng (2001b) instead and was estimated, if not given in the original source, from f_c' using (MacGregor 1988)

$$f_{ct} = 0.53\sqrt{f_c'} \tag{2.13}$$

Table 2.1 Test-to-predicted bond strength ratios

Source of model	FRP plates			Steel plates			All plates		
	Ave	Std	CoV	Ave	Std	CoV	Ave	Std	CoV
Hiroyuki and Wu (1997)	2.87	0.95	33 %	3.85	1.18	31 %	3.24	1.09	34 %
Tanaka (1996)	2.92	1.65	56 %	5.51	5.30	96 %	4.02	3.96	99 %
van Gemert (1980)	2.19	1.12	51 %	1.64	0.57	35 %	1.91	0.96	50 %
Chaallal *et al.* (1998)	1.81	0.89	49 %	1.68	0.70	42 %	1.71	0.79	46 %
Khalifa *et al.* (1998)	1.07	0.24	23 %	0.76	0.26	34 %	0.93	0.29	31 %
Neubauer and Rostásy (1997)	0.82	0.15	18 %	0.65	0.09	13 %	0.74	0.15	20 %
Chen and Teng (2001b)	1.05	0.18	17 %	0.94	0.11	12 %	1.00	0.16	16 %

Note: Ave = average, Std = standard deviation, CoV = coefficient of variation.

The first four models in Table 2.1 clearly do not fit the experimental data statistically. They hugely underestimate the bond strength, and, more importantly, lead to a very large scatter. The chief cause for the poor performance was considered by Chen and Teng (2001b) to be that the effective bond length is not considered in these models.

The models by Khalifa *et al.* (1998) and Neubauer and Rostásy (1997) show a reasonable performance. Khalifa *et al.*'s (1998) proposal was based on the model proposed by Maeda *et al.* (1997) from regression of test data of FRP-to-concrete joints. As a result, it agrees reasonably well with experimental data for FRP-to-concrete joints, but not so well for steel-to-concrete joints. The chief drawback of this model is that it may greatly overestimate the shear stress at failure and underestimate the effective bond length. For example, Equation 2.12 predicts an average shear stress at failure of about 60 MPa and Equation 2.3b predicts an effective bond length of about 11 mm for the set of steel-to-concrete joints reported by Täljsten (1997), compared with an observed effective bond length of about 300 mm. It thus cannot be used for safe practical design.

Table 2.1 also shows that on average the experimental observation is 0.82 of that predicted by the model of Neubauer and Rostásy (1997) for FRPs (i.e. predictions are $(1 - 0.82)/0.82 = 22\,\%$ higher than observations on average). For steel plates, the average test-to-predicted strength ratio is 0.74 (i.e. predictions are 55 % higher than test observations). For the whole data set including both FRP and steel plates, the predictions of their model are 35 % higher than experimental observations on average. Another drawback of this model is the use of concrete surface tensile strength, which requires special tests, whilst concrete compressive strength is readily available in most cases.

2.4 Chen and Teng's Model

Chen and Teng (2001b) recently proposed a bond strength model which does not suffer from the shortcomings of the previous models, as discussed in the preceding section. Their model was developed by combining fracture mechanics analysis with experimental evidence and provides accurate predictions. This model is described below.

Chen and Teng (2001b) noted that the shear–slip behaviour of plate to concrete bonded joints may be well represented by a triangular shear–slip model (Figure 2.3d) (Neubauer and Rostásy 1997, Yuan and Wu 1999, Yuan *et al.* 2001). For FRP-to-concrete joints, the typical slip values are $\delta_1 = 0.02$ mm at peak shear and $\delta_f = 0.2$ mm at failure; that is, δ_1 is small compared with δ_f (Yuan and Wu 1999). A linearly decreasing shear–slip model (Figure 2.3b) may thus be used. The NLFM solution for this case is (Yuan and Wu 1999)

$$P_u = \begin{cases} \tau_f b_p/\lambda & \text{if } L \geqslant L_e \\ (\tau_f b_p/\lambda)\sin(\lambda L) & \text{if } L < L_e \end{cases} \tag{2.14a}$$

where

$$L_e = \frac{\pi}{2\lambda} \tag{2.14b}$$

$$\lambda^2 = \frac{\tau_f}{\delta_f E_p t_p}(1+\alpha_Y) \tag{2.14c}$$

Chen and Teng also noted that the coefficient α_Y in Equation 2.14 is generally small for practical configurations compared with unity. This term is based on the assumption that the stress distribution is uniform over the whole cross-section of the concrete member as well as that of the bonded plate. Because of localised bond behaviour, this assumption is clearly invalid for the concrete member. Instead, the width ratio of the bonded plate to the concrete member, b_p/b_c, was shown to have a significant effect on the ultimate bond strength (Chen and Teng 2001b), in a form similar to the coefficient proposed by Holzenkämpfer (1994). If the width of the bonded plate is smaller than that of the concrete member, the force transfer from the plate to the concrete leads to a nonuniform stress distribution across the width of the concrete member. A smaller b_p compared with b_c may result in a higher shear stress at the interface at failure, due to the contribution from the concrete outside the bond area. Regression of the data collected from the literature showed that the ultimate bond strength is linearly related to β_p (Figure 2.4) in the form of (Chen and Teng 2001b)

$$\beta_p = \sqrt{\frac{2 - b_p/b_c}{1 + b_p/b_c}} \tag{2.15}$$

Accurate measurements of the shear–slip properties δ_f and τ_f are difficult and not practical for design use. It is thus desirable that the shear–slip properties in Equation 2.14 be related to simple measurements such as the concrete strength. The ultimate bond strength was related to the surface tensile strength of concrete by van Gemert (1980), Neubauer and Rostásy (1997), Holzenkämpfer (1994)

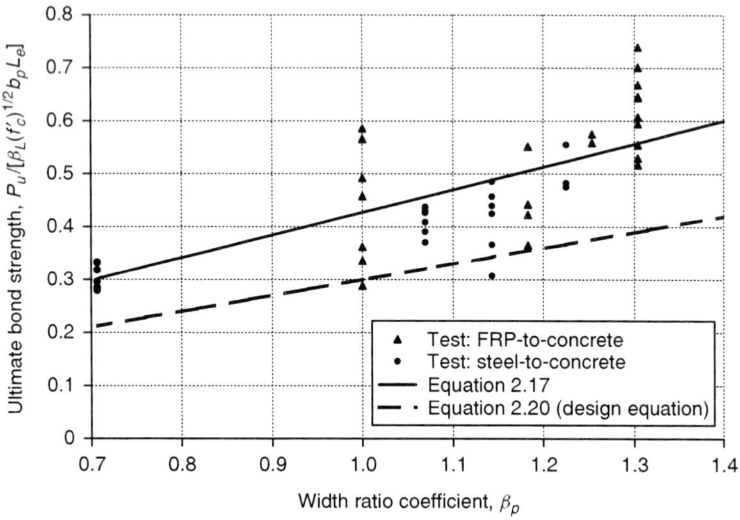

Figure 2.4 Effect of width ratio of bonded plate to concrete on ultimate bond strength

and to the shear strength of concrete by Maeda *et al.* (1997). However, various experimental observations (e.g. Chajes *et al.* 1996) showed that the ultimate bond strength may be related to $\sqrt{f'_c}$, similar to the bond strength of internal steel (e.g. BS 8110 1997) and FRP (e.g. Ehsani *et al.* 1996) reinforcement. Chen and Teng therefore approximated τ_f by the tensile strength of concrete which can be related to the compressive strength of concrete in the form of $\sqrt{f'_c}$ (e.g. through Equation 2.13). Noting that a typical value for δ_f is about 0.2 mm, the effective bond length L_e in Equation 2.14b was thus found to be approximately (Chen and Teng 2001b)

$$L_e = \sqrt{\frac{E_p t_p}{\sqrt{f'_c}}}, \quad \text{mm} \tag{2.16}$$

where $E_p t_p$ is in MPa mm and f'_c is in MPa.

By taking account of the above considerations, a simple ultimate bond strength model was proposed by Chen and Teng (2001b) based on Equation 2.14 and regression of test data:

$$P_u = 0.427\beta_p\beta_L\sqrt{f'_c}b_pL_e, \quad \text{N} \tag{2.17a}$$

where

$$\beta_L = \begin{cases} 1 & \text{if } L \geqslant L_e \\ \sin\left[\pi L/(2L_e)\right] & \text{if } L < L_e \end{cases} \tag{2.17b}$$

in which β_p and L_e are defined in Equation 2.15 and 2.16 respectively, and in Equation 2.17a L_e and b_p are in mm and f'_c is in MPa.

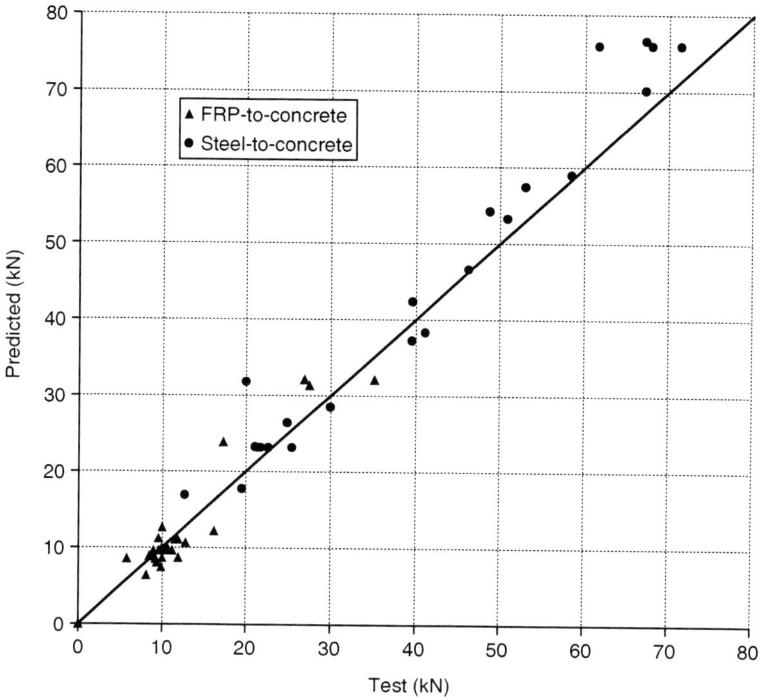

Figure 2.5 Test versus predicted bond strengths

This model is more convenient for practical application because it uses the cylinder compressive strength of concrete f_c' which is readily available in most cases. Table 2.1 and Figure 2.5 show that Equation 2.17 agrees well with test data for both FRP-to-concrete and steel-to-concrete bonded joints collected by Chen and Teng (2001b). The ratios of the test-to-predicted ultimate bond strength for the two types of joints have an overall average value of 1.00 and a corresponding standard deviation of 0.159.

Equation 2.16 shows that L_e increases linearly with $\sqrt{E_p t_p}$. The predictions by Chen and Teng's model for the effective bond length are in close agreement with the very limited experimental observations (Table 2.2). A comparison in Figure 2.6 shows that the empirical model proposed by Maeda *et al.* (1997) predicted the wrong trend for the effect of $E_p t_p$ on the effective bond length. Figure 2.7 shows the effect of bond length on the ultimate bond strength (Chen and Teng 2001b). The experimental data are nicely scattered around the curve predicted by Equation 2.17, statistically validating the important concept of effective bond length. Figure 2.8 shows the effect of the concrete strength on the bond strength. It may be noted that Figures 2.4 to 2.8 also show the bound of the relevant parameters within which Chen and Teng's model was verified.

Table 2.2 Effective bond length L_e (mm)

Data source	Test specimen	Measured (A)	Prediction			B/A
			Khalifa *et al.* (1998)	Neubauer and Rostásy (1997)	Chen and Teng (2001b) (B)	
Bizindavyi and Neale (1999)	BN1 (GFRP)	75	65.1	64.6	66.9	0.89
	BN2 (GFRP)	100	43.6	91.3	94.6	0.95
	BN3 (CFRP)	55	71.3	59.7	61.9	1.13
	BN4 (CFRP)	70	47.7	84.5	87.5	1.25
Täljsten (1997)	S100-40A to S800-80A (steel)	∼300	11.3	260–280	275–293	∼0.94

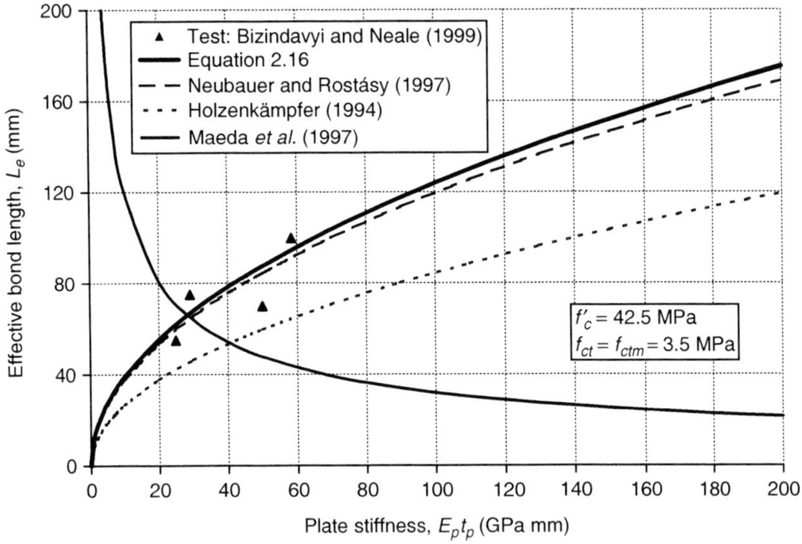

Figure 2.6 Effect of bonded plate stiffness on effective bond length

Substituting Equation 2.16 and $\sigma_{db} = P_{u/}(b_p t_p)$ into Equation 2.17a gives the stress in the bonded plate at failure:

$$\sigma_{db} = 0.427\beta_p\beta_L\sqrt{\frac{E_p\sqrt{f_c'}}{t_p}} = 0.4\beta_p\beta_L\sqrt{\frac{E_p\sqrt{f_{cu}}}{t_p}} \tag{2.18}$$

where the cube compressive strength $f_{cu} = 1.25f_c'$.

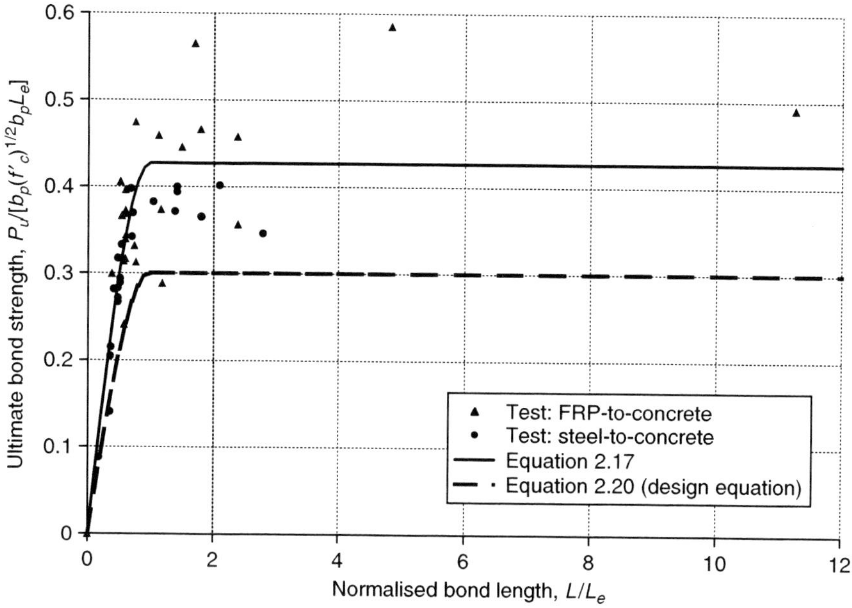

Figure 2.7 Effect of bond length on ultimate bond strength (Chen and Teng 2001b)

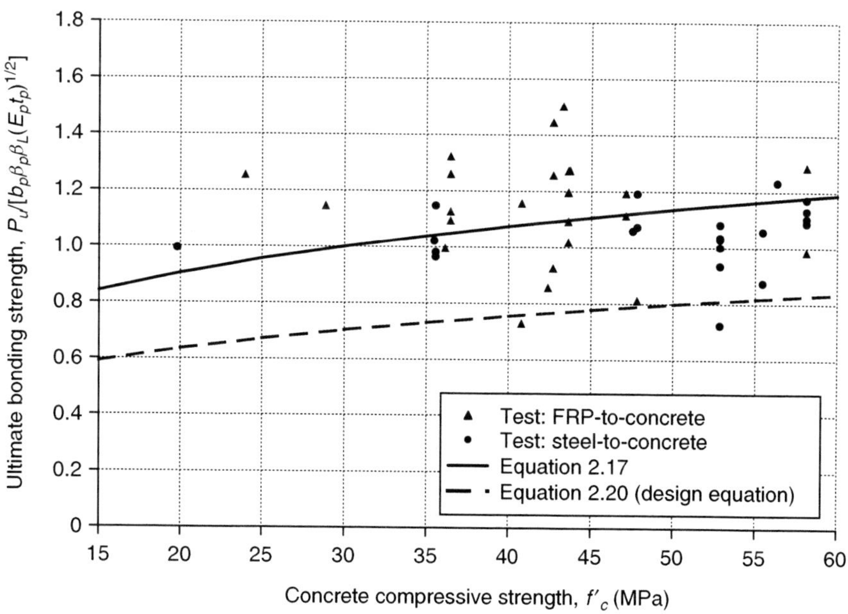

Figure 2.8 Effect of concrete strength on ultimate bond strength (Chen and Teng 2001b)

Equation 2.18 clearly shows that plates with a high elastic modulus and a small thickness should be used if a high stress needs to be reached in the bonded plate (concrete strength can hardly be changed in strengthening works). The ratio of the stress in the plate at bond failure to the plate tensile strength is given by

$$\frac{\sigma_{db}}{f_p} = \frac{0.427\beta_p\beta_L}{E_p\varepsilon_p}\sqrt{\frac{E_p\sqrt{f_c'}}{t_p}} = \frac{0.427\beta_p\beta_L}{\varepsilon_p}\sqrt{\frac{\sqrt{f_c'}}{E_pt_p}} \tag{2.19}$$

where f_p and ε_p are the ultimate strength and ultimate strain of FRP plates or yield strength and yield strain of steel plates respectively. Obviously, if two materials have similar ultimate strains (such as GFRP and CFRP), a thin plate made of the softer material should be used in order to utilise best the full tensile strength of the material, provided that the desired strength gain can still be achieved.

2.5 Design Recommendation

Based on their bond strength model, Chen and Teng (2001b) proposed a lower bound design equation. Their proposal is recommended here for use in practical design. They recommended that the coefficient in Equation 2.17 be reduced to the 95 percentile characteristic value of $0.427 \times (1 - 1.64 \times 0.159) = 0.315$ so that it can provide a conservative ultimate strength for design use:

$$P_u = \frac{0.315\beta_p\beta_L\sqrt{f_c'}b_pL_e}{\gamma_b} = \frac{0.3\beta_p\beta_L\sqrt{f_{cu}}b_pL_e}{\gamma_b} \tag{2.20}$$

where γ_b is the partial safety factor for bond strength as specified in Section 1.6.

In practical design, generally the designer needs to know the stress rather than the load carried by the FRP plate. The ultimate stress in the FRP plate can be found by replacing 0.427 with 0.315 in Equation 2.18:

$$\sigma_{db} = \frac{0.315\beta_p\beta_L}{\gamma_b}\sqrt{\frac{E_p\sqrt{f_c'}}{t_p}} = \frac{0.3\beta_p\beta_L}{\gamma_b}\sqrt{\frac{E_p\sqrt{f_{cu}}}{t_p}} \tag{2.21}$$

It is worth mentioning that the introduction of a separate partial safety factor γ_b instead of incorporating the partial safety factor for the compressive strength of concrete in Equation 2.21 is to achieve a consistent level of safety margin for both debonding and FRP tensile rupture.

2.6 Conclusions

In a number of cases of strengthening RC structures using externally bonded FRP plates, failure can occur owing to high interfacial stresses near a major crack away from the ends of the bonded plate. This problem has been idealised and studied as simple FRP-to-concrete bonded joints subject to tension. This chapter has been concerned with the bond behaviour and strength of such joints.

Existing test data suggest that the main failure mode is concrete failure under shear, leading to crack propagation parallel to the bonded plate adjacent to the adhesive-to-concrete interface, starting from the critically stressed position. The bond strength therefore depends on the concrete strength. In addition, the width ratio of the plate to the concrete member also has a significant effect. A very important aspect of bond behaviour is that there exists an effective bond length beyond which an extension of the bond length cannot increase the bond strength. This is a fundamental difference between the bond design of an externally bonded plate and that of an internal reinforcement for which a sufficiently long bond length can always be found so that the full tensile strength of the reinforcement can be achieved, provided a sufficient concrete cover is available. Thin plates with a low elastic modulus (e.g. GFRP plates) should be employed to make better use of the tensile strength of the bonded plate so long as the desired strength gain can still be achieved.

A number of bond strength models are available, but all of them except the recent model of Chen and Teng (2001b) do not predict existing test results closely. Deficiencies in these models are generally due to the omission of one or more of the important aspects mentioned above, such as the effective bond length limit. Chen and Teng's model, modified from an existing fracture mechanics model with suitable simplifications, captures all the main features of bond behaviour. Both the bond strength and effective bond length can be correctly predicted using this model. Design rules based on this model have been recommended for practical use.

It needs to be emphasised that Chen and Teng's bond strength model was developed from theoretical and experimental analysis of simple shear tests on idealised joints. When applied to a particular failure mode resembling the failure mode of such joints, such as intermediate crack-induced interfacial debonding in beams and slabs, some modification may be necessary because the stress states are not identical in the two problems. Such modification will be presented in Section 3.7.

References

Autocon (1994) *FORCA Tow Sheets Technical Notes*, Autocon Composites Incorporated, Ontario, Canada.

Bizindavyi, L. and Neale, K. W. (1999) 'Transfer lengths and bond strengths for composites bonded to concrete', *Journal of Composites for Constructions, ASCE*, Vol. 3, No. 4, pp. 153–160.

Blaschko, M., Niedermeier, R. and Zilch, K. (1998) 'Bond failure modes of flexural members strengthened with FRP', *Proceedings of the Second International Conference on Composites in Infrastructure (ICCI 1998)*, Tucson, Arizona, USA, edited by H. Saadatmanesh and M. R. Ehsani, pp. 315–327.

Brosens, K. and van Gemert, D. (1997) 'Anchoring stresses between concrete and carbon fibre reinforced laminates', *Non-Metallic (FRP) Reinforcement for Concrete Structures, Proceedings of the Third International Symposium*, Sapporo, Japan, pp. 271–278.

BS 8110 (1997) *Structural Use of Concrete, Part 1. Code of Practice for Design and Construction*, British Standards Institution, London, UK.

Chaallal, O., Nollet, M. J. and Perraton, D. (1998) 'Strengthening of reinforced concrete beams with externally bonded fibre-reinforced-plastic plates: design guidelines for shear and flexure', *Canadian Journal of Civil Engineering*, Vol. 25, No. 4, pp. 692–704.

Chajes, M. J., Januszka, T. F., Mertz, D. R., Thomson, T. A., Jr and Finch, W. W., Jr (1995) 'Shear strengthening of reinforced concrete beams using externally applied composite fabrics', *ACI Structural Journal*, Vol. 92, No. 3, pp. 295–303.

Chajes, M. J., Finch, W. W., Jr, Januszka, T. F. and Thomson, T. A., Jr (1996) 'Bond and force transfer of composite material plates bonded to concrete', *ACI Structural Journal*, Vol. 93, No. 2, pp. 295–303.

Chen, J. F. and Teng, J. G. (2001a) 'Shear capacity of FRP strengthened RC beams: FRP debonding', *to be published*.

Chen, J. F. and Teng, J. G. (2001b) 'Anchorage strength models for FRP and steel plates attached to concrete', *Journal of Structural Engineering, ASCE*, Vol. 127, No. 7, pp. 784–791.

Chen, J. F., Yang, Z. J. and Holt, G. D. (2001) 'FRP or steel plate-to-concrete bonded joints: effect of test methods on experimental bond strength', *Steel and Composite Structures–An International Journal*, Vol. 1, No. 2, pp. 231–244.

Deutsches Institut für Normung (1991) *DIN 1048, Augsgabe 6.91, Teil 2: Prüfverfahren für Beton, Festbeton in Bauwerken und Bauteilen*, Beuth Verlag, Berlin, Germany.

Ehsani, M. R., Saadatmanesh, H. and Tao, S. (1996) 'Design recommendations for bond of GFRP rebars to concrete', *Journal of Structural Engineering, ASCE*, Vol. 122, No. 3, pp. 247–254.

Fukuzawa, K., Numao, T., Wu, Z., Yoshizawa, H. and Mitsui, M. (1997) 'Critical strain energy release rate of interface debonding between carbon fibre sheet and mortar', *Non-Metallic (FRP) Reinforcement for Concrete Structures, Proceedings of the Third International Symposium*, Sapporo, Japan, pp. 295–302.

Hiroyuki, Y. and Wu, Z. (1997) 'Analysis of debonding fracture properties of CFS strengthened member subject to tension', *Non-Metallic (FRP) Reinforcement for Concrete Structures, Proceedings of the Third International Symposium*, Sapporo, Japan, pp. 287–294.

Holzenkämpfer, O. (1994) *Ingenieurmodelle des Verbundes geklebter Bewehrung für Betonbauteile*, Dissertation, TU Braunschweig.

Horiguchi, T. and Saeki, N. (1997) 'Effect of test methods and quality of concrete on bond strength of CFRP sheet', *Non-Metallic (FRP) Reinforcement for Concrete Structures, Proceedings of the Third International Symposium*, Sapporo, Japan, pp. 475–482.

Khalifa, A., Gold, W. J., Nanni, A. and Aziz, A. (1998) 'Contribution of externally bonded FRP to shear capacity of RC flexural members', *Journal of Composites for Construction, ASCE*, Vol. 2, No. 4, pp. 195–203.

Kobatake, Y., Kimura, K. and Katsumata, H. (1993) 'A retrofitting method for reinforced concrete structures using carbon fibre', *Fibre-Reinforced-Plastic (FRP) Reinforcement for Concrete Structures: Properties and Applications*, edited by A. Nanni, Elsevier, Amsterdam, The Netherlands, pp. 435–450.

Kong, F. K. and Evans, R.H. (1987) *Reinforced and Prestressed Concrete*, Third Edition, Chapman & Hall, London, UK.

MacGregor, J. G. (1988) *Reinforced Concrete: Mechanics and Design*, Prentice Hall, Englewood Cliff, New Jersey, USA.

Maeda, T., Asano, Y., Sato, Y., Ueda, T. and Kakuta, Y. (1997) 'A study on bond mechanism of carbon fiber sheet', *Non-Metallic (FRP) Reinforcement for Concrete Structures, Proceedings of the Third International Symposium*, Sapporo, Japan, pp. 279–285.

Neubauer, U. and Rostásy, F. S. (1997) 'Design aspects of concrete structures strengthened with externally bonded CFRP plates', *Proceedings of the Seventh International Conference on Structural Faults and Repairs*, edited by M. C. Forde, Engineering Technics Press, Edinburgh, UK, pp. 109–118.

Niedermeier, R. (1996) 'Stellungnahme zur Richtlinie für das Verkleben von Betonbauteilen durch Ankleben von Stahllaschen–Entwurf März 1996', *Schreiben Nr 1390 vom 30.10.1996 des Lehrstuhls für Massivbau*, TU München.

Roberts, T. M. (1989) 'Approximate analysis of shear and normal stress concentrations in the adhesive layer of plated RC beams', *The Structural Engineer*, Vol. 67, No. 12, pp. 229–233.

Sato, Y., Ueda, T., Kakuta, Y. and Tanaka, T. (1996) 'Shear reinforcing effect of carbon fibre sheet attached to side of reinforced concrete beams', *Advanced Composite Materials in Bridges and Structures, Proceedings of the Second International Conference*, Quebec, Canada, edited by M. M. El-Badry, pp. 621–627, Canadian Society for Civil Engineering, Montreal, Canada.

Swamy, R. N., Jones, R. and Charif, A. (1986) 'Shear adhesion properties of epoxy resin adhesives', *Adhesion Between Polymers and Concrete, Proceedings of an International Symposium*, edited by H. R. Sasse, Paris, France, pp. 741–755.

Täljsten, B. (1994) *Plate Bonding. Strengthening of Existing Concrete Structures with Epoxy Bonded Plates of Steel or Fibre Reinforced Plastics*, Doctoral Thesis, Luleå University of Technology.

Täljsten, B. (1997) 'Defining anchor lengths of steel and CFRP plates bonded to concrete', *International Journal of Adhesion and Adhesives*, Vol. 17, No. 4, pp. 319–327.

Tanaka, T. (1996) *Shear Resisting Mechanism of Reinforced Concrete Beams with CFS as Shear Reinforcement*, Graduation Thesis, Hokkaido University.

Triantafillou, T. C. and Plevris, N. (1992) 'Strengthening of RC beams with epoxy-bonded fibre-composite materials', *Materials and Structures*, Vol. 25, pp. 201–211.

van Gemert, D. (1980) 'Force transfer in epoxy-bonded steel-concrete joints', *International Journal of Adhesion and Adhesives*, No. 1, pp. 67–72.

Varastehpour, H. and Hamelin, P. (1996). 'Analysis and study of failure mechanism of RC beams strengthened with FRP plate', *Advanced Composite Materials in Bridges and Structures, Proceedings of the Second International Conference*, Quebec, Canada, edited by M. M. El-Badry, pp. 527–537, Canadian Society for Civil Engineering, Montreal, Canada.

Volnyy, V. A. and Pantelides, C. P. (1999) 'Bond length of CFRP composites attached to precast concrete walls', *Journal of Composites for Construction, ASCE*, Vol. 3, No. 4, pp. 168–176.

Yuan, H. and Wu, Z. (1999) 'Interfacial fracture theory in structures strengthened with composite of continuous fiber', *Proceedings of Symposium of China and Japan, Science and Technology of 21st Century*, Tokyo, Japan, pp. 142–155.

Yuan, H., Wu, Z. S. and Yoshizawa, H. (2001) 'Theoretical solutions on interfacial stress transfer of externally bonded steel/composite laminates', *Journal of Structural Mechanics and Earthquake Engineering, JSCE*, No. 675/1–55, pp. 27–39.

Ziraba, Y. N., Baluch, M. H., Basunbul, A. M., Azad, A. K., Al-Sulaimani, G. J. and Sharif, I. A. (1995) 'Combined experimental–numerical approach to characterization of steel-glue-concrete interface', *Materials and Structures*, Vol. 28, pp. 518–525.

3 Flexural Strengthening of Beams

3.1 Introduction

Traditional methods which are commonly used for the flexural strengthening of RC beams include external post-tensioning and the bonding of steel plates. These methods suffer from inherent disadvantages ranging from difficult application procedures to lack of durability. In recent years, the bonding of FRP plates or sheets has become a very popular method for the flexural strengthening of RC beams. As a matter of fact, the application of FRPs to the strengthening of structures was first researched in the middle 1980s for the flexural strengthening of RC beams using CFRP plates at the Swiss Federal Laboratory for Materials Testing and Research (EMPA) (Meier *et al.* 1993). An RC beam with an FRP plate bonded to its soffit is shown in Figure 3.1.

Most existing research on FRP plate bonding for flexural strengthening has been carried out in the last decade (e.g. Ritchie *et al.* 1991, Saadatmanesh and Ehsani 1991, Triantafillou and Plevris 1992, Chajes *et al.* 1994, Sharif *et al.* 1994, Meier 1995, Quantrill *et al.* 1996a, Shahawy *et al.* 1996, Takeda *et al.* 1996, Arduini and Nanni 1997, Garden *et al.* 1997, 1998, Varastehpour and Hamelin 1997, Buyukozturk and Hearing 1998, GangaRao and Vijay 1998, Garden and Hollaway 1998a, Grace *et al.* 1998, Spadea *et al.* 1998, Grace *et al.* 1999, Ross *et al.* 1999, Swamy and Mukhopadhyaya 1999, Vecchio and Bucci 1999, Bonacci and Maalej 2000, Ramana *et al.* 2000, Spadea *et al.* 2000, Mukhopadhyaya and Swamy 2001, Nguyen *et al.* 2001, Rahimi and Hutchinson 2001), with the research effort experiencing an explosive growth in recent years as a result of great worldwide needs for structural strengthening works and reductions in the cost of FRPs. Prior to this research on FRP plate bonding, considerable research had already accumulated on steel plate bonding for flexural strengthening (e.g.

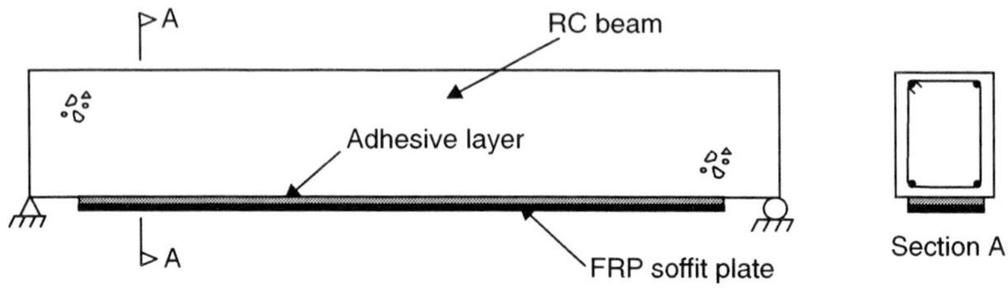

Figure 3.1 RC beam bonded with an FRP soffit plate

Jones *et al.* 1980, 1988, Macdonald and Calder 1982, Swamy *et al.* 1982, 1987, 1989) and this effort has continued through the 1990s (e.g. Basunbul *et al.* 1990, Oehlers and Moran 1990, Oehlers 1992, Hussain *et al.* 1995, Ashrafuddin *et al.* 1999). Much of what was learnt in research on steel plate bonding is relevant to FRP plate bonding.

This chapter is concerned with the behaviour and strength of RC beams strengthened with FRP soffit plates. The focus of the chapter is on RC beams strengthened with unstressed plates without mechanical anchors, as these are more common in practical applications, although a brief discussion is also given of prestressed plates and different anchorage systems. The chapter starts with a summary of different methods of FRP plate bonding, followed by a description of the different failure modes and typical behaviour of such beams. Existing research is then summarised, with a particular emphasis on suitable strength models for various failure modes. A rational design procedure is finally given, catering for various possible failure modes, including those involving premature debonding between the FRP plate and the RC beam. It should be mentioned that all discussions are on simply supported beams which have been the predominant concern of existing research, although the proposed design procedure can also be applied to indeterminate beams.

3.2 Methods of Strengthening

3.2.1 General

Flexural strengthening of a simply supported RC beam using FRP composites is generally by bonding an FRP plate to the soffit of the beam. Prior to the application of the FRP plate, the soffit of the RC beam must be prepared. The purpose of adequate surface preparation is to remove the weak surface layer of the concrete, expose the concrete aggregate to improve the bond with the FRP, and provide an even surface. The FRP plate may be a prefabricated (e.g. pultruded) plate, in which case some preparation of the bonding surface of the FRP plate may be necessary. On the other hand, the FRP plate may be constructed on site in a wet lay-up process.

There are a number of variations of the basic procedure. These include prestressing of the plate, and the provision of anchors such as U strips at the plate ends to reduce the risk of debonding failures. Further details of these variations are given in the rest of this section.

3.2.2 Unstressed Soffit Plates

The bonding of unstressed FRP plates to the soffit of RC beams is the most common and has received the greatest amount of theoretical and experimental research to date. Three schemes exist for the adhesion of unstressed FRP plates to the soffit of an RC beam. They consist of (a) adhesive bonding of prefabricated/pultruded FRP plates, (b) wet lay-up, and (c) resin infusion.

In the first approach, prefabricated or pultruded FRP plates are cut to the required size and bonded to the soffit of an RC beam. Bonding of pultruded plates ensures the highest degree of material uniformity and quality control and has been used by Saadatmanesh and Ehsani (1991), Quantrill *et al.* (1996a), Garden *et al.* (1997, 1998), Ross *et al.* (1999) and Nguyen *et al.* (2001) to name a few.

Wet lay-up is perhaps the most commonly used and gives the greatest flexibility for field use. It is also the cheapest (Karbhari and Zhao 1998). Resin is applied to the concrete surface and layers of fabric are impregnated in place using rollers. Here the adhesive also forms the matrix of the FRP and this creates a strong bond with the RC beam. This method is, however, sensitive to unevenness of the RC beam soffit and such unevenness can lead to debonding (Meier 1995) as discussed in Section 3.3.7.1. This plating technique has been used in many experimental studies including those by Chajes *et al.* (1994), Arduini and Nanni (1997) and Ahmed and van Gemert (1999).

The resin infusion method (Varastehpour and Hamelin 1997, Karbhari and Zhao 1998) has similar characteristics to those of wet lay-up but is the least commonly used.

3.2.3 End Anchorage for Unstressed Soffit Plates

To prevent debonding at the ends of the soffit plate, mechanical anchors can be installed. Prefabricated steel or FRP U strips (Figure 3.2), a common type

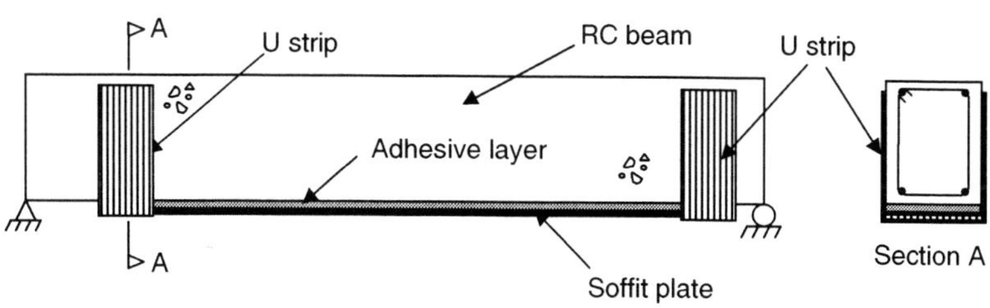

Figure 3.2 Plated RC beam with FRP U strip end anchorage

of end anchorage, can be bonded or bolted (Ritchie *et al.* 1991, Swamy and Mukhopadhyaya 1999) to the ends of the plate. The use of steel bolts to anchor the ends of the soffit plate directly to the concrete has also been reported (Garden and Hollaway 1998a). FRP end anchorage strips can also be formed by wet lay-up and they can be completely wrapped (GangaRao and Vijay 1998) or partially wrapped (van Gemert *et al.* 1999) around the RC beam near the ends of the plate.

Experimental studies on mechanical anchors have been carried out by Ritchie *et al.* (1991), Sharif *et al.* (1994), Spadea *et al.* (1998), van Gemert *et al.* (1999), Swamy and Mukhopadhyaya (1999) and Smith and Teng (2001a) to name a few. A more detailed summary of different anchorage schemes and experimental investigations is given in Hollaway and Mays (1999).

Experimental investigations have established that mechanical anchors delay, if not prevent, the onset of debonding (Sharif *et al.* 1994, Garden and Hollaway 1998a) although the increase in the plate-end debonding load appears to be limited (Smith and Teng 2001a). Tentative models have been proposed (e.g. van Gemert *et al.* 1999) to predict the increase in strength offered by mechanical anchors. Such models are semi-empirical in nature and were calibrated with a limited number of tests. Considerably more research is required in this area before reliable strength models emerge for such anchors.

3.2.4 Prestressed Soffit Plates

A pultruded FRP plate can be bonded to the RC beam soffit with prestressing. The main advantage of prestressing the FRP plate is that the bonded plate contributes to the load-carrying capacity before additional loading is applied on the structure. The prestressing also contributes to the reduction of crack widths if this is an important issue. Since FRPs have high tensile strengths, prestressing leads to more efficient use of them as a prestressed FRP plate is more likely to reach its ultimate tensile strength at failure. Prestressing, however, is labour intensive, with special care required for the design and construction of the anchorage system. This technique is still in its infancy as there have been limited experimental studies or field applications. A detailed review of existing research on RC beams strengthened with prestressed FRP soffit plates can be found in Hollaway and Leeming (1999).

Figure 3.3 shows part of the sequence of application of a prestressed FRP soffit plate to an RC beam in a laboratory environment (Garden and Hollaway 1998b, Garden and Mays 1999). Following the preparation of the beam soffit, the plate is stressed to the required level. The adhesive is then applied to both the plate and the RC beam soffit. The RC beam is lifted until it is pressed against the stressed plate. The adhesive is now left to cure with a small amount of pressure being applied to the plate. After the adhesive has cured, steel clamps are installed near each end of the soffit plate to ensure adequate anchorage of the plate prior to the release of the pretension by cutting the FRP just outside the end clamps. Thick FRP plates are then bonded on top of the soffit plate just inside the steel clamps.

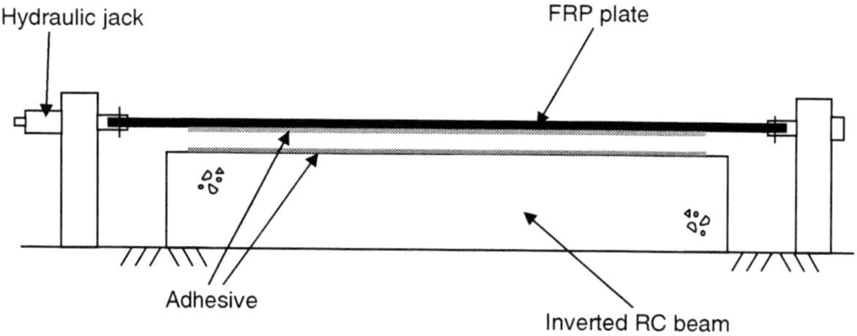

Figure 3.3 Procedure of prestressing an FRP plate. Reprinted from Garden and Holl-away, Copyright (1998b), with permission from Elsevier Science

After this adhesive has cured, holes are drilled through the end plates and the soffit plate into the RC beam where bolts are then inserted into the holes prefilled with epoxy. Upon curing of the bolt adhesive, the steel clamps are removed, and the excess length of the soffit plate just outside the end plates is also removed.

Garden and Mays (1999) described the strengthening of a series of prestressed beams with stressed soffit plates *in situ*. Here the soffit plate was cut to a predefined length and predrilled anchor plates were bonded to the ends. The plate was then positioned against the RC beam soffit and a prestressing device mounted onto the concrete beam. Once the required load was applied to the plate, holes were drilled into the RC beam and the soffit plate using the anchor plates as templates and bolts then fitted. The plate was then pressed against a freshly spread adhesive layer on the beam soffit. A small upward camber was initially present in the RC beam since it was already prestressed, and this allowed room for the easy addition of an adhesive layer. The pressure was maintained until the adhesive had cured, upon which the prestressing device was removed. An alternative form of prestressing was used by Saadatmanesh and Ehsani (1991) where an RC beam was cambered prior to the application of an unstressed FRP soffit plate. The plate became stressed when the camber was released.

3.2.5 Additional Remarks

In the remainder of the chapter, only RC beams bonded with unstressed FRP plates without mechanical anchors are discussed. This restriction is adopted because they are the dominant concern of existing research and are widely used in practice. Prestressed FRP plates are excluded from further discussion as there have been limited existing research on and applications of prestressed FRP plates. Mechanical end anchors are an important topic, but again the limited amount of research available does not allow an in-depth treatment at present. Nevertheless, observations of failures made in recent tests on RC beams bonded with unstressed soffit plates anchored at their ends by U strips (Smith and Teng 2001a) are summarised in Section 3.3.8 to provide some insight into the behaviour of such beams. In practical applications, mechanical end anchorage should be considered whenever necessary and convenient, and its effectiveness should be verified by prototype testing. In many applications, mechanical end anchorage in the form of U strips may not be possible or effective, such as in the case of a wide beam or slab bonded with narrow strips in the middle. In such cases, alternative forms of mechanical end anchorage should be investigated. One effective alternative is the fibre anchors used by Teng *et al.* (2000a) in the strengthening of cantilever slabs. The use of such anchors in the flexural strengthening of RC beams does not appear to have been studied.

3.3 Failure Modes and Typical Behaviour

3.3.1 Classification of Failure Modes

A number of failure modes for RC beams bonded with FRP soffit plates have been observed in numerous experimental studies to date (e.g. Ritchie *et al.* 1991, Saadatmanesh and Ehsani 1991, Triantafillou and Plevris 1992, Chajes *et al.* 1994, Sharif *et al.* 1994, Heffernan and Erki 1996, Shahawy *et al.* 1996, Takeda *et al.* 1996, Arduini and Nanni 1997, Garden *et al.* 1997, 1998, Grace *et al.* 1998, Ross *et al.* 1999, Bonacci and Maleej 2000, Nguyen *et al.* 2001, Rahimi and Hutchinson 2001). A schematic representation of typical failure modes observed in tests is shown in Figure 3.4. The failure modes are classified into seven main categories in the figure and are termed (a) flexural failure by FRP rupture, (b) flexural failure by crushing of compressive concrete, (c) shear failure, (d) concrete cover separation, (e) plate-end interfacial debonding, (f) intermediate flexural crack-induced interfacial debonding, and (g) intermediate flexural shear crack-induced interfacial debonding. Collectively, failure modes (d) and (e) are referred to as plate-end debonding failures, while failure modes (f) and (g) are referred to as intermediate crack-induced interfacial debonding failures.

All the failure modes identified, with the exception of intermediate crack-induced interfacial debonding with a long process of debonding propagation, occur in a brittle manner. In particular, debonding at the plate ends occurs with little or no indication given of failure. Whether a particular mode is critical for a given beam depends on numerous parameters such as the amount of internal flexural and shear reinforcement, and geometric and material properties of the FRP plate, the adhesive layer and the RC beam. Each of the failure modes is discussed in some detail in the rest of this section.

3.3.2 Flexural Failure

If the ends of the plate are properly anchored, the ultimate flexural capacity of the beam is reached when either the FRP plate fails by tensile rupture (Figures 3.4a

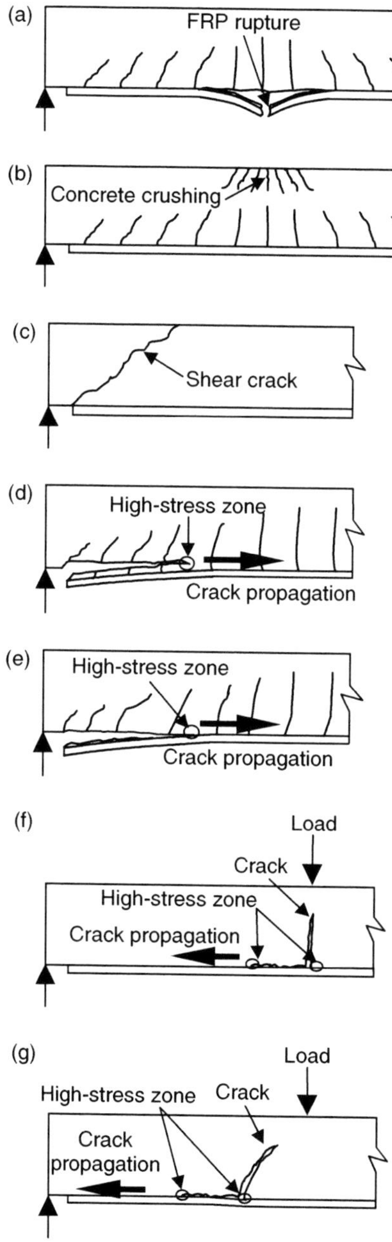

Figure 3.4 Failure modes of FRP-plated RC beams: (a) FRP rupture; (b) crushing of compressive concrete; (c) shear failure; (d) concrete cover separation; (e) plate-end interfacial debonding; (f) intermediate flexural crack-induced interfacial debonding; (g) intermediate flexural shear crack-induced interfacial debonding

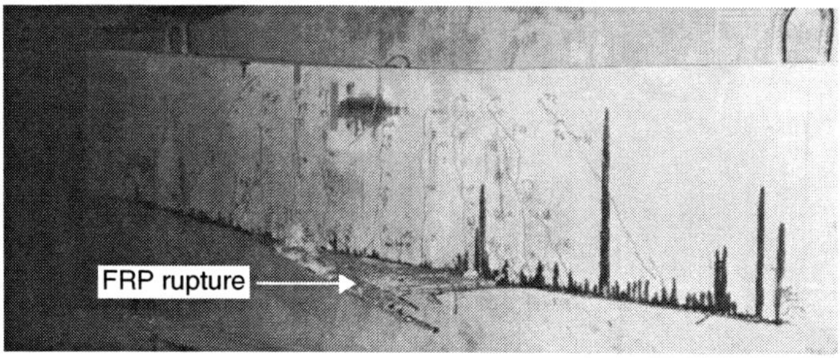

Figure 3.5 FRP-plated RC beam: FRP rupture

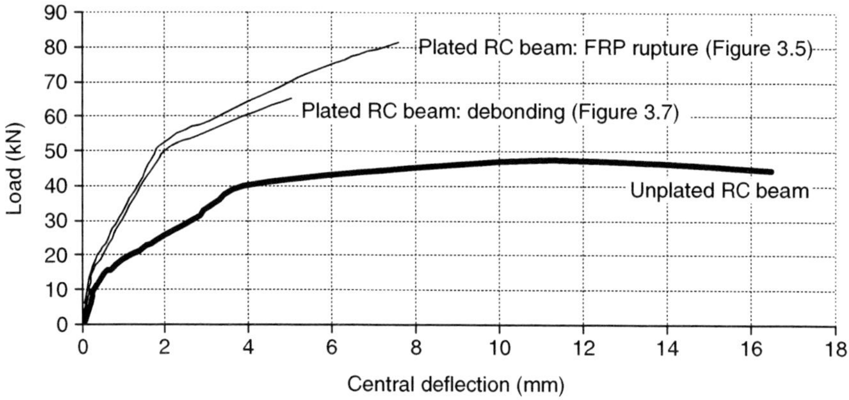

Figure 3.6 Typical load–deflection curves of plated and unplated RC beams

and 3.5) or the compressive concrete is crushed (Figure 3.4b). This is very similar to the classical flexural failure of RC beams, except for small differences due to the brittleness of the bonded FRP plate. FRP rupture generally occurs following the yielding of the longitudinal steel bars, although steel yielding may not have been reached if the steel bars are located quite far away from the tension face.

Figure 3.6 shows a typical load versus mid-span deflection response of a simply supported RC beam bonded with an FRP plate subjected to four-point bending (Hau 1999). For this particular beam, the plate was terminated very close to the support and premature debonding did not occur. Instead, the beam failed in flexure by FRP rupture. Compared with the corresponding response of an unplated control RC beam (Figure 3.6), the plated beam achieved a strength gain of 76 %, but showed much reduced ductility. The strength gain and the ductility reduction are the two main consequences of flexural strengthening of RC beams using FRP plates. Beams which fail by concrete crushing when a large amount of FRP is used also show much reduced ductility (Buyukozturk and Hearing 1998).

3.3.3 Shear Failure

The strengthened beam can fail brittlely in shear (Buyukozturk and Hearing 1998, Lopez *et al.* 1999) as shown in Figure 3.4c. While normal RC beams are designed to fail in flexure rather than in shear which is a brittle mode, the shear failure mode can be made critical by flexural strengthening. In such cases, the shear capacity of the RC beam alone dictates its strength in shear as the FRP soffit plate provides little contribution to shear resistance, as discussed in Section 4.1. In such situations, shear strengthening of the RC beam should be carried out simultaneously to ensure that the required flexural strength is not compromised by shear failure and that flexural failure still precedes shear failure. The latter is desirable as the flexural failure mode of a plated RC beam, although brittle, is still more ductile than its shear failure mode.

3.3.4 Plate-end Debonding Failures

Premature failure may occur before the ultimate flexural capacity of the beam is reached owing to debonding. The most commonly reported mode of debonding in experimental studies has been the separation of the concrete cover (Figure 3.4d) that initiates at or near one of the two ends of the soffit plate (referred to as the end of the soffit plate, the end of the plate or the plate end for simplicity hereafter, unless when the two ends need to be differentiated from each other). A less commonly observed debonding mode that also initiates at or near the plate end is the separation of the FRP plate from the RC beam soffit (Figure 3.4e). These two modes of debonding are discussed in the following. In real beams, the two may be mixed together as seen in the beam tested by Hau (1999) and shown in Figure 3.7. This test beam was bonded with an FRP plate of the same width

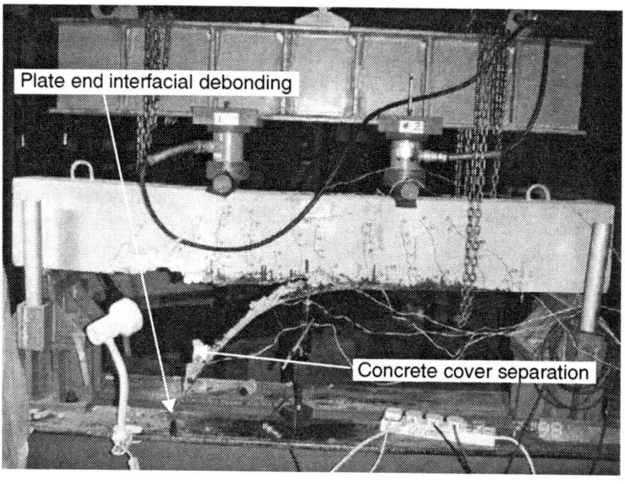

Figure 3.7 FRP-plated RC beam: debonding in a mixed mode (plate-end debonding in combination with concrete cover separation)

and thickness as the other beam shown in Figure 3.5, but the length of the FRP plate was reduced (Hau 1999). The load–deflection response of this beam shows little ductility (Figure 3.6), which is typical of plate-end debonding failures.

3.3.4.1 Concrete Cover Separation

This is the most commonly reported premature failure mode (e.g. Ritchie *et al.* 1991, Sharif *et al.* 1994, Garden *et al.* 1997, 1998, Tumialan *et al.* 1999a, Nguyen *et al.* 2001). This mode of failure has been referred to elsewhere as 'end-of-plate failure through concrete' (Ritchie *et al.* 1991), 'concrete rip-off failure' (Sharif *et al.* 1994, Nguyen *et al.* 2001), 'debond at rebar layer' (Buyukozturk and Hearing 1998), 'local shear failure' (Saadatmanesh and Malek 1998), and 'concrete cover delamination' (Tumialan *et al.* 1999a). A typical picture of such failure is shown in Figure 3.8. As the failure occurs away from the bondline, this

(a)

(b)

Figure 3.8 FRP-plated RC beam, concrete cover separation: (a) overall view; (b) close-up view

is not a debonding failure mode in strict terms, although it does stem from stress concentration near the ends of the bonded plate. The term *concrete cover separation* previously used by Garden and Hollaway (1998a) and Juvandes *et al.* (1998) appears to describe this mode of failure most accurately and is recommended for use here (Figure 3.4d).

It is generally believed that failure of the concrete cover is initiated by the formation of a crack at or near the plate end, due to high interfacial shear and normal stresses caused by the abrupt termination of the plate here (e.g. Saadatmanesh and Malek 1998, Tumialan *et al.* 1999b, Mukhopadhyaya and Swamy 2001). A detailed discussion of these interfacial stresses is given in Section 3.5. Once a crack forms in the concrete at or near the end of the plate, the crack propagates to the level of the tension reinforcement and then progresses horizontally, along the level of the steel tension reinforcement, resulting in the separation of the concrete cover (Garden and Hollaway 1998a). Figure 3.8(b) shows a close-up view of the detached plate end, where the flexural tension reinforcement of the beam can be clearly seen.

A careful study on concrete cover separation has recently been undertaken by Smith and Teng (2001a) through tests on beams in three-point bending, where one of the issues examined was the location of the plate end. Figure 3.9 shows a series of photographs taken at failures of these beams, where a is the distance between the plate end and the nearer support while B is the shear span. It was found from these tests that as the plate end moved closer to the support, the crack at the plate end became more inclined (Figure 3.9), the failure process became more brittle and the failure load increased. These changes are consequences of the shear force becoming more important relative to the bending moment at the plate end.

3.3.4.2 Plate-end Interfacial Debonding

Debonding between the adhesive and the beam (Figure 3.4e) that propagates from the end of the plate has been observed in experiments (e.g. Saadatmanesh and Ehsani 1991, Quantrill *et al.* 1996a, Täljsten 1999) and is herein referred to as *plate-end interfacial debonding*. The load–deflection response for this mode of failure is similar to that of failure by concrete cover separation.

The general consensus among researchers is that debonding failures of this form are initiated by high interfacial shear and normal stresses near the end of the plate that exceed the strength of the weakest element, generally the concrete (e.g. Varastehpour and Hamelin 1997, Täljsten 1999). Upon debonding, a thin layer of concrete generally remains attached to the plate (Täljsen 1999). This suggests that failure generally occurs in the concrete adjacent to the concrete-to-adhesive interface.

The risk of plate-end interfacial debonding is increased by inferior adhesives and uneven concrete surfaces. These issues are discussed in Section 3.3.7.

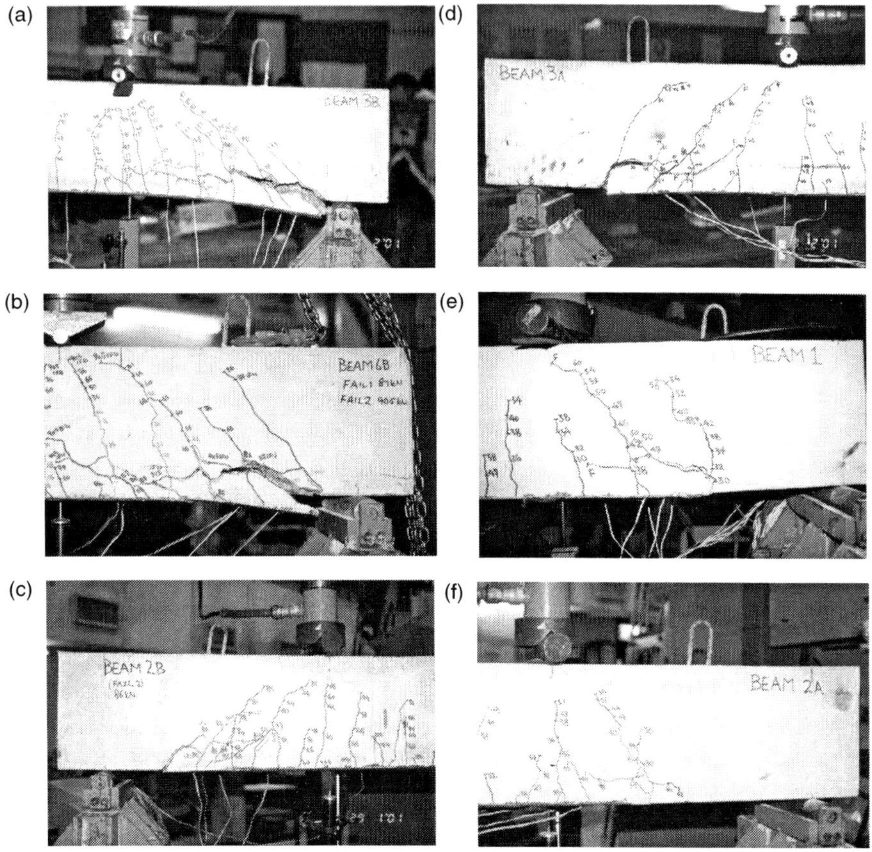

Figure 3.9 Cover separation failures of FRP-plated RC beams with varying plate termination-to-support distance for various values of a/B: (a) 0.10; (b) 0.15; (c) 0.25; (d) 0.35; (e) 0.50; (f) 0.75

3.3.5 Intermediate Crack-induced Interfacial Debonding

Debonding may initiate at a flexural or a mixed flexural shear crack away from the plate ends and then propagates towards one of the plate ends as shown in Figures 3.4f and 3.4g respectively. Debonding generally occurs in the concrete, adjacent to the adhesive-to-concrete interface, and is referred to herein as *intermediate flexural crack-induced interfacial debonding* or *intermediate flexural shear crack-induced interfacial debonding*. A typical picture of flexural crack-induced interfacial debonding is shown in Figure 3.10. A thin layer of concrete remained attached to the plate (Figure 3.10) and this suggests that failure occurred in the concrete, adjacent to the adhesive-to-concrete interface. It should be noted that for this beam, plate separation was forced to propagate towards the right end of the plate by the U jacket on the other side of the mid-span.

Figure 3.10 FRP-plated RC beam: intermediate flexural crack-induced interfacial debonding

Intermediate flexural crack-induced interfacial debonding has been observed in a limited number of experimental studies (e.g. Saadatmanesh and Ehsani 1991, Lopez *et al.* 1999, Tumialan *et al.* 1999a, Bonacci and Maleej 2000) as well as intermediate flexural shear crack-induced interfacial debonding (e.g. Triantafillou and Plevris 1992, Garden *et al.* 1998, Rahimi and Hutchinson 2001). Intermediate crack-induced interfacial debonding failures appear to be more likely in shallow beams and more ductile than plate-end debonding failures.

3.3.5.1 Intermediate Flexural Crack-induced Interfacial Debonding

The mechanism of intermediate flexural crack-induced debonding may be summarised as follows. When a major crack is formed in the concrete, the tensile stresses released by the cracked concrete are transferred to the FRP plate. As a result, high local interfacial stresses between the FRP plate and the concrete are induced near the crack. As the applied loading increases further, the tensile stresses in the plate and hence the interfacial stresses between the FRP plate and the concrete near the crack also increase. When these stresses reach critical values, debonding initiates at the crack and then propagates towards one of the plate ends, generally the nearer end.

3.3.5.2 Intermediate Flexural Shear Crack-induced Interfacial Debonding

For debonding induced by a flexural crack, the widening of the crack is the driving force for the propagation of debonding. However, in an intermediate flexural shear crack-induced debonding failure, the relative vertical displacement between the two faces of the crack produces peeling stresses on the FRP plate-to-concrete interface (Triantafillou and Plevris 1992, Meier 1995, Swamy and Mukhopadhyaya 1999, Rahimi and Hutchinson 2001). Nevertheless, it is believed by the present authors that the widening of the crack is normally the more important factor, with the relative movement of the crack faces being secondary.

3.3.6 Other Debonding Failure Modes

Debonding failures involving a mixture of the distinct modes described above can also occur. Hau (1999) observed a mixed mode of debonding which was a combination of concrete cover separation and plate-end interfacial debonding (Figure 3.7). Tumialan *et al.* (1999a) reported a mixed mode of failure where concrete cover separation occurred at an intermediate crack, but at a small distance away from the crack, interfacial debonding occurred between the adhesive and the concrete. Interlaminar shear failure within the FRP plate is another possible mode of failure (Leeming and Darby 1999) but this mode does not appear to have been reported in tests to date.

3.3.7 Other Aspects of Debonding

The risk of debonding is increased by a number of factors associated with the quality of on-site application. These include poor workmanship and the use of inferior adhesives. The effects of these factors can be minimised if due care is exercised in the application process.

3.3.7.1 Concrete Surface Preparation

The concrete surface should be properly prepared to avoid failure at the adhesive-to-concrete interface by following a suitable procedure (Hutchinson and Quinn 1999). An important requirement for good concrete surface preparation is that uneven concrete surfaces are evened out (Meier 1995). When a thin FRP plate bonded to an uneven concrete surface is subject to tension, it tries to straighten itself, which results in interfacial peeling stresses that can cause debonding in the concrete adjacent to the adhesive-to-concrete interface (Meier 1995). FRPs formed in a wet lay-up process are more sensitive to surface unevenness, as their profile follows the uneven surface.

3.3.7.2 Adhesive

Strong adhesives are available for FRP plate bonding, and their strength generally exceeds that of concrete so failure in the adhesive is rare. However, if substandard adhesives are used or if adhesives are not properly applied, adhesive failure may occur within the adhesive layer, adjacent to the adhesive-to-concrete interface, and adjacent to the FRP plate-to-adhesive interface (Meier 1995, Karbhari and Zhao 1998, Leeming and Darby 1999).

3.3.7.3 FRP Surface Preparation

For pultruded plates, failure can also occur at the FRP-to-adhesive interface if the surface of the FRP plate is not properly prepared (e.g. inadequate removal of impurities such as grease from the FRP plate surface) (Leeming and Darby 1999).

3.3.8 U Strip End Anchorage

As mentioned in Section 3.2.3, mechanical anchors can be installed at the plate ends to delay or prevent plate-end debonding failures. Smith and Teng (2001a) recently reported a series of tests on FRP-plated beams with end U strip anchorage such as that shown in Figure 3.11(a). All five tests reported by Smith and Teng (2001a) were for RC beams of the same cross-sectional geometry, reinforcement, loading, and length of FRP plate, but the width and the placement of

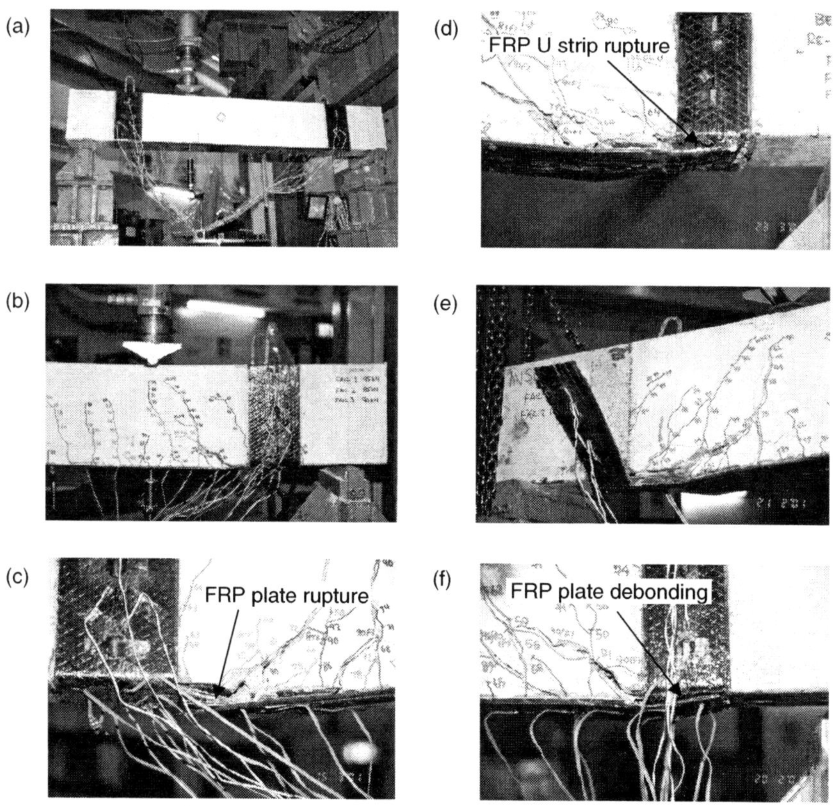

Figure 3.11 Failure modes of FRP-plated RC beams with FRP U strip end anchorage: (a) prior to test; (b) failure in concrete; (c) FRP plate rupture; (d) FRP U strip rupture; (e) FRP U strip debonding; (f) FRP plate debonding

the U strips (either wrapped around the FRP plate or the FRP plate positioned outside the U strips) were varied. Tan (1999) suggested that when both flexural strengthening by soffit plating and shear strengthening by either U jacketing or wrapping are required, the FRP for shear strengthening should be applied first. The variation in U strip placement considered in Smith and Teng (2001a) was prompted by this suggestion. Four different failure modes were observed by Smith and Teng (2001a) for beams with U strips wrapped around plate ends. In all beams, the first peak load was reached when the concrete near the U strip developed a major crack and the FRP plate debonded nearby (Figure 3.11b). With further loading, a number of other secondary modes were observed on different beams by Smith and Teng (2001a). These include rupture of the FRP plate near the U strip (Figure 3.11c), rupture along a bottom edge of the U strip (Figure 3.11d), or debonding of the U strip (Figure 3.11e). For beams with the FRP plate located outside the U strip, failure was by the development of a major crack in the concrete near the U strip and the associated debonding of the FRP plate from the base of the U strip (Figure 3.11f).

The presence of the U strip offered only a slight improvement to the load-carrying capacity when compared with a beam with the same length of FRP plate without U strips that failed prematurely by concrete cover separation. A U strip wrapped around the FRP soffit plate was found to be more effective than one placed above the FRP plate. More detailed studies on such end anchorage systems should be carried out in the future.

3.4 Flexural Strength

3.4.1 General

Existing research suggests that the ultimate flexural strength of FRP-strengthened RC beams can be predicted using existing RC beam design approaches with appropriate modifications to account for the brittle nature of FRPs (An *et al.* 1991, Chajes *et al.* 1994). Consequently, design equations for evaluating the flexural strength of RC beams or one-way slabs flexurally strengthened with FRPs have been presented following the approaches of the ACI code (ACI 318-95 1999) by An *et al.* (1991), Chajes *et al.* (1994), Nanni *et al.* (1998), Saadatmanesh and Malek (1998), El-Mihilmy and Tedesco (2000) and Lam and Teng (2001), the Canadian code (CSA 1994) by Chaallal *et al.* (1998) and the British code (BS 8110 1997) by Teng *et al.* (2000b). The following presentation is based on the work of Teng *et al.* (2000b).

These design equations consider only flexural failure of the plated beam by either FRP rupture (Figure 3.4a) or concrete crushing (Figure 3.4b) without premature debonding failure. The preferred modes of failure to be designed for are concrete crushing following yielding of steel reinforcement and FRP rupture following yielding of steel reinforcement, with the former being the most favourable (El-Mihilmy and Tedesco 2000). In both modes, yielding of the steel reinforcement precedes failure by either concrete crushing or rupture of the FRP, which ensures that failure will occur after the formation of large flexural cracks. Such flexural cracks will give some warning of failure despite the fact that these failure modes generally show limited ductility. Failure by FRP rupture without yielding of steel reinforcement is expected to display roughly the same degree of ductility as that of FRP rupture following yielding of steel reinforcement, but this is unlikely to occur unless the steel reinforcement is located near the neutral axis of the beam. Failure by concrete crushing without yielding of steel reinforcement should be avoided as far as possible.

A key assumption in developing flexural strength design equations is that a plane section remains plane. The equations presented in the rest of this section are based on BS 8110 (1997) and in developing these equations an important point to note is that the partial safety factor for FRP composites γ_{frp} is not readily available in BS 8110 (1997). A value of 1.25 for γ_{frp}, as suggested in Section 1.6, is believed to be appropriate.

3.4.2 Design Equations

Following BS 8110 (1997), the ultimate strain at the extreme concrete compression fibre is taken to be 0.0035. The stress–strain curve for concrete adopted by BS 8110 (1997) and more accurately defined in Kong and Evans (1987) is given in Figure 3.12. As the behaviour of the FRP composite is brittle, this ultimate strain of concrete may not have been reached when the FRP ruptures. Thus, the simplified rectangular stress block of the code (BS 8110 1997) for the compression concrete is no longer valid. The strains and stresses over the depth of a plated beam are shown in Figure 3.13.

Compressive strains are taken to be positive here, so the strains in the compression concrete are positive, while the strains in the FRP are negative. The sign of

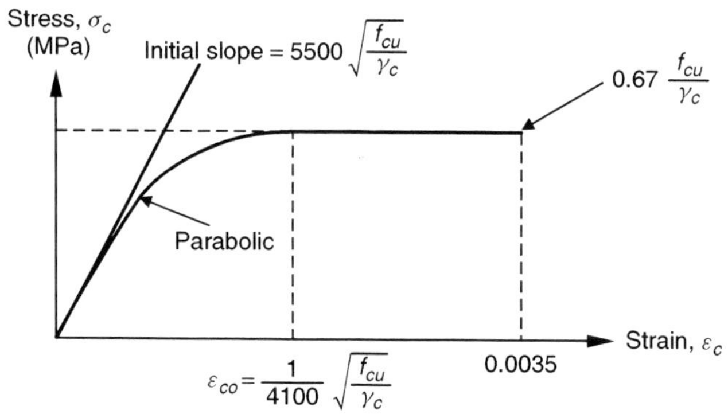

Figure 3.12 Stress–strain curve of concrete for design use. Reprinted from Kong and Evans (1987). Copyright CRC Press, Boca Raton, Florida

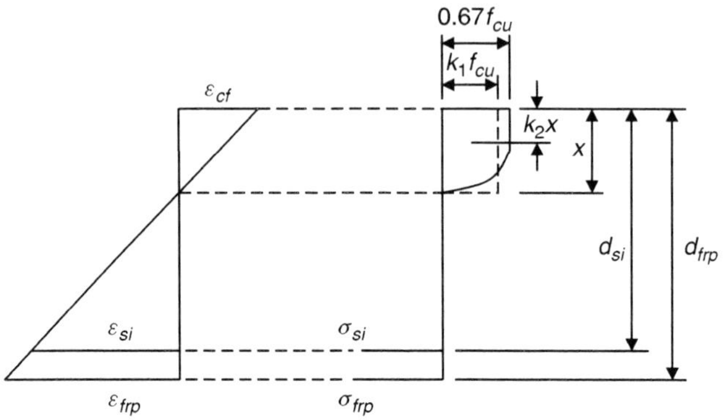

Figure 3.13 Strains and stresses over beam depth

strains in the reinforcing bars depends on their position in the beam. Based on the plane section assumption, the strains in the FRP, ε_{frp}, and those in the steel bars, ε_{si}, are related to the extreme compression fibre strain of concrete, ε_{cf}, as follows:

$$\varepsilon_{frp} = \varepsilon_{cf}\frac{x - d_{frp}}{x} \tag{3.1}$$

$$\varepsilon_{si} = \varepsilon_{cf}\frac{x - d_{si}}{x} \tag{3.2}$$

where x, d_{si} and d_{frp} are distances from the extreme concrete compression fibre to the neutral axis (i.e. neutral axis depth), the centroid of steel bars in layer i and the centroid of the FRP respectively. Similarly, when the strain in the FRP, ε_{frp}, is known, the value of ε_{cf} can be easily found as

$$\varepsilon_{cf} = \varepsilon_{frp}\frac{x}{x - d_{frp}} \tag{3.3}$$

According to the stress–strain curve of Figure 3.12, compressive stresses in the concrete are given by (Kong and Evans 1987)

$$\sigma_c = 5500\left(\sqrt{\frac{f_{cu}}{\gamma_c}}\varepsilon_c - \frac{4100}{2}\varepsilon_c^2\right) \quad \text{if } 0 \leqslant \varepsilon_c \leqslant \varepsilon_{co} \tag{3.4}$$

and

$$\sigma_c = 0.67\frac{f_{cu}}{\gamma_c} \quad \text{if } \varepsilon_{co} \leqslant \varepsilon_c \leqslant 0.0035 \tag{3.5}$$

where σ_c is the compressive concrete stress, ε_c is the compressive concrete strain which has a value of

$$\varepsilon_{co} = \frac{1}{4100}\sqrt{\frac{f_{cu}}{\gamma_c}}$$

at the peak stress of concrete, f_{cu} is the cube compressive strength of concrete, and γ_c is the partial safety factor for concrete.

For any given ε_{cf}, the total concrete compression force C is

$$C = k_1 \frac{f_{cu}}{\gamma_c} b_c x \tag{3.6}$$

where b_c is the beam width and k_1 is the mean stress factor defined by

$$k_1 = \frac{\int_0^{\varepsilon_{cf}} \sigma_c \, \mathrm{d}\varepsilon_c}{(f_{cu}/\gamma_c)\,\varepsilon_{cf}} \tag{3.7}$$

Substituting Equations 3.4 and 3.5 into Equation 3.7 yields

$$k_1 = 0.67\left(\frac{\varepsilon_{cf}}{\varepsilon_{co}} - \frac{\varepsilon_{cf}^2}{3\varepsilon_{co}^2}\right) \quad \text{if } 0 \leqslant \varepsilon_{cf} \leqslant \varepsilon_{co} \tag{3.8a}$$

and

$$k_1 = 0.67\left(1 - \frac{\varepsilon_{co}}{3\varepsilon_{cf}}\right) \quad \text{if } \varepsilon_{co} \leqslant \varepsilon_{cf} \leqslant 0.0035 \tag{3.8b}$$

The depth of the neutral axis x can be determined by solving the following force equilibrium equation:

$$k_1 \frac{f_{cu}}{\gamma_c} b_c x + \sum_{i=1}^{n} \sigma_{si} A_{si} + \sigma_{frp} A_{frp} = 0 \tag{3.9}$$

where σ_{si} and σ_{frp} are the stresses in the steel bars and the FRP respectively, A_{si} is the total area of steel in layer i, n is the total number of steel layers, and A_{frp} is the area of the FRP. σ_{si} and σ_{frp} are given by

$$\sigma_{si} = E_s \varepsilon_{si} \quad \text{if } |\varepsilon_{si}| < \frac{f_y}{\gamma_s E_s} \tag{3.10}$$

$$\sigma_{si} = \frac{\varepsilon_{si}}{|\varepsilon_{si}|}\frac{f_y}{\gamma_s} \quad \text{if } |\varepsilon_{si}| \geqslant \frac{f_y}{\gamma_s E_s} \tag{3.11}$$

and

$$\sigma_{frp} = E_{frp}\varepsilon_{frp} \geqslant \frac{-f_{frp}}{\gamma_{frp}} \tag{3.12}$$

where E_s and E_{frp} are moduli of elasticity of the steel bars and the FRP respectively, f_y is the yield strength of the steel, f_{frp} is the tensile strength of the FRP, and γ_s and γ_{frp} are the partial safety factors for steel and FRP respectively. It should be noted that f_y and f_{frp}, being material properties, are always positive.

The position of concrete compression force C is defined by D which is the distance from the extreme concrete compression fibre to the line of action of concrete compression force. D can be related to the height of the compression zone through

$$D = k_2 x \tag{3.13}$$

where the centroid factor k_2 of the compression force is given by

$$k_2 = 1 - \frac{\int_0^{\varepsilon_{cf}} \varepsilon_c \sigma_c \, d\varepsilon_c}{\varepsilon_{cf} \int_0^{\varepsilon_{cf}} \sigma_c \, d\varepsilon_c} \tag{3.14}$$

Evaluating the right-hand side of Equation 3.14 results in

$$k_2 = \frac{\frac{1}{3} - \varepsilon_{cf}/(12\varepsilon_{co})}{1 - \varepsilon_{cf}/(3\varepsilon_{co})} \quad \text{if } 0 \leqslant \varepsilon_{cf} \leqslant \varepsilon_{co} \tag{3.15a}$$

and

$$k_2 = \frac{\varepsilon_{cf}/2 + \varepsilon_{co}^2/(12\varepsilon_{cf}) - \varepsilon_{co}/3}{\varepsilon_{cf} - \varepsilon_{co}/3} \quad \text{if } \varepsilon_{co} \leqslant \varepsilon_{cf} \leqslant 0.0035 \tag{3.15b}$$

The moment capacity of the beam M_u is finally determined by

$$M_u = k_1 \frac{f_{cu}}{\gamma_c} b_c x \left(\frac{h}{2} - k_2 x \right) + \sum_{i=1}^{n} \sigma_{si} A_{si} \left(\frac{h}{2} - d_{si} \right) + \sigma_{frp} A_{frp} \left(\frac{h}{2} - d_{frp} \right) \tag{3.16}$$

where h is the depth of the RC beam. In Equation 3.16, the three terms on the right-hand side represent the contributions of the compression concrete, the steel bars and the FRP respectively.

The beam is deemed to have reached failure when either the concrete compression strain attains the maximum usable strain 0.0035 according to BS 8110 (1997) and/or the FRP reaches its rupture strain $\varepsilon_{frp,rup} = f_{frp}/(\gamma_{frp} E_{frp})$ (An *et al.* 1991). In practical design, the mode of failure should first be determined. Assuming that in a balanced design both the compression concrete and the FRP reach their respective failure states, then $\varepsilon_{cf} = 0.0035$ and $\varepsilon_{frp} = -\varepsilon_{frp,rup}$. The critical FRP ratio $\rho_{frp,cr}$ for a balanced beam can be found from the equilibrium equation

(Equation 3.9) to be

$$\rho_{frp,cr} = -\frac{k_1 (f_{cu}/\gamma_c)(x_{cr}/h) + \sum_{i=1}^{n} \sigma_{si}\rho_{si}}{f_{frp}/\gamma_{frp}} \tag{3.17}$$

where ρ_{si} is the steel reinforcement ratio of layer i, and x_{cr} is the critical depth of the neutral axis, given by

$$x_{cr} = \frac{0.0035}{0.0035 + f_{frp}/\left(\gamma_{frp} E_{frp}\right)} d_{frp} \tag{3.18}$$

If $\rho_{frp} > \rho_{frp,cr}$ the beam fails by concrete crushing, but if $\rho_{frp} < \rho_{frp,cr}$ the beam fails by FRP rupture. It should be noted that both ρ_{si} and ρ_{frp} are defined here using the gross area of the cross-section.

In the case of concrete crushing, $\varepsilon_{cf} = 0.0035$ and ε_{frp} is found from Equation 3.1. The mean stress factor k_1 is then directly obtained from Equation 3.8b. The depth of the neutral axis x can be determined by solving Equation 3.9 making use of Equations 3.1, 3.2 and 3.10–3.12.

In the case of FRP rupture, $\varepsilon_{frp} = -\varepsilon_{frp,rup}$. The depth of the neutral axis x is first assumed a value so that Equation 3.3 leads to an ε_{cf} value between ε_{co} and 0.0035. The mean stress factor k_1 is then obtained using Equation 3.8b, which is then substituted into Equation 3.9 to see if it is satisfied. The process of finding x is thus iterative and ends with an x value which satisfies Equation 3.9. If Equation 3.9 cannot be satisfied with $\varepsilon_{co} < \varepsilon_{cf} < 0.0035$, smaller values of x should be assumed to give $\varepsilon_{cf} \leqslant \varepsilon_{co}$. The mean stress factor k_1 is then calculated using Equation 3.8a.

Hand calculations using the above design equations can be a lengthy process, so it is best to do these calculations using a computer spreadsheet program like Microsoft Excel.

3.4.3 Presence of Preloading

The preceding design equations are for beams without preloading, but preloading including self-weight is likely to exist in practical applications. Preloading leads to an initial tensile strain ε_{ini} at the tension face of the beam, where the FRP is to be bonded. Lam and Teng (2001) have examined the effect of preloading on slabs and their conclusions are equally applicable to RC beams. These conclusions, adapted for beams, are: (a) the effect of preloading due to self-weight and service loads is generally beneficial if the beam fails by FRP rupture, but this effect is generally insignificant, and (b) the effect of preloading is more significant and detrimental if the beam fails by concrete crushing and should be investigated in design calculations. The effect of preloading should therefore be considered if the strengthened section is found to fail by concrete crushing. It should be noted that the failure mode of FRP-plated RC beams may change from FRP rupture to concrete crushing as a result of preloading. For such beams, the effect of preloading should also be considered in design.

The effect of preloading can be considered in strengthening design by modifying the strain value in the FRP as has been suggested by Saadatmanesh and Malek (1998). An alternative interpretation of their procedure is suggested here which employs a modified stress–strain curve of the FRP. This procedure involves two steps: (a) a cracked section analysis based on the assumptions of the design equations presented above is carried out to determine the strain ε_{ini} of the tension face of the beam without the bonded FRP at the critical section, and (b) the strength of the beam with the bonded FRP is then evaluated using the design equations presented above except that the modified stress–strain curve of the FRP given in Figure 3.14 should be used in calculations, which requires Equation 3.12 to be replaced by

$$\sigma_{frp} = 0 \quad \text{if } \varepsilon_{frp} \geqslant \varepsilon_{ini} \tag{3.19a}$$

$$\sigma_{frp} = E_{frp}\left(\varepsilon_{frp} - \varepsilon_{ini}\right) \geqslant \frac{-f_{frp}}{\gamma_{frp}} \tag{3.19b}$$

Tensile rupture of the FRP now occurs at the following modified rupture strain:

$$\varepsilon_{frp,rup} = \frac{f_{frp}}{\gamma_{frp} E_{frp}} - \varepsilon_{ini} \tag{3.20}$$

which is greater than the original rupture strain as ε_{ini} is negative. The effect of preloading is therefore accounted for by a modified stress–strain curve for the FRP without affecting the rest of the design equations presented earlier. This procedure is conceptually very simple, as the design is basically the same as for a beam without preloading. Physically, the 'modified' beam is one which is bonded with FRP without any preloading and the FRP has no stiffness during the initial stage of straining up to a strain of ε_{ini} at the tension face of the beam.

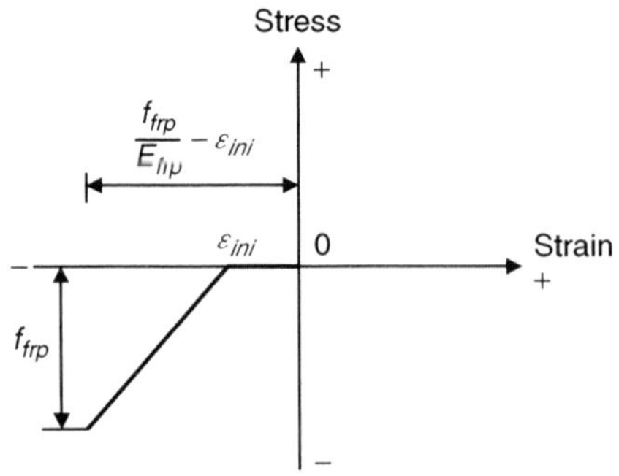

Figure 3.14 Modified FRP stress–strain curve for flexural design with preloading

3.5 Interfacial Stresses

3.5.1 General

As discussed earlier, the most commonly reported debonding failure mode is the separation of the concrete cover, with the soffit plate still attached, which propagates from the end of the plate. Initiation of this debonding failure mode is believed to be due to high interfacial stresses at the plate cut-off. These high interfacial stresses can also cause the less common mode of plate-end interfacial debonding which also propagates from the plate end. This section presents a summary of existing knowledge on these interfacial stresses.

3.5.2 Approximate Analytical Solutions

Figure 3.15 shows the stresses acting on an element of concrete adjacent to the end of the adhesive layer where τ and σ_y denote the shear and normal (peeling) stresses respectively and σ_x is the longitudinal stress. As debonding failures are believed to be controlled by the interfacial shear and normal stresses, much research effort has been expended in the last decade on the determination of interfacial stresses in plated beams. Seven simple approximate closed-form solutions suitable for direct use in design have been formulated (Vilnay 1988, Roberts 1989, Roberts and Haji-Kazemi 1989, Liu and Zhu 1994, Täljsten 1997, Malek *et al.* 1998, Smith and Teng 2001b). The earlier efforts (Vilnay 1988, Roberts 1989, Roberts and Haji-Kazemi 1989, Liu and Zhu 1994) were concerned with steel-plated beams, while more recent studies (Täljsten 1997, Malek *et al.* 1998) were concerned with FRP-plated beams. As elastic behaviour is generally assumed in interfacial stress analysis, steel differs from FRPs only in the elastic modulus. All existing solutions can thus be applied to FRP-plated RC beams, and no distinction is made between the two types of plated beams in the recent study by Smith and Teng (2001b). All these solutions are based on the assumption that the shear and normal stresses in the adhesive layer are uniform across the adhesive layer thickness. Apart from other differences, different solutions have different restrictions in loading conditions (Smith and Teng 2001b).

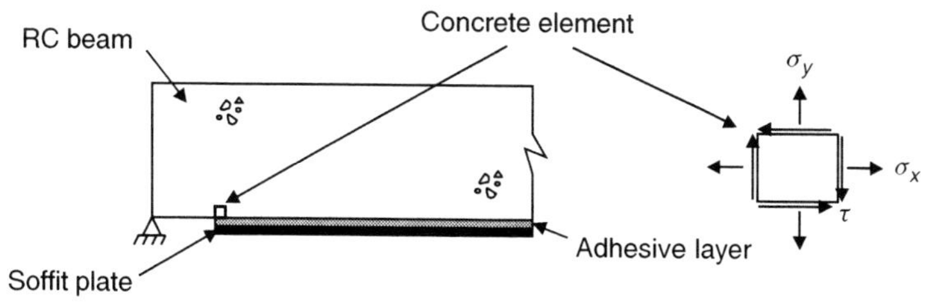

Figure 3.15 Stresses acting on a concrete element adjacent to the plate end

A comparison of the interfacial shear and normal stresses in plated RC beams, given by all the above solutions, shows that their predictions are close to each other (Smith and Teng 2001b) for RC beams strengthened with a thin plate. A summary of the analytical solution derived by Smith and Teng (2001b) is presented in Appendix A of this chapter (Section 3.10). This solution is relatively simple and is one of the most accurate of the closed-form solutions mentioned above (Teng *et al.* 2001a).

The interfacial shear and normal stress distributions predicted by Smith and Teng's (2001b) solution are shown in Figure 3.16 for a simply supported beam under a uniformly distributed load previously analysed by Roberts and Haji-Kazemi (1989). A positive normal stress at the end of the plate signifies tension between the RC beam and the plate. Both the maximum normal stress and the maximum shear stress are predicted to occur at the end of the plate, and this is the region where plate-end debonding failures are expected to occur.

3.5.3 Finite-element Results

While limited finite-element (FE) results for interfacial stresses were presented in some existing studies (e.g. Täljsten 1997, Malek *et al.* 1998, Rabinovich and Frostig 2000), the first comprehensive FE study of interfacial stresses in plated RC beams has only been conducted recently by Teng *et al.* (2001a) to (a) investigate the complex stress distributions in the adhesive layer near the end of the plate, and (b) assess the accuracy of the simple approximate analytical solutions. Figures 3.17 and 3.18 compare results from the analytical solution of Smith and Teng (2001b) with FE results, again for the example beam of Roberts and Haji-Kazemi (1989).

The FE results show that stresses vary strongly across the adhesive layer, with the stresses along the adhesive-to-concrete (AC) interface being very different from those along the plate-to-adhesive (PA) interface and those at the middle of the adhesive layer (MA). In particular, near the end of the plate, the interfacial

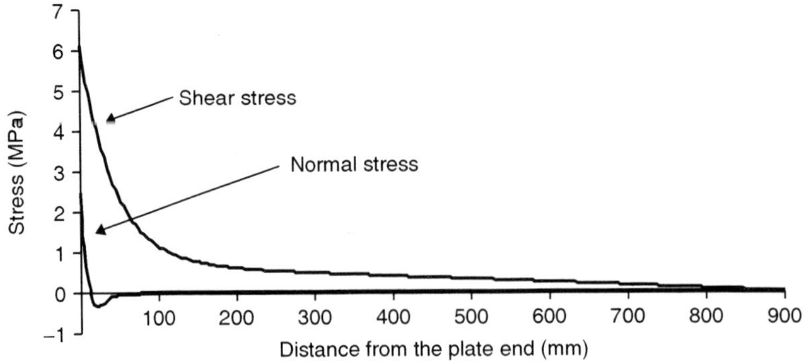

Figure 3.16 Typical interfacial shear and normal stress distributions

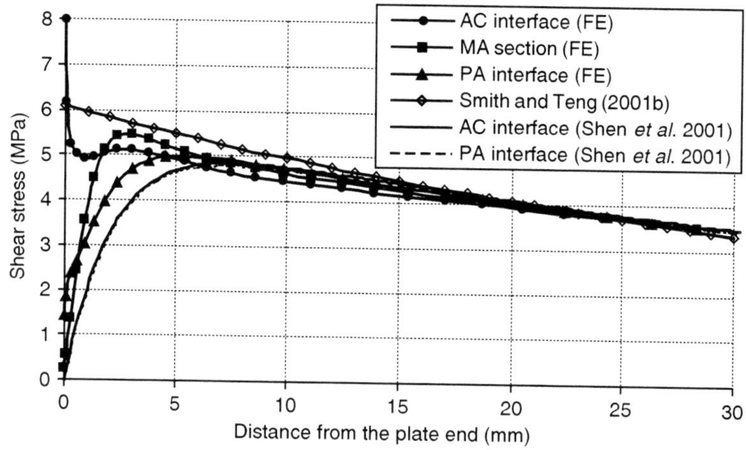

Figure 3.17 Predicted interfacial shear stresses in a plated RC beam

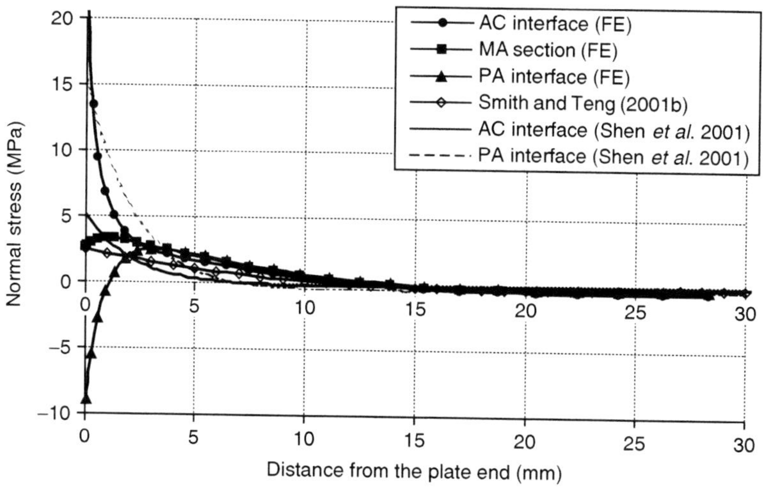

Figure 3.18 Predicted interfacial normal stresses in a plated RC beam

normal stress (Figure 3.18) is tensile along the AC interface but compressive along the PA interface, offering an explanation for the fact that PA interfacial failure in tests has rarely been – if at all – reported. This point was previously noted by Rabinovich and Frostig (2000).

The closed-form solution of Smith and Teng (2001b), based on the assumption of uniform stresses across the thickness of the adhesive layer, is incapable of predicting such complex stresses, but nevertheless provides reasonably close predictions of stresses along the middle-thickness section of the adhesive layer, particularly when the adhesive layer is not very thin (Teng *et al.* 2001a). For the current problem, the thickness of the adhesive layer is 4 mm. From a distance of about two times the thickness of the adhesive layer from the end of the plate, the agreement between the results of the FE analysis and the analytical solution is close for both the shear and normal stresses.

3.5.4 Higher Order Analytical Solutions

It should be noted that approximate closed-form solutions of the type discussed above do not satisfy the zero shear stress condition at the end of the adhesive layer. For this condition to be satisfied, a higher order analysis is required, which considers stress variation across the adhesive layer thickness (Rabinovich and Frostig 2000, Shen *et al.* 2001). The recent higher order solution of Rabinovich and Frostig (2000) does not provide explicit expressions for the interfacial stresses. Furthermore, its correctness is in doubt (Shen *et al.* 2001). The solution given in Shen *et al.* (2001) does provide explicit expressions for the interfacial stresses although the expressions are complex.

The interfacial stresses predicted by Shen *et al.*'s (2001) higher order solution are also plotted in Figures 3.17 and 3.18 for comparison with those from Smith and Teng's (2001b) closed-form solution and the FE predictions of Teng *et al* (2001a). The interfacial shear stresses predicted by Shen *et al.*'s solution are almost identical for the two interfaces (i.e. the AC and PA interfaces), and agree closely with those from Smith and Teng's (2001b) approximate solution and the FE predictions of Teng *et al.* (2001a) except near the plate end (Figure 3.17). Smith and Teng's solution predicts increasing interfacial shear stresses towards the plate end, while FE results near the plate end differ for the two interfaces and the MA section, with only the shear stresses in the middle-thickness section of the adhesive layer reducing to zero at the plate end. The interfacial shear stresses of Shen *et al.*'s solution reduce to zero at the plate end, satisfying the free surface condition.

The interfacial normal stresses predicted by Shen *et al.*'s solution are significantly different for the two interfaces, indicating that the distribution of normal stresses across the adhesive thickness is not uniform (Figure 3.18). At the AC interface, Shen *et al.*'s solution is capable of predicting the high normal stresses from FE analysis, although some differences in distributions still exist. These high tensile stresses are consistent with the observation that interfacial debonding at the PA interface in tests has rarely been reported, as noted by Rabinovich and Frostig (2000). At the PA interface, the predictions of Shen *et al.*'s solution for the small region near the plate end are very different from the FE results. Since this interface is not the critical one for debonding failures except when it is weakened by poor workmanship, this inaccuracy is considered to be of secondary importance.

3.5.5 Remarks

The preceding discussions of interfacial stresses in RC beams bonded with a soffit plate allow some useful comments to be made. First, the distributions of these stresses are very complex near the end of the plate, as revealed by the careful FE study of Teng *et al.* (2001a). In particular, the FE results show that high peeling stresses exist at the AC interface rather than the PA interface, making the AC interface the critical interface for debonding failures. This explains why PA interfacial debonding has been rarely reported in tests. An important point to note is that the end points of both the AC and the PA interfaces are points of stress singularity because of different material properties of the two adjacent materials, where the peak values of stresses depend on mesh refinement (Teng *et al.* 2001a). The FE results of Teng *et al.* (2001a) presented above were obtained using a very fine mesh so that any further refinement would not appreciably affect the results when plotted except within 2 mm from the plate end. In real beams, these high stresses cannot be achieved owing to factors such as a spew fillet (Teng *et al.* 2001a) and local damage at high stresses.

Simple closed-form solutions assuming uniform stresses across the adhesive layer thickness are capable only of giving reasonable predictions of the stresses along the MA section, as these solutions do not differentiate between the two interfaces. These solutions violate the zero shear stress condition, but this violation is not important for practical applications, as it affects only stress predictions in a tiny zone near the plate end. The inability of such solutions in predicting the high peeling stresses at the AC interface is of more concern. Nevertheless, they provide a useful but simple tool for understanding the interfacial stress behaviour and for use in developing a design rule. As discussed in the next section, several debonding strength models rely on such a closed-form solution for predicting the interfacial stresses. In these models, debonding failure is deemed to occur when these stresses satisfy certain failure criteria. These debonding strength models are assessed together with other debonding strength models against test data in the next section.

Higher order solutions such as that of Shen *et al.* (2001) are capable of satisfying the zero shear stress condition, and predicting the high peeling stresses. They are therefore a definite improvement over the simple approximate solutions in providing a more accurate picture of interfacial behaviour. Their disadvantage lies in their complexity, as the solution is complex and numerical results are not easily obtainable. This means that they are difficult to exploit in developing a design rule, although their potential for such purposes is yet to be fully explored.

3.6 Plate-end Debonding Strength Models

3.6.1 General

Plate-end debonding failures, including both concrete cover separation and plate-end interfacial debonding, have been subjected to extensive studies as they are commonly encountered in tests. Various strength models have also been developed for plate-end debonding failures. Smith and Teng (2001c, d) recently presented a review of existing debonding strength models and then assessed them with experimental data collected from the literature. The relative strengths and weaknesses of each model were then revealed. This section is based on the work of Smith and Teng (2001c, d).

3.6.2 Review of Debonding Strength Models

Several strength models for plate-end debonding have been developed for FRP-plated RC beams in the last decade (Varastehpour and Hamelin 1997, Saadatmanesh and Malek 1998, Wang and Ling 1998, Ahmed and van Gemert 1999, Tumialan *et al.* 1999b, Raoof and Hassanen 2000, Smith and Teng 2001d). In addition, a number of strength models have also been developed for plate-end debonding in steel-plated beams (Oehlers 1992, Ziraba *et al.* 1994, Jansze 1997, Raoof and Zhang 1997). Plate-end debonding occurs either by concrete cover separation or in the concrete adjacent to the adhesive-to-concrete interface. Although the different mechanical properties of the bonded plate are expected to play a significant role, it can be expected that strength models are applicable to different types of plate if the geometrical and material properties of the plate are properly included. It is thus not unreasonable to expect that models developed for steel-plated beams may well be applicable to FRP-plated beams, with modifications if necessary. Therefore, all the above-mentioned models, except the recent model of Smith and Teng (2001d) are reviewed and assessed based on the studies of Smith and Teng (2001c, d). Smith and Teng's (2001d) model, which is recommended for use in design later in the chapter, is described in detail in Section 3.6.5.

In terms of approaches, existing debonding strength models can be divided into three categories, namely (a) shear-capacity-based models (Oehlers 1992, Jansze 1997, Ahmed and van Gemert 1999), (b) concrete tooth models (Raoof and Zhang 1997, Wang and Ling 1998, Raoof and Hassanen 2000), and (c) interfacial stress-based models (models I and II of Ziraba *et al.* 1994, Varastehpour and Hamelin 1997, Saadatmanesh and Malek 1998, Tumialan *et al.* 1999b). Model II of Ziraba *et al.* (1994) in fact combines approaches (a) and (c). In addition, it should be mentioned that Oehlers' (1992) model is not purely based on the shear capacity, as it also takes into account the interaction between shear and bending. The main concepts of these models are described below whilst a more detailed review is given in Appendix B of this chapter (Section 3.11). It is worth noting that all the models developed specifically for FRP-strengthened RC beams (Varastehpour and Hamelin 1997, Saadatmanesh and Malek 1998, Tumialan *et al.* 1999b) follow the interfacial stress-based approach. On the other hand, three of the five models specifically developed for steel-plated beams follow either the shear capacity approach (Oehlers 1992, Jansze 1997) or a combined approach (model II of Ziraba *et al.* 1994).

3.6.2.1 Shear-capacity-based Models

Oehlers and Moran (1990) and Oehlers (1992) investigated simply supported steel-plated beams subjected to four- and three-point bending respectively and developed a strength model by first considering two extreme positions of soffit plate termination. For a plate terminated in the constant moment region, an expression for the flexural debonding moment $M_{db,f}$ at the end of the plate was proposed and calibrated with steel-plated beam tests. It was suggested that in design, this moment needs only to be compared with the additional moment applied to the beam at the plate end after the plate is bonded. For a plate terminated near the support, it was proposed that shear debonding occurs when the shear force at the plate end $V_{db,s}$ reaches the shear capacity of the concrete in the RC beam alone, without the contribution from the steel shear reinforcement, based on experimental observations. For general cases where both the shear force and the moment at the plate end are important, an interaction equation was suggested based on test results (Oehlers 1992).

Jansze (1997) also proposed a debonding strength model for steel-plated beams but the background information to the model was not available to Smith and Teng (2001c). They instead referred to Ahmed and van Gemert (1999) where the final form of the model with general comments was given. It was stated in Ahmed and van Gemert (1999) that Jansze's (1997) model was based on the initiation of shear cracking in an RC beam without the contribution of shear reinforcement. Ahmed and van Gemert later modified Jansze's model and calibrated the modified model with test results of FRP-plated RC beams reported to have failed by concrete cover separation. Ahmed and van Gemert did not explain in detail the basis of their modification, but Smith and Teng (2001c) noted that the increase in shear strength offered by shear reinforcement was included by an inspection of their debonding strength equation.

3.6.2.2 Concrete Tooth Models

Raoof and Zhang (1997) developed yet another strength model to predict concrete cover separation failures in steel-plated beams. The model makes use of the concept of a concrete 'tooth' between two adjacent cracks deforming like a cantilever under the action of horizontal shear stresses at the base of the beam (Figure 3.19). Debonding is deemed to occur when these shear stresses lead to

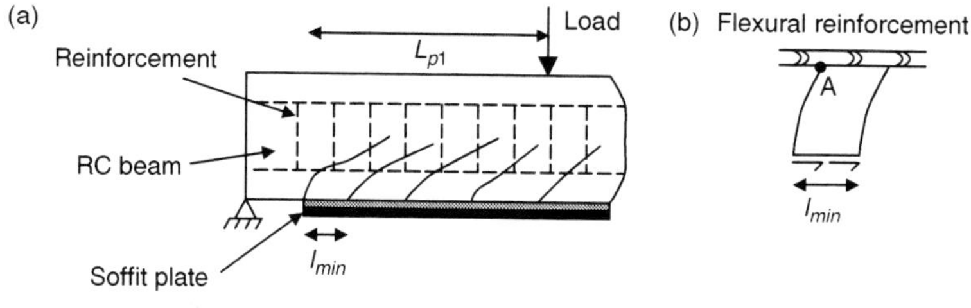

Figure 3.19 Concrete tooth model: (a) cracked beam; (b) concrete tooth between two adjacent flexural cracks. From Zhang *et al.* (1995). Reproduced by permission of Thomas Telford Publishing/Institute of Civil Engineers

tensile stresses at the root of the 'tooth' that exceed the tensile strength of the concrete. The stress in the soffit plate at debonding can then be determined by defining an effective length for the plate for end anchorage over which a uniform shear stress is assumed. Once the stress in the plate is known, the moment to cause separation of the plate can be calculated using a conventional section analysis with the assumption of plane section bending as described in Zhang *et al.* (1995). A lower bound and an upper bound bending moment are determined as a consequence of the assumed minimum (l_{min}) and maximum stabilised crack spacings differing by a factor of two.

Raoof and Hassanen (2000) recently modified Raoof and Zhang's (1997) model for application to FRP-plated beams in which two expressions were presented for the effective length of the FRP plate for end anchorage, with two corresponding values for the ultimate bond shear stress between the FRP and the concrete. These expressions were calibrated in the same manner as that described in Raoof and Zhang (1997) but with FRP-plated RC beam test results. Earlier, Wang and Ling (1998) proposed a modification to the 1995 version of the 'tooth' model (Zhang *et al.* 1995) by slightly redefining the minimum crack spacing formula for application to FRP-plated RC beams.

3.6.2.3 Interfacial Stress-based Models

A logical assumption is that concrete cover separation or plate-end interfacial debonding is due to high interfacial stresses at the end of the soffit plate (Figure 3.15) (Smith and Teng 2001c). Interfacial stress-based debonding strength models are generally based on interfacial stresses from an existing closed-form solution (e.g. Roberts 1989) and a concrete failure criterion. Models of this type include model I of Ziraba *et al.* (1994), and those of Varastehpour and Hamelin (1997), Saadatmanesh and Malek (1998) and Tumialan *et al.* (1999b).

An alternative use of the interfacial stresses was made in model II of Ziraba *et al.* (1994) which assumed that the shear force resisted at debonding consists of the shear resistance of concrete and a part of the shear resistance of steel stirrups as determined by an efficiency factor. The stirrup efficiency factor was related to the peak interfacial normal stress at the end of the soffit plate and calibrated with experimental data. This model therefore combines the shear capacity approach discussed above with interfacial stresses.

3.6.2.4 Model Limitations

Of the 12 debonding strength models reviewed above, model I of Ziraba *et al.* (1994) and Varastehpour and Hamelin's (1997) model are for plate-end interfacial debonding only. All the concrete tooth models and the remainder of the interfacial stress-based models appear to be for cover separation only, while all the shear-capacity-based models appear to be suitable for both cover separation and plate-end interfacial debonding. These limitations should be borne in mind when they are applied to make predictions.

3.6.3 Experimental Data

The complexity of behavior of FRP-strengthened RC structures means that test databases are still the sole verification tool for the reliability of a strength model. As a result, a number of attempts have been made to develop test databases for simply supported RC beams strengthened with a soffit FRP plate (Smith and Teng 2001d). The database recently compiled by Smith and Teng (2001d), containing 59 data points collected from 14 studies based on clear selection criteria, is much larger and more reliable than any previous database. All test results in this database are for beams reported to have failed by plate-end debonding which consisted of 44 concrete cover separation results, 8 plate-end interfacial debonding results, and 7 mixed concrete cover and plate-end interfacial debonding results. This database was used by Smith and Teng (2001d) to assess the performance of existing debonding strength models. Before describing their findings, selection criteria used and assumptions made by Smith and Teng in developing their database are given first.

3.6.3.1 Selection Criteria for Smith and Teng's Database

For inclusion in Smith and Teng's database, the following requirements had to be met: (a) failure of the beam was by plate-end debonding; (b) the FRP plate was neither prestressed nor anchored in any form at its ends; (c) the beam never experienced loading before being loaded to debonding failure; and (d) sufficient details for various geometric and material parameters were provided to enable the results to be used with confidence. Criterion (c) above was used for consistency in the database as the vast majority of tests were conducted on beams without any preloading. Limited research on beams preloaded to cracking and then completely unloaded before plate bonding appears to suggest that such preloading has little effect on the plate-end debonding load (Arduini and Nanni 1997), but further research is required on this aspect before firm conclusions can be drawn. A few tests on beams with FRP plates bonded under sustained loading have also been carried out (Arduini and Nanni 1997), but no conclusion can yet be drawn from these tests.

3.6.3.2 Assumptions for Geometric and Material Properties

Many of the references on FRP soffit-plated RC beams reviewed for possible inclusion in Smith and Teng's database could not be used. Often the failure mode was poorly reported and/or important geometric and material properties of the RC beam, the FRP plate or the adhesive layer were omitted. Only those tests with the failure mode sufficiently well described and important material and geometric properties adequately reported could be included in the database. The only geometric and material properties assumed in the database, if not specified in the original source, were depth of concrete cover, elastic modulus of adhesive, thickness of adhesive layer, and modulus of elasticity of steel reinforcement (200 GPa). The depth of concrete cover, if not reported, was assumed to be 10 % of the overall beam depth, which was similar to the average of the different cover depths used for other beams in the database. The ACI code (ACI 318-95 1999) expressions for the elastic modulus of concrete E_c and the splitting tensile strength of concrete f_{ct} were adopted which are related to the concrete cylinder compressive strength f_c' as follows:

$$E_c = 4730\sqrt{f_c'}, \quad \text{MPa} \tag{3.21}$$

$$f_{ct} = 0.53\sqrt{f_c'}, \quad \text{MPa} \tag{3.22}$$

For beams with only the concrete cube compressive strength f_{cu} reported, the concrete cylinder strength was taken to be $0.8 f_{cu}$ (Kong and Evans 1987).

For beams with pultruded plates, the adhesive layer thickness was sometimes not reported. In such cases, the thickness of the adhesive layer was assumed to be 2 mm, a value similar to the average of those reported for the remaining beams with pultruded plates. This appears to be the most rational approach that one can take without having to exclude these test data from the database, and should have minimised errors due to this assumed adhesive layer thickness. The assumption may be further justified by noting that any model which is overly sensitive to adhesive layer thickness is unlikely to be suitable as design models because adhesive layer thicknesses are difficult to control on site and have been found to have a limited effect on test debonding loads (Swamy *et al.* 1987, Quantrill *et al.* 1996b, Oehlers and Ali 1998).

For beams strengthened with FRP plates constructed in a wet lay-up process, more uncertainty existed. For many of the tests reported on such beams, the thickness of the FRP plate versus the thickness of the adhesive was generally not available. Although in a small number of cases, both the fibre sheet thickness alone, t_{sheet}, and the gross thickness of the fibre sheet plus the applied adhesive, $t_{frp,g}$, were given, only the sheet thickness t_{sheet} was given in most cases. For the former cases, the adhesive layer thickness and FRP plate thickness were worked out using the following formulae for a single layer of FRP:

$$t_a = \frac{t_{frp,g} - t_{sheet}}{2} \tag{3.23}$$

$$t_{frp} = t_a + t_{sheet} \tag{3.24}$$

based on the assumptions that the sheet sits in the middle of the adhesive, and the sheet and the outside adhesive form the plate (Figure 3.20). For the latter

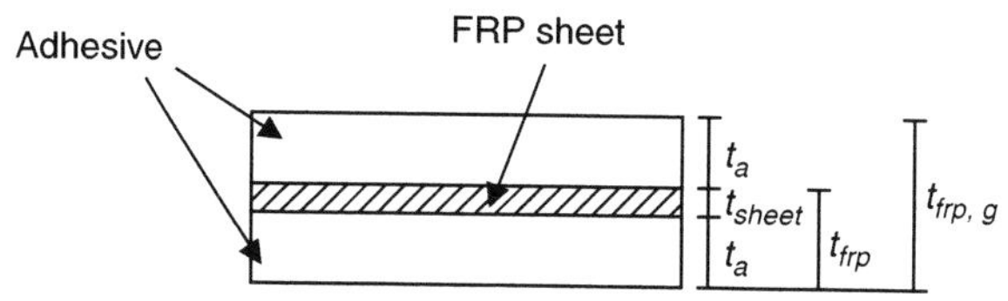

Figure 3.20 Wet lay-up plate thickness

cases, the adhesive layer thickness t_a was assumed to be 0.42 mm based on measurements of plate samples formed from carbon fibre sheets by wet lay-up (Smith and Teng 2001d), and the FRP plate thickness t_{frp} was worked out from Equation 3.24 for a single layer of FRP. For cases where the FRP plate was formed from a number of layers, the total plate thickness was found by multiplying the single layer thickness by the number of layers by assuming that the adhesive thickness between two sheets is t_a. The above rational process adopted by Smith and Teng (2001d) is believed to have minimised the effect of uncertainty in the existing test data. Furthermore, any debonding strength model that is overly sensitive to this uncertainty is unlikely to be suitable for design use for a wet lay-up process as adhesive and plate thicknesses are difficult to control on site or to define for calculations. Owing to this additional uncertainty with wet lay-up plates, test data of beams with wet lay-up plates and those with pultruded plates were appropriately differentiated from each other by Smith and Teng (2001d) in comparisons.

In addition, the elastic modulus of the adhesive was not reported in many studies. For these cases, the elastic modulus was assumed to be 8500 MPa, a value close to the average of the values reported for the other cases of the database (Smith and Teng 2001d). Again, this is a rational approach, and is not expected to compromise the usefulness of the database.

3.6.4 Performance of Debonding Strength Models

Using their test database, Smith and Teng (2001d) carried out a systematic assessment of the debonding strength models reviewed above. Statistical comparisons were carried out using either the cover separation alone or the entire database. In this section, a detailed discussion of the former is presented, followed by some brief comments on the latter. A limited discussion of direct graphical comparisons is also given.

Tables 3.1–3.3 show the statistical characteristics of the test-to-predicted debonding strength ratios for all 12 debonding strength models generated by Smith and Teng (2001d) using the cover separation data of their test database. Tables 3.1 is for pultruded and wet lay-up plates combined, Table 3.2 for pultruded plates alone, and Table 3.3 for wet lay-up plates only.

It should be noted that the models of Jansze (1997), Ahmed and van Gemert (1999), Saadatmanesh and Malek (1998) and Tumialan *et al.* (1999b) cannot be

Table 3.1 Test-to-predicted debonding strength ratios for failure by concrete cover separation: pultruded and wet lay-up plates

Category	Model	Average	Standard deviation	Coefficient of variation	Percentage of exceedence	No. of tests
Shear-capacity-based models	Oehlers (1992)	2.35	0.49	21 %	0.3 %	44
	Jansze (1997)	1.50	0.45	30 %	13 %	39
	Ahmed and van Gemert (1999)	1.15	0.25	21 %	27 %	39
	Smith and Teng's model ($V_{db,end} = 1.4\,V_c$)	1.67	0.35	21 %	2.8 %	44
	Smith and Teng's model ($V_{db,end} = 1.5\,V_c$)	1.56	0.33	21 %	4.4 %	44
Concrete tooth models	Raoof and Zhang: lower (1997)	1.47	0.52	36 %	19 %	44
	Raoof and Zhang: upper (1997)	1.06	0.31	29 %	43 %	44
	Wang and Ling: lower (1998)	1.20	0.35	29 %	28 %	44
	Wang and Ling: upper (1998)	0.88	0.16	18 %	78 %	44
	Raoof and Hassanen I: lower (2000)	1.61	0.65	41 %	18 %	44
	Raoof and Hassanen I: upper (2000)	1.13	0.43	38 %	38 %	44
	Raoof and Hassanen II: lower (2000)	1.57	0.60	38 %	17 %	44
	Raoof and Hassanen II: upper (2000)	1.09	0.35	32 %	40 %	44
Interfacial stress-based models	Ziraba *et al.* I (1994)	4.89	3.64	74 %	14 %	44
	Ziraba *et al.* II (1994)	1.63	0.55	34 %	13 %	44
	Varastehpour and Hamelin (1997)	0.59	0.20	34 %	98 %	44
	Saadatmanesh and Malek (1998)	3.24	2.83	87 %	21 %	39
	Tumialan *et al.* (1999b)	1.83	1.28	70 %	26 %	39

Table 3.2 Test-to-predicted debonding strength ratios for failure by concrete cover separation: pultruded plates only

Category	Model	Average	Standard deviation	Coefficient of variation	Percentage of exceedence	No. of tests
Shear-capacity-based models	Oehlers (1992)	2.09	0.27	13 %	0 %	25
	Jansze (1997)	1.21	0.35	29 %	27 %	23
	Ahmed and van Gemert (1999)	1.06	0.23	22 %	40 %	23
	Smith and Teng's model ($V_{db,end} = 1.4\,V_c$)	1.48	0.18	12 %	0.4 %	25
	Smith and Teng's model ($V_{db,end} = 1.5\,V_c$)	1.38	0.17	12 %	1.2 %	25
Concrete tooth models	Raoof and Zhang: lower (1997)	1.32	0.38	28 %	19 %	25
	Raoof and Zhang: upper (1997)	0.95	0.26	27 %	58 %	25
	Wang and Ling: lower (1998)	1.12	0.27	24 %	32 %	25
	Wang and Ling: upper (1998)	0.81	0.13	16 %	93 %	25
	Raoof and Hassanen I: lower (2000)	1.41	0.36	25 %	12 %	25
	Raoof and Hassanen I: upper (2000)	0.96	0.21	22 %	57 %	25
	Raoof and Hassanen II: lower (2000)	1.43	0.42	30 %	16 %	25
	Raoof and Hassanen II: upper (2000)	0.96	0.23	24 %	57 %	25
Interfacial stress-based models	Ziraba *et al.* I (1994)	3.54	3.25	92 %	22 %	25
	Ziraba *et al.* II (1994)	1.54	0.45	29 %	11 %	25
	Varastehpour and Hamelin (1997)	0.57	0.14	25 %	100 %	25
	Saadatmanesh and Malek (1998)	2.99	3.30	110 %	27 %	23
	Tumialan *et al.* (1999b)	1.63	1.53	94 %	34 %	23

Table 3.3 Test-to-predicted debonding strength ratios for failure by concrete cover separation: wet lay-up plates only

Category	Model	Average	Standard deviation	Coefficient of variation	Percentage of exceedence	No. of tests
Shear-capacity-based models	Oehlers (1992)	2.69	0.51	19 %	0 %	19
	Jansze (1997)	1.91	0.18	9.7 %	0 %	16
	Ahmed and van Gemert (1999)	1.28	0.21	16 %	9.3 %	16
	Smith and Teng's model ($V_{db,end} = 1.4\,V_c$)	1.92	0.36	19 %	0.6 %	19
	Smith and Teng's model ($V_{db,end} = 1.5V_c$)	1.79	0.34	19 %	1.0 %	19
Concrete tooth models	Raoof and Zhang: lower (1997)	1.66	0.63	38 %	15 %	19
	Raoof and Zhang: upper (1997)	1.20	0.33	27 %	28 %	19
	Wang and Ling: lower (1998)	1.30	0.41	32 %	24 %	19
	Wang and Ling: upper (1998)	0.97	0.15	15 %	57 %	19
	Raoof and Hassanen I: lower (2000)	1.87	0.85	46 %	15 %	19
	Raoof and Hassanen I: upper (2000)	1.35	0.54	40 %	26 %	19
	Raoof and Hassanen II: lower (2000)	1.76	0.74	42 %	15 %	19
	Raoof and Hassanen II: upper (2000)	1.25	0.41	33 %	27 %	19
Interfacial stress-based models	Ziraba *et al.* I (1994)	6.67	3.43	51 %	4.9 %	19
	Ziraba *et al.* II (1994)	1.74	0.66	38 %	13 %	19
	Varastehpour and Hamelin (1997)	0.60	0.26	43 %	93 %	19
	Saadatmanesh and Malek (1998)	3.59	2.02	56 %	9.9 %	16
	Tumialan *et al.* (1999b)	2.10	0.76	36 %	7.4 %	16

used for plates terminated at the support. This limitation accounts for the smaller number of test data used to assess these models (Tables 3.1–3.3).

It should also be noted that many of the upper bound predictions and a few of the lower bound predictions of debonding moments by the tooth models (Raoof and Zhang 1997, Wang and Ling 1998, Raoof and Hassanen 2000) exceeded the theoretical moment capacities of beams without considering the possibility of debonding. For such beams, these theoretical moment capacities instead of the predicted debonding moments were used by Smith and Teng (2001d) for comparisons with test data, as previously done for these models (e.g. Raoof and Hassanen 2000).

3.6.4.1 Concrete Cover Separation: Better Models

Smith and Teng (2001d) found the statistically better performing models for concrete cover separation failures, based on the results given in Tables 3.1–3.3 and in terms of providing safe and close predictions, to be given by Oehlers (1992), Ziraba *et al.* (1994) (model II), Jansze (1997), Raoof and Zhang (1997) (lower bound predictions) and Raoof and Hassanen (2000) (lower bound predictions of both of their models).

Table 3.1 shows that for the combined data of pultruded plates and wet lay-up plates, the most unconservative of the better performing models is Raoof and Zhang's model which provides an 81 percentile lower bound (i.e. 19 % of exceedence), while the most conservative is that given by Oehlers which has a percentage of exceedence of 0.3 %. Oehlers' model also shows the least scatter based on the coefficient of variation, followed by Jansze's model. The predictions of both models of Raoof and Hassanen (lower bound predictions) are seen to be the most scattered, with coefficients of variation being about 40 %.

If pultruded plates are examined alone (Table 3.2), the same six models already identified above remain the better ones. Predictions by Oehlers' model remain the most conservative and least scattered. For these beams, the predictions of this model are conservative in all cases. The degree of scatter now does not vary too much among the other better performing models, with coefficients of variation in the range of 22 % to 30 %. Jansze's model is the most unconservative for these beams, with a percentage of exceedence of 27 %.

Table 3.3 shows that for wet lay-up plates alone, Jansze's model becomes the best performing model, with the lowest coefficient of variation (9.7 %) and conservative predictions for all cases. Oehlers' model is also conservative for all cases and has the second-lowest coefficient of variation. The rest of the better performing models all show a high degree of scatter with coefficients of variation in the range of 38 % to 46 %, with the percentages of exceedence being either 13 % or 15 %.

Based on the statistical characteristics given in Tables 3.1–3.3, Smith and Teng (2001d) concluded that Oehlers' model is the best performing model in terms of providing safe and reliable predictions, although it may be considered overly conservative as indicated by its relatively high average ratios between the test and predicted debonding loads (Tables 3.1–3.3).

3.6.4.2 Concrete Cover Separation: Poorer Models

The models of Raoof and Zhang (1997) (upper bound predictions), Varastehpour and Hamelin (1997), Saadatmanesh and Malek (1998), Wang and Ling (1998) (both lower and upper bound predictions), Ahmed and van Gemert (1999), Tumialan *et al.* (1999b), Raoof and Hassanen (2000) (upper bound predictions), and model I of Ziraba *et al.* (1994) have been identified by Smith and Teng (2001d) to be the worse performing models in terms of giving close and safe predictions (Tables 3.1–3.3).

For the combined data for the two types of plates, Varastehpour and Hamelin's model suffers from great unconservativeness with the percentage of exceedence being 98 %, which means that its predictions are unconservative in almost all cases. On the other hand, Ziraba *et al.*'s model I suffers from a great scatter, with its coefficient of variation being 74 %. These two models developed for plate-end interfacial debonding are thus not suitable for cover separation failures as can be expected. The upper bound predictions of all tooth models are unsafe as they are expected to be. The interfacial stress-based cover separation models of Saadatmanesh and Malek and Tumialan *et al.* suffer from great scatters, with Saadaatmanesh and Malek's model having the highest coefficient of variation of all models (87 %). The model of Ahmed and van Gemert and the lower bound predictions of Wang and Ling's model show more reasonable performance.

The worse performing models are seen to perform differently for the two types of plates (Tables 3.2 and 3.3). Ahmed and van Gemert's model based on the shear capacity approach, which was modified from Jansze's model, is rather unconservative for pultruded plates (Table 3.2) as evidenced by the high percentages of exceedence, but is satisfactory for wet lay-up plates alone (Table 3.3). The worse performing tooth models generally show greater scatters (except the upper bound predictions of Wang and Ling's model) and reduced percentages of exceedence for wet lay-up plates than for pultruded plates, but they are still too unconservative for design use for wet lay-up plates. Of the interfacial stress-based models, Varastehpour and Hamelin's model performs worse for wet lay-up plates than for pultruded plates as the percentages of exceedence are similar for the two types of plates, but the degree of scatter is greater for the former. The other interfacial stress based models show much better performance for wet lay-up plates, with both the coefficient of variation and the percentage of exceedence substantially reduced from those for pultruded plates. In fact for wet lay-up plates, Tumialan *et al.*'s model and Ziraba *et al.*'s model I are comparable with some of the better performing models.

3.6.4.3 All Test Results

The statistical performance of all 12 models as assessed using all the test results in Smith and Teng's database, which includes eight results for plate-end interfacial debonding and another seven results for mixed-mode debonding, can be found in Smith and Teng (2001d). Of these 15 results 12 are for pultruded plates with only 3 for wet lay-up plates. The inclusion of these additional results has no or very limited effect on the statistical characteristics of the models, with some showing a small deterioration in performance (e.g. Oehlers' model). The better and worse performing models for all plate-end debonding failures combined, based on the statistics given in Smith and Teng (2001d), are the same as those for concrete cover separation (Tables 3.1–3.3).

3.6.4.4 Graphical Comparisons of Debonding Strength Models

Smith and Teng (2001d) also carried out graphical comparisons of the reviewed debonding strength models with their test data. Figures 3.21 and 3.22 show such comparisons for Oehlers' and Raoof and Zhang's models, where the test data are differentiated based on both plate types and failure modes. In these figures, M_u is the ultimate moment capacity of the beam section without the influence of debonding, M_{exp} is the maximum experimental moment in the beam at debonding, and M_{db} is the corresponding debonding moment predicted by the strength models. It is worth noting that Oehlers' model is more conservative for concrete cover separation than for plate-end interfacial debonding, which is not seen in the case of Raoof and Zhang's model.

Figure 3.23 shows a comparison of the predictions of Tumialan *et al.*'s model with the test data reported in Smith and Teng (2001d). The predictions of this

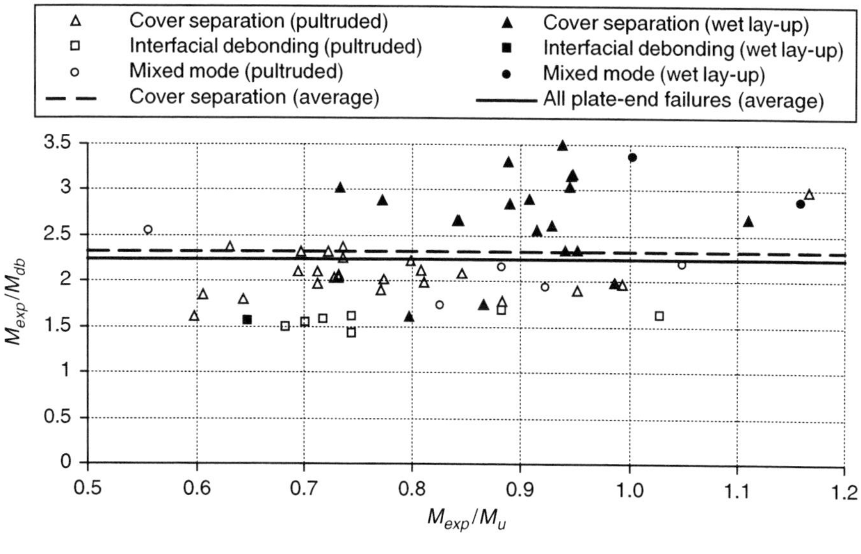

Figure 3.21 Test results versus predictions of Oehlers' model

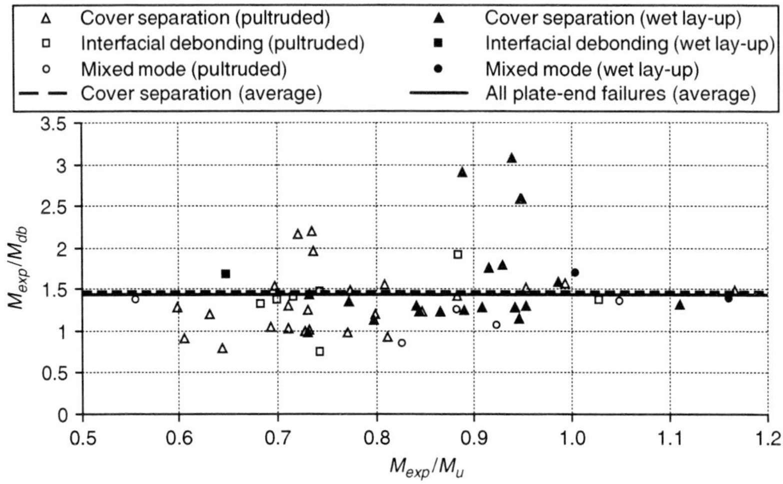

Figure 3.22 Test results versus predictions of Raoof and Zhang's lower bound model

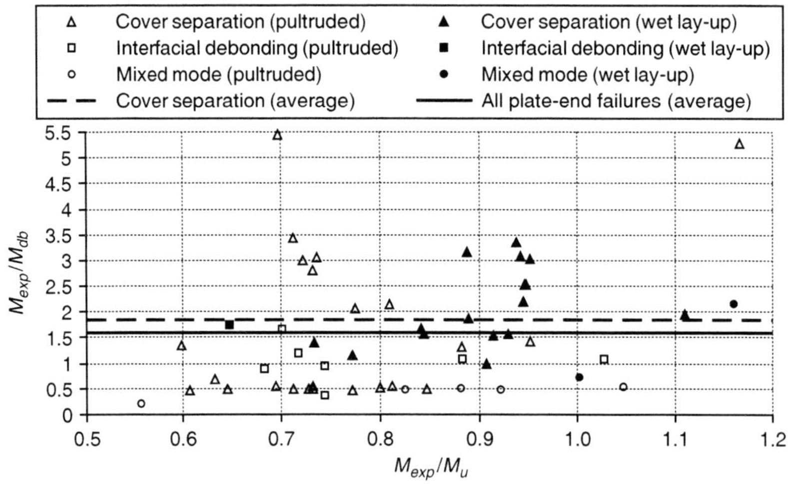

Figure 3.23 Test results versus predictions of Tumialan *et al.*'s model

model are widely scattered but are conservative when only beams with wet lay-up plates are considered. For beams with pultruded plates, about half of the predictions from this model are unconservative.

3.6.4.5 Additional Remarks on Debonding Strength Models

The discussions above show that, of the six better performing debonding strength models, Oehlers' model is the only one which gives safe but slightly conservative predictions for design use when either concrete cover separation failures are examined alone or all plate-end failure modes are considered. It also has the smallest scatter for the combined data (Table 3.1). Oehlers' model does have a drawback for direct application to FRP-strengthened beams as discussed in Section 3.6.5 where Smith and Teng's model (2001d), which was modified from Oehlers' model, is presented.

It is worth noting that among the six better performing models, three of them (Oehlers', Ziraba *et al.*'s model II and Jansze's) relate the debonding strength to the shear strength of concrete in an RC beam, and the two models based purely on the shear capacity approach (Oehlers' and Jansze's) show smaller scatters in predictions. This shear-capacity-based approach thus appears to be a promising approach, which is another reason for building a new model on the basis of Oehlers' model.

3.6.5 Smith and Teng's Model

Of the six better performing models identified above, Oehlers' model is the only one which gives safe but overly conservative predictions for design use when either cover separation data are examined alone or when all test data are considered together (Tables 3.1–3.3), despite the fact that it was developed for steel-plated RC beams. It also generally has the smallest scatter. Smith and Teng (2001d) therefore proposed a more accurate debonding strength model by modifying Oehlers' model.

It is also worth noting that of these six better performing models, three of them (Oehlers', Ziraba *et al.*'s model II and Jansze's) relate the debonding strength to the shear strength of concrete in an RC beam, while the other three are concrete tooth models from the same group with only small differences from each other. The shear-capacity-based approach independently proposed by three different groups appears to be a promising approach. This was another reason for Smith and Teng to base their model on Oehlers' model.

3.6.5.1 Oehlers' Model

Oehlers' model is given by

$$\frac{M_{db,end}}{M_{db,f}} + \frac{V_{db,end}}{V_{db,s}} = 1.17 \tag{3.25a}$$

and

$$M_{db,end} \leqslant M_{db,f}, \quad V_{db,end} \leqslant V_{db,s} \tag{3.25b}$$

where $M_{db,end}$ and $V_{db,end}$ are the plate-end bending moment and shear force respectively at debonding. The bending moment is based on the load applied after plating while the shear force includes the part from loading before plating.

Oehlers (1992) equated $V_{db,s}$ to the shear capacity of the concrete in the RC beam, V_c, and used the following expression for V_c which was taken from the Australian concrete code (AS 3600 1988):

$$V_c = [1.4 - (d/2000)]b_c d \left[\rho_s f_c'\right]^{1/3} \tag{3.26}$$

where it is required that $1.4 - (d/2000) \geqslant 1.1$. In Equation 3.26, $\rho_s = A_s/b_c d$ is the ratio of steel tension reinforcement, f_c' is the concrete cylinder compressive strength (in MPa), A_s is the area of steel tension reinforcement (in mm^2), b_c is the section width (in mm) and d is the effective depth (in mm) of the section.

3.6.5.2 Smith and Teng's Model

Smith and Teng (2001d) plotted Equation 3.25a against the test data of FRP-plated beams and noted that no interaction appears to exist between the moment and the shear force, and the shear force at failure is always greater than $V_{db,s}(V_c)$ predicted by Equation 3.26 (Figure 3.24). As a result, the following simple debonding strength model was proposed by Smith and Teng (2001d):

$$V_{db,end} = \eta V_c \tag{3.27}$$

This equation with $\eta = 1.5$ and with V_c given by Equation 3.26 provides a 95.6 percentile lower bound prediction to the cover separation data (Table 3.1). When all plate-end failures are considered, $\eta = 1.4$ provides a 94.2 percentile lower bound prediction (Smith and Teng 2001d) and is shown as the solid line in

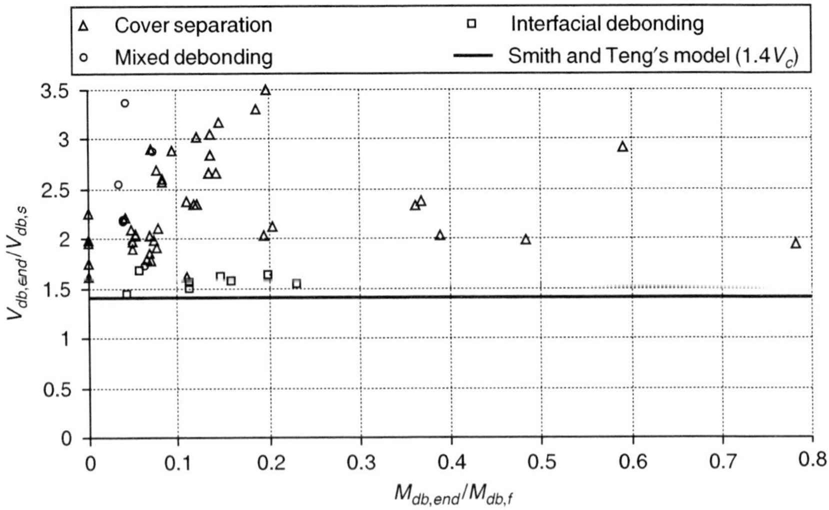

Figure 3.24 Comparison of Smith and Teng's model with test data

Figure 3.24. In practical design, since it is difficult to ascertain whether cover separation will be the failure mode, Equation 3.27 with $\eta = 1.4$ was recommended by Smith and Teng (2001d) for use to design against all forms of plate end debonding failures.

Smith and Teng also suggested that the range of applicability of their model be limited to that of the collected data in terms of the magnitude of the moment at the plate end. That is, the model should only be used when $M_{db,end}/M_u \leqslant 0.67$ as this approximates the upper limit of the test data. The choice of this ratio as a parameter means that little extra work is required as M_u is expected to be calculated in any flexural strengthening design. The use of the alternative ratio of $M_{db,end}/M_{db,f}$ may be considered with its value limited to 0.78 (Figure 3.24), but this would require the extra work of evaluating $M_{db,f}$. The use of $M_{db,end}/M_u$ thus appears to be preferable unless evidence to the contrary emerges.

3.6.5.3 Advantages of Smith and Teng's Model

The model proposed by Smith and Teng has a number of advantages over all the other debonding strength models reviewed in this chapter and is recommended for use in design. These include: (a) the model was calibrated with a large and reliable test database of FRP-strengthened beams and provides a suitable lower bound for design use; (b) the model is simpler to apply as the shear capacity of the concrete in the RC beam is either known or has to be evaluated in flexural strengthening design, so little extra work is required for checking plate-end debonding; (c) in this model, V_c may be evaluated according to any national or international design code, and this facilitates the inclusion of the model in any national code on FRP strengthening; and (d) the model gives designers a better intuitive feel of the possibility of plate-end debonding as it is directly related to the shear capacity of the concrete in the beam, particularly when steel shear reinforcement is not present as is often the case for slabs.

Although Smith and Teng's model is the best proposal available for design use, the predicted debonding strength, being only 40 % higher than the shear capacity of the concrete, can become a limiting factor to FRP flexural strengthening of beams. Specialist advice should be sought and/or tests conducted if a more liberal design is desired. Further research is urgently needed to develop a more accurate debonding strength model to reduce the conservativeness of Smith and Teng's model.

3.7 Strength Model for Intermediate Crack-induced Interfacial Debonding

3.7.1 General

Intermediate crack-induced interfacial debonding failures, described in detail in Section 3.3.5, have much in common with debonding failures observed in simple shear tests (Figure 2.2), particularly when these are induced by flexural cracks. Therefore, the simple strength model of Chen and Teng (2001) presented in Chapter 2 for steel and FRP plate-to-concrete bonded joints is expected to be useful for predicting at least intermediate flexural crack-induced debonding.

The potential applicability of Chen and Teng's debonding strength model to intermediate flexural crack-induced debonding in RC beams was initially noted in Chen and Teng (2001). A simple modification to their model to enable the prediction of intermediate crack-induced debonding failures was first explored in Smith *et al.* (2001). In that study, test results for flexural crack-induced debonding and flexural–shear crack-induced debonding were not differentiated from each other. A more refined study has recently been presented by Teng *et al.* (2001b) where this differentiation was made. The new study included test data for simply supported RC beams as well as test data for cantilever slabs and beams. Intermediate crack-induced interfacial debonding in cantilever slabs/beams occurred in a similar manner to that of simply supported beams, with debonding initiating at a flexural crack near the fixed support end and then propagating towards the free end of the slab.

This section reports the results of Teng *et al.*'s (2001b) study. Although this chapter is concerned with beams, reference is also made to one-way slabs, because beams and slabs show similar behaviour with regard to at least flexural crack-induced debonding and test data for both beams and slabs were made use of by Teng *et al.* in developing a modified Chen and Teng (2001) model. This failure mode is shown in Chapter 5 to be a particularly important failure mode for slabs.

3.7.2 Intermediate Flexural Crack-induced Debonding

Chen and Teng's (2001) bond strength model was developed from simple lap shear tests. Although there are many similarities between the behaviour of intermediate flexural crack-induced debonding and the failure behaviour of a simple lap shear specimen, a recent finite-element study (Chen *et al.* 2001) revealed that the stresses, especially the peak values in the adhesive near an intermediate flexural crack, can be different from those in a simple lap shear specimen. Direct application of Chen and Teng's model to intermediate flexural crack-induced debonding leads to conservative predictions. Teng *et al.* (2001b) therefore proposed a simple modification to Chen and Teng's model in which the following general form of their equation is retained:

$$\sigma_{dbic} = \alpha\beta_p\beta_L\sqrt{\frac{E_{frp}\sqrt{f'_c}}{t_{frp}}} \tag{3.28}$$

but the coefficient α in the equation is recalibrated against the test data of beams and slabs they collected. The values for the coefficient α are given in

Tables 3.4–3.6 for different test data sets and different degrees of conservativeness in prediction, and β_p and β_L are the width and bond length coefficients respectively as defined in Section 2.4.

In arriving at these values, the assumption was made by Teng *et al.* (2001b) that plane sections remain plane so that the moment at debonding failure at a

Table 3.4 Test-to-predicted debonding strength ratios: beams

Ratio	α	Average	Standard deviation	Coefficient of variation	Percentage of exceedence	No. of tests
$\sigma_{exp}/\sigma_{dbic}$	0.427	2.58	0.79	31 %	2.3 %	7
	1.100 (average)	1.00	0.31	31 %	50 %	7
	0.544 (95 %)	2.02	0.62	31 %	5.0 %	7
	0.4	2.75	0.85	31 %	1.9 %	7
M_{exp}/M_{db}	0.427	1.56	0.57	37 %	16 %	8
	1.100 (average)	1.04	0.16	15 %	41 %	8
	0.544 (95 %)	1.38	0.42	30 %	18 %	8
	0.4	1.61	0.61	38 %	16 %	8

Table 3.5 Test-to-predicted debonding strength ratios: slabs

Ratio	α	Average	Standard deviation	Coefficient of variation	Percentage of exceedence	No. of tests
$\sigma_{exp}/\sigma_{dbic}$	0.427	1.69	0.35	21 %	2.3 %	9
	0.720 (average)	1.00	0.21	21 %	50 %	9
	0.478 (95 %)	1.51	0.31	21 %	5.0 %	9
	0.4	1.80	0.37	21 %	1.5 %	9
M_{exp}/M_{db}	0.427	1.24	0.23	19 %	15 %	9
	0.720 (average)	1.00	0.15	15 %	50 %	9
	0.478 (95 %)	1.18	0.19	16 %	18 %	9
	0.4	1.27	0.26	20 %	15 %	9

Table 3.6 Test-to-predicted debonding strength ratios: combined beams and slabs

Ratio	α	Average	Standard deviation	Coefficient of variation	Percentage of exceedence	No. of tests
$\sigma_{exp}/\sigma_{dbic}$	0.427	2.08	0.72	35 %	6.8 %	16
	0.887 (average)	1.00	0.35	35 %	50 %	16
	0.379 (95 %)	2.34	0.81	35 %	5.0 %	16
	0.4	2.22	0.77	35 %	5.7 %	16
M_{exp}/M_{db}	0.427	1.39	0.44	32 %	19 %	17
	0.887 (average)	1.02	0.20	19 %	46 %	17
	0.379 (95 %)	1.47	0.51	35 %	18 %	17
	0.4	1.43	0.48	33 %	18 %	17

critical section is related to the stress in the bonded plate by a conventional section analysis using equations given in Section 3.4. Although this assumption is not strictly valid in zones where the FRP has debonded from the beam and should be carefully examined in future research, it does allow a simple way to relate the stress in the FRP plate to the applied moment. This assumption made it possible to find the strain in the FRP plate at debonding using the known debonding moment or to calculate the debonding moment capacity once the plate stress at the critical section is predicted using Equation 3.28. For the cantilever slab tests (Yao *et al.* 2001), strains measured on the FRP plates were also reported. Teng *et al.* (2001b) found that the use of these strains led to less scatter in the deduced value for α but this option was not available for the other test data. Therefore, in finding a suitable value of α for design use, the plate stresses in the slabs at the critical section were found from the measured strains while those for the other tests were found from a section analysis using the debonding moment (Teng *et al.* 2001b). The critical sections where debonding is expected to initiate include the position of maximum moment (e.g. the mid-span in three-point bending tests or the fixed end of a cantilever beam or slab), the end of the maximum moment region (e.g. in simply supported beams/one-way slabs subjected to four-point bending), and where the amount of flexural reinforcement or bonded FRP changes.

The application of Equation 3.28 also requires the definition of the bond length, being the distance between the critical section and the nearer plate end. In design, as recommended later in the chapter, it should be ensured that the FRP plate extends for a distance greater than the effective bond length beyond the critical section so that the bond length coefficient β_L in Equation 3.28 is equal to 1.0.

Tables 3.4–3.6 detail the statistical performance of Chen and Teng's model with various values for the coefficient α based on both the debonding plate stresses and the debonding moments as assessed using the collected test data presented in Teng *et al.* (2001b). These test and predicted stresses and moments are denoted by σ_{exp}, σ_{dbic}, M_{exp} and M_{db} respectively. Tables 3.4–3.6 are for beam, cantilever slab, and combined beam and cantilever slab results respectively. One beam data point was treated as a statistical outlier and thus excluded in the comparison using plate stresses (Teng *et al.* 2001b). It was, however, included in all statistical analyses for the moment capacity which is less sensitive to the stress in the plate because of the presence of steel flexural reinforcement.

The stresses obtained from Chen and Teng's model, using a value of 0.427 for α (i.e. Chen and Teng's original best-fit model) are assessed in the first line of each table. The second line in the tables is for an α value from a regression analysis of the respective data set. The third α value corresponds to the 5 % exceedence (i.e. 95 percentile lower bound) while the fourth α value of 0.4 is for use in design. Also reported in these tables are the corresponding test-to-predicted debonding moment statistics and these are given in the bottom half of the tables. In evaluating the theoretical moment capacities, all limiting conditions including FRP rupture and concrete crushing were considered (Teng *et al.* 2001b). In a few cases of the slab tests, when the values of α from regression analyses were used, some of the predicted stresses exceeded the rupture strength of the FRP. In these cases, the actual predicted stress was used in the statistical analyses for stresses but the rupture strength of the FRP was used when determining the moment in the section for the corresponding moment-based statistics.

A value of 0.4 for α is seen to correspond to a 5.7 % exceedence in terms of the stresses in the plate. Although this corresponds to an 18 % exceedence in terms of the debonding moments (Table 3.6) due to the scatter in the data, all but one test result are conservatively predicted (Figure 3.25), and for this one case the prediction is also close. This value of α is therefore appropriate for providing lower bound predictions of stresses/moments at intermediate flexural crack-induced debonding. This modified Chen and Teng model (Teng *et al.* 2001b) is therefore recommended for design use.

3.7.3 Intermediate Flexural Shear Crack-induced Debonding

Intermediate flexural shear crack-induced debonding differs from flexural crack-induced debonding in that in the former a relative vertical displacement between the crack faces will impose peeling stresses on the FRP plate-to-concrete interface. Test data for flexural shear crack-induced debonding gathered by Teng *et al.* (2001b) are also shown in Figure 3.25. Surprisingly, all such failures are conservatively predicted by the modified Chen and Teng model described above. It is therefore recommended that the modified Chen and Teng model be used to design against both intermediate crack-induced debonding modes until further studies are carried out.

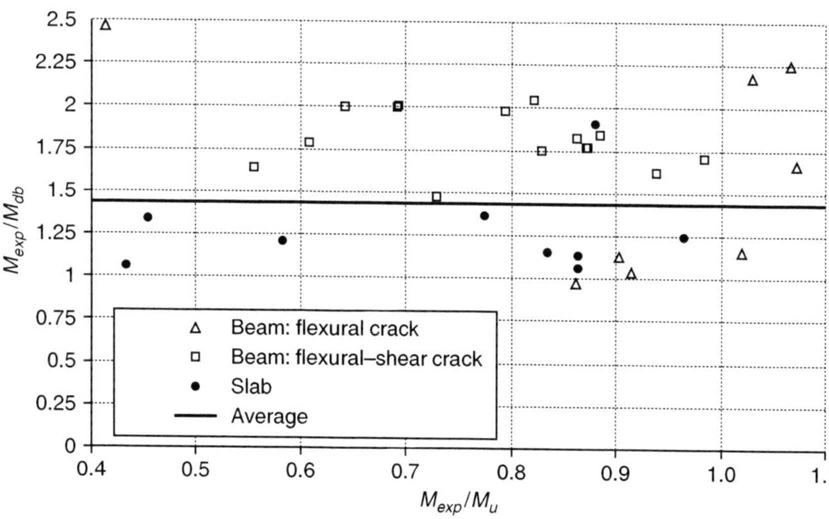

Figure 3.25 Test results versus predictions of the modified Chen and Teng model: $\alpha = 0.4$

3.8 Design Recommendation

3.8.1 Critical Sections and Plate-end Anchorage

For a strengthened RC beam, the ultimate moment capacity, taking into account the possibility of debonding, needs to be checked for two critical sections: (a) the maximum moment section; and (b) the section just outside the effectively strengthened region, provided that the moment capacity of the original beam does not change along the beam length. If the moment capacity changes, it may be necessary to check it for all sections at a change. Moment capacity changes can be due to changes in section size, the nonuniformity of the original steel reinforcement, or the nonuniformity of the external FRP reinforcement. The design recommendations described below are applicable to both situations, but specific reference is only made to beams of uniform original moment capacity for ease of presentation.

The effectively strengthened region L_{eff} is taken to be the plated region L_p minus an anchorage length L_a of the plate at each end (Figure 3.26). The effective bond length determined from lap shear tests as given in Section 2.4 with an additional safety factor of two, is recommended as the anchorage length. This ensures that in the case of debonding induced by intermediate cracking, the plate stress sustained is maximised. Following BS 8110 (1997), this anchorage length (Equation 2.16b) is now given in terms of the concrete cube compressive strength as

$$L_a = 2.11\sqrt{\frac{E_{frp}t_{frp}}{\sqrt{f_{cu}}}} \approx 2\sqrt{\frac{E_{frp}t_{frp}}{\sqrt{f_{cu}}}} \tag{3.29}$$

The capacity of the RC beam alone at the ends of the effectively strengthened region needs to be checked to ensure that it exceeds the applied moment here for the strengthened beam. This check can be done using any existing codes of practice such as BS 8110 (1997). For checking the strength of the maximum moment section, the following approach is recommended.

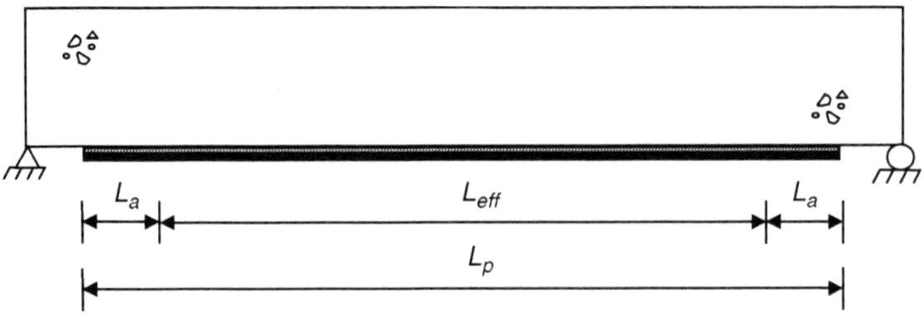

Figure 3.26 Definitions of L_a, L_{eff} and L_p

3.8.2 Strength Check

Smith and Teng's (2001d) model is recommended for the plate-end debonding strength, while the modified Chen and Teng (2001) model is recommended for the intermediate crack-induced debonding strength.

The strength check for the maximum moment section of the beam consists of the following five steps:

(a) Determination of the ultimate plate stress σ_{dbic} for intermediate crack-induced debonding according to the modified Chen and Teng model with a value of 0.4 for α in Equation 3.28. In terms of the concrete cube compressive strength instead of the cylinder compressive strength, the plate stress at debonding is given by

$$\sigma_{dbic} = 0.38\beta_p\beta_L\sqrt{\frac{E_{frp}\sqrt{f_{cu}}}{t_{frp}}} \tag{3.30}$$

(b) The reduced tensile strength $f_{frp,r}$ of the FRP plate is worked out as the smaller of the ultimate tensile strength f_{frp} and σ_{dbic}, that is

$$f_{frp,r} = \min\left(f_{frp}, \sigma_{dbic}\right) \tag{3.31}$$

(c) The flexural strength M_u of the maximum moment section is then evaluated using the design equations presented in Section 3.4 with f_{frp} replaced by $f_{frp,r}$.

(d) The plate-end debonding strength $V_{db,end}$ is now determined using Smith and Teng's model (Equation 3.32) presented in Section 3.6 as

$$V_{db,end} = 1.4V_c \tag{3.32}$$

and the moment at the end of the plate $M_{db,end}$ is checked to see if it satisfies

$$M_{db,end} \leqslant 0.67M_u \tag{3.33}$$

where M_u is that evaluated in step (c). If Equation 3.33 is not satisfied, the plate-end position needs to be adjusted.

(e) The lower load-carrying capacity from steps (c) and (d) is the ultimate strength of the beam.

In Equation 3.33, the value of M_u is evaluated with intermediate crack-induced debonding taken into account. This differs from that initially recommended in Section 3.6.6. As the new M_u is smaller, this deviation from the original recommendation leads to safe designs and avoids the extra work of evaluating another value for M_u solely for use with Equation 3.33.

3.8.3 Partial Safety Factors

The reduced tensile strength of the FRP, $f_{frp,r}$, is calculated as the minimum stress of either the tensile capacity of the FRP or the stress determined from the modified Chen and Teng (2001) model. This reduced tensile strength is the characteristic strength of the FRP taking into account the effect of intermediate crack-induced debonding and should replace f_{frp} in the design equations presented in Section 3.4 where an FRP partial safety factor is included. It may be noted that this treatment does not lead to any inconsistency in the adopted safety margin in design if the partial safety factor for the tensile strength of the FRP is the same as that for the bond strength, as recommended in Section 1.6.

3.8.4 Application of the Design Procedure to Indeterminate Beams

The above design recommendation made for simply supported beams can also be applied to the design of FRP strengthening measures for indeterminate beams. This requires simply treating each part of the beam between points of contraflexure as a simply supported beam. In addition, if a clamped support exists, then the portion of the beam near the clamped support needs to be treated as a cantilever beam. For a cantilever beam, it is important to anchor the FRP strips into the clamped support. Proper anchorage for this purpose is discussed in detail in Section 5.2 for cantilever slabs. One interesting and important situation of this kind arises in the strengthening of coupling beams in coupled shear walls. Here, the coupling beams are subject to double bending, so for effective strengthening, the beam needs to be dealt with as two cantilever beams. Of course, such coupling beams are subject to reversed cyclic loading from either wind or earthquake, so flexural strengthening is required for bending in both directions.

3.9 Conclusions

Flexural strengthening of RC beams using bonded FRP soffit plates has become popular over the last few years. The basic strengthening procedure involves the bonding of FRP plates to the soffit of the beam. Variations from this basic procedure include additional end anchorage and prestressing of the plates. This chapter has been concerned chiefly with the behaviour and strength of RC beams bonded with unanchored and unstressed FRP plates, as such beams have been extensively researched and widely used in practice.

These FRP-plated beams can fail in various modes, which can be broadly classified into normal flexural failures and debonding failures. Both types of failures are brittle, but there exists limited ductility in flexural failures, and in intermediate crack-induced debonding when there is a sufficient distance between the crack and the plate end. If debonding failure occurs, the ultimate tensile strength of the FRP cannot be fully utilised.

For normal flexural failures, design equations have been presented based on the design approach of BS 8110 (1997), with modifications to take into account the brittle nature of the FRP. A method of accounting for preloading has also been presented.

In plated beams, debonding failures often govern the strength. Many different debonding failure modes are possible and have been discussed in the chapter. For design, they can be lumped into two modes: (a) plate-end debonding due to high interfacial shear and normal stresses at the end of the soffit plate, and (b) intermediate crack-induced debonding due to high interfacial stresses near a major crack away from the plate ends. Recommendations have been made about suitable debonding strength models for design use in this chapter. For the plate-end debonding mode, Smith and Teng's (2001d) model provides a close lower bound, while the modified Chen and Teng model (Teng *et al.* 2001b) gives close predictions for the intermediate crack-induced debonding mode.

Finally, the design recommendations for flexural strengths and debonding strengths are tied together to form a design approach. Smith and Teng's model and the modified Chen and Teng model are recommended for plate-end debonding and intermediate crack-induced debonding respectively. The flexural strength of the beam is evaluated using a reduced value of the FRP tensile strength, being the smaller of the ultimate tensile stress and the ultimate plate stress at debonding as given by the modified Chen and Teng model. In addition, the plate-end debonding strength from Smith and Teng's model needs to be checked.

3.10 Appendix A

The analytical solution derived by Smith and Teng (2001b) for the interfacial shear and normal stresses is given here. Linear elastic behaviour is assumed, with deformations due to both axial forces and bending moments included for both the RC beam and the soffit plate. Shear deformations of both components are neglected, as their effect is small and their inclusion couples the two governing equations. The adhesive layer is assumed to be subject to stresses invariant across its thickness. The governing differential equation for the interfacial shear stress τ is given by

$$\frac{d^2\tau(x)}{dx^2} - \frac{G_a b_p}{t_a}\left(\frac{(y_c + y_p)(y_c + y_p + t_a)}{E_c I_c + E_p I_p} + \frac{1}{E_c A_c} + \frac{1}{E_p A_p}\right)\tau(x)$$
$$= -\frac{G_a}{t_a}\left(\frac{y_c + y_p}{E_c I_c + E_p I_p}\right) V(x) \qquad \text{(A3.1)}$$

where b_p is the width of the soffit plate, t_a is the adhesive layer thickness and A, E, I and G are the cross-sectional area, elastic modulus, second moment of area and shear modulus respectively. The subscripts c, p and a refer to the RC beam, the soffit plate and the adhesive layer respectively. The distance from the centroid of the RC beam to its base and the distance from the centroid of the soffit plate to its top are denoted by y_c and y_p respectively. The coordinate x is the distance along the soffit plate from its end and $V(x)$ is the applied shear force at any point along the soffit plate.

The governing differential equation for the interfacial normal stress σ_y is given by

$$\frac{\mathrm{d}^4\sigma_y(x)}{\mathrm{d}x^4} + \frac{E_a b_p}{t_a}\left(\frac{1}{E_c I_c} + \frac{1}{E_p I_p}\right)\sigma_y(x) = -\frac{E_a b_p}{t_a}\left(\frac{y_c}{E_c I_c} - \frac{y_p}{E_p I_p}\right)\frac{\mathrm{d}\tau(x)}{\mathrm{d}x} - \frac{E_a}{t_a}\frac{1}{E_c I_c}q \quad \text{(A3.2)}$$

where q is a uniformly distributed load. A general expression for the interfacial shear stress for all load cases, and applicable over the whole length of the soffit plate, is given as

$$\tau(x) = m_1 V(x) + \frac{m_2}{\lambda} M(0)\mathrm{e}^{-\lambda x} + \tau^*(x) \quad \text{(A3.3)}$$

where

$$\lambda^2 = \frac{G_a b_p}{t_a}\left(\frac{(y_c + y_p)(y_c + y_p + t_a)}{E_c I_c + E_p I_p} + \frac{1}{E_c A_c} + \frac{1}{E_p A_p}\right)$$

$$m_1 = \frac{G_a}{t_a}\frac{1}{\lambda^2}\left(\frac{y_c + y_p}{E_c I_c + E_p I_p}\right), \quad m_2 = \frac{G_a}{t_a}\frac{y_c}{E_c I_c}$$

and $M(0)$ is the applied bending moment at the end of the soffit plate. The term $\tau^*(x)$ varies for each load case and is given in Table A3.1, where $k = \lambda(B - a)$, a is the distance from the support to the nearer end of the soffit plate, and B is the distance from the support to the nearest applied concentrated load P (i.e. shear span). The region of the soffit plate, for which the additional shear stress term is applicable, is also indicated in Table A3.1 by double-arrowed lines.

The solution for interfacial normal stress is given by

$$\sigma_y(x) = \mathrm{e}^{-\beta x}\left[C_1 \cos(\beta x) + C_2 \sin(\beta x)\right] - n_1 \frac{\mathrm{d}\tau(x)}{\mathrm{d}x} - n_2 q \quad \text{(A3.4)}$$

where

$$\beta = \sqrt[4]{\frac{E_a b_p}{4t_a}\left(\frac{1}{E_c I_c} + \frac{1}{E_p I_p}\right)}, \quad n_1 = \left(\frac{y_c E_p I_p - y_p E_c I_c}{E_c I_c + E_p I_p}\right),$$

$$n_2 = \frac{E_p I_p}{b_p\left(E_c I_c + E_p I_p\right)}$$

Table A3.1 Interfacial shear stress – additional shear stress terms

Load case	$\tau^*(x)$
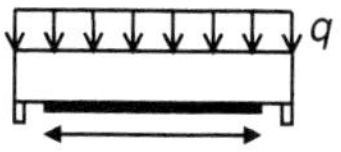	$-\frac{m_1}{\lambda}qe^{-\lambda x}$
	$-m_1 P\cosh(\lambda x)e^{-k}$
	$m_1 P\sinh(k)e^{-\lambda x}$
	0

$$C_1 = \frac{E_a}{2\beta^3 t_a}\frac{1}{E_c I_c}[V(0)+\beta M(0)] - \frac{n_3}{2\beta^3}\tau(0) + \frac{n_1}{2\beta^3}\left(\left.\frac{d^4\tau(x)}{dx^4}\right|_{x=0} + \beta\left.\frac{d^3\tau(x)}{dx^3}\right|_{x=0}\right)$$

and

$$C_2 = -\frac{E_a}{2\beta^2 t_a}\frac{1}{E_c I_c}M(0) - \frac{n_1}{2\beta^2}\left.\frac{d^3\tau(x)}{dx^3}\right|_{x=0}$$

3.11 Appendix B

Details of all 12 models reviewed in Section 3.6.2 are given here. These models are presented in the following three categories based on their approaches: (a) shear-capacity-based models (Oehlers 1992, Jansze 1997, Ahmed and van Gemert 1999), (b) concrete tooth models (Raoof and Zhang 1997, Wang and Ling 1998, Raoof and Hassanen 2000), and (c) interfacial stress-based models (models I and II of Ziraba *et al.* 1994, Varastehpour and Hamelin 1997, Saadatmanesh and Malek 1998, Tumialan *et al.* 1999b). All models are presented here using consistent notation and in a form that best facilitates comparisons between them, based on the work of Smith and Teng (2001c). It should be noted that some of the models are dimensionally inconsistent, and that all models presented in this book are based on the following set of consistent units: forces in Newtons (N), lengths in millimetres (mm) and hence stresses in Megapascals (MPa).

3.11.1 Shear-capacity-based Models

The common feature of these models is that the debonding failure strength is related to the shear strength of the concrete with no or only partial contribution of the steel shear reinforcement. The debonding strength is generally given as the shear force acting at the plate end, with or without taking into account the effect of any coexistent moment. As interfacial stresses between the plate and the beam need not be evaluated, the required calculations are generally simple.

3.11.1.1 Oehlers' Model

Oehlers and Moran (1990) and Oehlers (1992) investigated simply supported steel-plated beams subjected to either four- or three-point bending and developed a strength model by first considering two extreme positions of soffit plate termination. For a plate terminated in the constant moment region, the following expression for the flexural debonding moment $M_{db.f}$ at the end of the plate was proposed and calibrated with test results of steel-plated beams:

$$M_{db.f} = \frac{E_c I_{trc,c} f_{ct}}{0.901 E_{frp} t_{frp}} \tag{B3.1}$$

where E_c and E_{frp} are the moduli of elasticity of the concrete and the FRP respectively, $I_{trc,c}$ the cracked second moment of area of the plated section transformed to concrete, f_{ct} the cylinder splitting tensile strength of concrete, and t_{frp} the plate thickness. If the splitting tensile strength is not determined from tests, f_{ct} (in MPa) is taken as $0.5\sqrt{f_c'}$ where f_c' is the concrete cylinder compressive strength in MPa (Oehlers 1992). It was suggested that in design, this moment needs only to be compared with the additional moment applied to the beam at the plate end after the plate is bonded.

For a plate terminated near the support, it was proposed that debonding occurs when the shear force at the plate end, $V_{db,s}$, reaches the shear capacity of the concrete in the RC beam alone, without the contribution from the steel shear reinforcement, based on experimental observations (Oehlers 1992). That is,

$$V_{db,s} = V_c = [1.4 - (d/2000)] b_c d \left[\rho_s f_c'\right]^{1/3} \tag{B3.2}$$

where $\rho_s = A_s/b_c d$ is the ratio of steel tension reinforcement, A_s the area of steel tension reinforcement, b_c the section width and d the effective depth of the section. The above expression for V_c, the concrete shear capacity in the beam, is that given by the Australian concrete code (AS 3600 1988) and it is a requirement that $1.4 - (d/2000) \geqslant 1.1$.

For general cases where both the shear force and the moment at the plate end are important, the following interaction equation was suggested based on test results (Oehlers 1992):

$$\frac{M_{db,end}}{M_{db.f}} + \frac{V_{db,end}}{V_{db,s}} = 1.17 \tag{B3.3}$$

and

$$M_{db,end} \leqslant M_{db,f}, \quad V_{db,end} \leqslant V_{db,s} \tag{B3.4}$$

In terms of the shear force at the plate end, Equation B3.3 can be rearranged to read

$$V_{db,end} = \frac{1.17}{a/M_{db,f} + 1/V_{db,s}} \tag{B3.5}$$

and

$$V_{db,end}a \leqslant M_{db,f}, \quad V_{db,end} \leqslant V_{db,s} \tag{B3.6}$$

where a is the distance from the support to the nearer end of the soffit plate.

3.11.1.2 Jansze's Model

Jansze (1997) also proposed a debonding strength model for steel-plated beams but the background information to the model was not available to Smith and Teng (2001c). They instead referred to Ahmed and van Gemert (1999) where the final form of the model with general comments were given. It was stated in Ahmed and van Gemert (1999) that Jansze's model was based on the initiation of shear cracking in an RC beam without the contribution of shear reinforcement. The critical shear force in the RC beam at the plate end to cause debonding $V_{db,end}$ is given as follows:

$$V_{db,end} = \tau_{PES} b_c d \tag{B3.7}$$

where

$$\tau_{PES} = 0.18\sqrt[3]{3\frac{d}{B_{mod}}}\left(1 + \sqrt{\frac{200}{d}}\right)\sqrt[3]{100\rho_s f_c'} \tag{B3.8}$$

$$B_{mod} = \sqrt[4]{\frac{\left(1 - \sqrt{\rho_s}\right)^2}{\rho_s} da^3} \tag{B3.9}$$

Here, B is the shear span and B_{mod} a modified shear span as given by Equation B3.9. If B_{mod} as defined by Equation B3.9 is greater than the actual shear span B of the beam then the modified shear span should be given by $(B_{mod} + B)/2$. Jansze's model appears to be invalid for soffit plates terminated at the support as B_{mod} becomes zero and Equation B3.8 predicts that debonding is never possible.

3.11.1.3 Ahmed and Van Gemert's Model

Ahmed and van Gemert (1999) modified Jansze's (1997) model to be suitable for use with FRP-plated RC beams. Ahmed and van Gemert's model is given as follows:

$$V_{db,end} = (\tau_{PES} + \Delta\tau_{mod})\, b_c d \tag{B3.10}$$

where

$$\Delta\tau_{MOD} = \tau_{PES} b_c d \left(\frac{S_s}{I_s b_{frp}} - \frac{S_{frp}}{I_{trc,c} b_a} \right) + 6188.5 \left(\frac{\tau - 4.121}{b_c d} \right) \tag{B3.11}$$

$$\tau = \left(0.15776\sqrt{f_c'} + \frac{17.2366 \rho_s d}{B} \right) + 0.9 \frac{A_{sv} f_{yv}}{s b_c} \tag{B3.12}$$

where τ_{PES} is given by Equation B3.8, S_{frp} and S_s are respectively the first moment of area of the FRP plate and that of an equivalent steel plate about the neutral axis of a cracked plated section transformed to concrete, where the equivalent steel plate is one that has the same total tensile capacity and width as that of the FRP plate, but with an equivalent thickness determined assuming that the yield stress of steel is 550 MPa, $I_{trc,c}$ and I_s are the second moments of area of a cracked plated section transformed to concrete with an FRP plate and an equivalent steel plate respectively, and b_{frp} and b_a are the widths of the FRP and adhesive respectively. The stirrup spacing is denoted by s, whilst A_{sv} and f_{yv} are the cross-sectional area and the yield stress of the steel stirrups respectively. For all practical purposes b_a is equal to b_{frp}.

The modifications given in Equations B3.11 and B3.12 appear to rely on determining the difference in the interfacial shear stress between the RC beam and the plate as a result of replacing a steel plate with an FRP plate. This difference, $\Delta\tau_{mod}$, is then added to the shear stress τ_{PES} in Equation B3.8 where the shear stress is determined over a fictitious shear span of length B_{mod}. The increase in shear strength offered by the shear reinforcement is also included as seen in Equation B3.12.

3.11.2 Concrete Tooth Models

Concrete tooth models make use of the concept of a concrete 'tooth' between two adjacent cracks deforming like a cantilever under the action of horizontal shear stresses at the base of the beam (Figure 3.19). Debonding is deemed to occur when these shear stresses lead to tensile stresses at the root of the 'tooth' that exceed the tensile strength of the concrete. The stress in the soffit plate at debonding can then be determined by defining an effective length for the plate for end anchorage over which a uniform shear stress is assumed.

3.11.2.1 Raoof and Zhang's Model

The concept of a concrete tooth was first described in Zhang *et al.* (1995). Based on this concept, Zhang *et al.* (1995) and Raoof and Zhang (1997) developed a strength model to predict concrete cover separation failures in steel-plated beams. Minimum and maximum crack widths (l_{min} and l_{max} respectively) are determined which lead to a lower and an upper bound stress in the plate to cause debonding respectively and the width of these cracks differs by a factor of two ($l_{max} = 2l_{min}$). The minimum width of a crack l_{min}, termed the minimum stabilised crack spacing (Zhang *et al.* 1995), is given by

$$l_{min} = \frac{A_e f_{ct}}{u\left(\sum O_{bars} + b_{frp}\right)} \tag{B3.13}$$

where A_e is the area of concrete in tension, u the steel-to-concrete average bond strength, and $\sum O_{bars}$ the total perimeter of the tension reinforcing bars. It is assumed that $u = 0.28\sqrt{f_{cu}}$ (in MPa) and $f_{ct} = 0.36\sqrt{f_{cu}}$ (in MPa) where f_{cu} is the concrete cube compressive strength. For the case of an RC beam with a single layer of steel tension reinforcement, A_e is twice the distance from the centroid of the tension reinforcement to the base of the RC beam multiplied by the width of the RC beam.

Ignoring interaction between adjacent teeth and assuming elastic behaviour, the required shear stress to cause failure of a tooth can be determined. Failure of the tooth occurs when the stress at point A (Figure 3.19b) exceeds the tensile strength of the concrete. The tensile stress at point A, σ_A, can be determined from

$$\sigma_A = \frac{M_A}{I_A}\left(\frac{l}{2}\right) \tag{B3.14}$$

where $M_A = \tau l b_{frp} h'$ and $I_A = b_{frp} l^3/12$. Here, l is the crack spacing (minimum or maximum), h' the net height of concrete cover measured from the base of the steel tension reinforcement to the base of the concrete beam, τ the shear stress at the interface between the concrete and the soffit plate, I_A the second moment of area of the tooth, and M_A the moment at the base of the tooth.

Substitution of M_A and I_A into Equation B3.14 and assuming that at the instant of plate debonding $\sigma_A = f_{ct}$, the shear stress at the interface between the concrete and the steel plate based on a minimum stabilised crack spacing can be determined as follows. In this tooth theory, all teeth in the end anchorage zone are assumed to fail simultaneously at debonding.

$$\tau_{min} = \frac{f_{ct} l_{min}}{6h'} \frac{b_c}{b_{frp}} \tag{B3.15}$$

The minimum stress in the soffit plate, $\sigma_{s,min}$, required to cause flexural cracking and failure of a tooth covering the minimum stabilised crack spacing can then be determined as follows (Zhang *et al.* 1995):

$$\sigma_{s,min} = 0.154 \frac{L_p h_1 b_c^2 \sqrt{f_{cu}}}{h' b_{frp} t_{frp}\left(\sum O_{bars} + b_{frp}\right)} \tag{B3.16}$$

where L_p is an effective length of soffit plate for end anchorage, and h_1 is the distance from the centroid of the tension reinforcement to the base of the RC beam. This model has been applied only to simply supported beams subject to three- or four-point bending, for which the limiting stress of Equation B3.16 was compared with the stress in the soffit plate directly under a point load to check the possibility of debonding. It is not clear how this model should be applied to beams subject to other loading conditions (e.g. uniformly distributed load), but it is not unreasonable to choose the section of maximum moment in the beam for an assessment.

In Zhang *et al.* (1995), the effective length for end anchorage was taken as the length of the soffit plate in the shear span (in Figure 3.19a this is indicated as L_{p1}). In Raoof and Zhang (1997), it was recommended that the effective length for end anchorage be the smaller of the length of the soffit plate in the shear span L_{p1} and the following lengths L_{p2} which were calibrated against test data of steel-plated RC beams that failed by plate-end debonding:

$$L_{p2} = l_{min}(21 - 0.25 l_{min}) \quad l_{min} \leqslant 72\,\text{mm} \tag{B3.17a}$$

$$L_{p2} = 3 l_{min} \quad l_{min} > 72\,\text{mm} \tag{B3.17b}$$

Only the modified effective length for plate-end anchorage as proposed by Raoof and Zhang (1997) was considered by Smith and Teng (2001c, d) as this is the more recent tooth model derived by this research group for steel-plated RC beams.

Once the stress in the plate is known, the moment to cause separation of the plate, at the same location as the stress, can be calculated using a conventional section analysis with the assumption of plane section bending in accordance with Zhang *et al.* (1995) with the tensile strength of concrete taken into account. A lower and an upper bound bending moment are determined as a consequence of the assumed minimum (l_{min}) and maximum (l_{max}) stabilised crack spacings differing by a factor of two.

For a simply supported beam under four-point bending with the plate positioned in the constant moment region, Raoof and Zhang (1997) specified that the effective length should be obtained from Equation B3.17 as the length of the plate in the shear span L_{p1} is zero.

3.11.2.2 Wang and Ling's Model

Wang and Ling (1998) proposed a modification to Zhang *et al.*'s (1995) tooth model to make it suitable for FRP-strengthened beams. This involved modifying the average bond strength between the concrete and the plate, leading to the following formula for the minimum crack spacing:

$$l_{min} = \frac{A_e f_{ct}}{u_s \sum O_{bars} + u_{frp} b_{frp}} \tag{B3.18}$$

where $u_s = 0.313\sqrt{f_c'}$ is the average bond strength between the steel tension reinforcement and concrete and is the same as that used by Zhang *et al.* (1995) assuming $f_c' = 0.8 f_{cu}$, while u_{frp} is the average bond shear strength between the FRP and the concrete and was taken as 1.96 MPa.

Once the minimum stabilised crack spacing is determined the remainder of the analysis is essentially the same as that given in Zhang *et al.* (1995), although the effective length for end anchorage was taken by Wang and Ling (1998) to be the total plate length in the shear span. In Wang and Ling's model, the tensile strength of concrete was not included in the section analysis for relating the bending moment in the beam to the stress in the plate. Similar to the model of Zhang *et al.* (1995), an upper bound solution of Wang and Ling's model is obtainable by replacing their minimum crack spacing (Equation B3.18) with a maximum crack spacing which is twice the former.

3.11.2.3 Raoof and Hassanen's Models

Raoof and Hassanen (2000) also modified Zhang *et al.*'s (1995) model for application to FRP-plated beams. Two expressions were presented by them for the effective length of the FRP plate for end anchorage with two corresponding values for the bond strength between the FRP and the concrete. Two new models thus resulted from these modifications. They are referred to as Raoof and Hassanen's (2000) model I and model II.

In Raoof and Hassanen's model I, Equation B3.13 is retained to calculate the minimum stabilised crack spacing. The effective length of the plate for end anchorage is taken to be the smaller of the plate length in the shear span (L_{p1} in Figure 3.19a) and the following lengths which were calibrated with test data of FRP-plated RC beams reported to have failed by plate-end debonding (Raoof and Hassanen 2000):

$$L_{p2} = l_{min}(24 - 0.5l_{min}) \quad l_{min} \leqslant 40\,\text{mm} \tag{B3.19a}$$

$$L_{p2} = 4l_{min} \quad l_{min} > 40\,\text{mm} \tag{B3.19b}$$

In Raoof and Hassanen's model II, the bond strength between the FRP plate and the concrete is specified as 0.8 MPa, whilst the bond strength between the steel tension reinforcement and the concrete is still the same as that originally specified in Zhang *et al.* (1995). The effective length of the FRP plate for end anchorage is defined as the smaller of the plate length in the shear span and the following lengths which were calibrated with test data of FRP-plated RC beams reported to have failed by plate-end debonding (Raoof and Hassanen 2000) using the new value of FRP-to-concrete bond strength:

$$L_{p2} = l_{min}(11.6 - 0.17l_{min}) \quad l_{min} \leqslant 56.5\,\text{mm} \tag{B3.20a}$$

$$L_{p2} = 2l_{min} \quad l_{min} > 56.5\,\text{mm} \tag{B3.20b}$$

Once the stabilised crack spacing and the effective length of FRP plate for end anchorage are determined, the remainder of the analysis is the same as that given in Zhang *et al.* (1995).

3.11.3 Interfacial Stress-based Models

A popular and logical assumption is that concrete cover separation or plate-end interfacial debonding is due to high interfacial stresses at the end of the soffit plate. Figure 3.15 shows the stresses acting on an element of concrete adjacent to the end of the adhesive layer where τ and σ_y denote the shear and transverse normal (peeling) stresses respectively, and σ_x is the longitudinal stress.

Interfacial stress-based debonding strength models generally make use of interfacial stresses from an existing closed-form solution (e.g. Roberts 1989) and a concrete failure criterion. Models of this type include model I of Ziraba *et al.* (1994), and those of Varastehpour and Hamelin (1997), Saadatmanesh and Malek (1998) and Tumialan *et al.* (1999b).

An alternative use of the interfacial stresses was made in developing model II of Ziraba *et al.* (1994), who assumed that the shear force resisted at debonding consists of the shear resistance of concrete and a part of the shear resistance of steel stirrups as determined by an efficiency factor. The stirrup efficiency factor is related to the peak interfacial stresses at the plate end and was calibrated with experimental data. This model therefore combines the shear capacity approach discussed above with interfacial stresses.

3.11.3.1 Ziraba *et al.*'s Models

Ziraba *et al.* (1994) proposed two debonding strength models for steel-plated RC beams, one for predicting plate-end interfacial debonding and the other for predicting concrete cover separation, which are termed Ziraba *et al.*'s model I and II respectively.

Model I: Plate-end Interfacial Debonding: In this model, the Mohr–Coulomb failure criterion is used to define the critical stress state at plate-end interfacial debonding:

$$\tau + \sigma_y \tan\phi \leqslant C \tag{B3.21}$$

where τ and σ_y are the peak interfacial shear and normal stresses at the plate end, C the coefficient of cohesion, and ϕ the angle of internal friction. The peak interfacial shear and normal stresses are given by

$$\tau = \alpha_1 f_{ct} \left(\frac{C_{R1} V_0}{f_c'} \right)^{5/4} \tag{B3.22}$$

$$\sigma_y = \alpha_2 C_{R2} \tau \tag{B3.23}$$

where

$$C_{R1} = \left[1 + \left(\frac{K_s}{E_{frp} b_{frp} t_{frp}}\right)^{1/2} \frac{M_0}{V_0}\right] \frac{b_{frp} t_{frp}}{I_{trc,frp} b_a} \left(d_{frp} - x_{trc,frp}\right) \quad \text{(B3.24)}$$

$$C_{R2} = t_{frp} \left(\frac{K_n}{4E_{frp} I_{frp}}\right)^{1/4} \quad \text{(B3.25)}$$

where C_{R1} and C_{R2} are obtained from Roberts' (1989) analytical solution for interfacial shear and normal stresses, while α_1 and α_2 are empirical multipliers calibrated from numerical studies (Ziraba *et al.* 1994) for RC beams retrofitted with steel plates. The shear stiffness K_s and the normal stiffness K_n of the adhesive layer are given by

$$K_s = \frac{G_a b_a}{t_a} \quad \text{(B3.26)}$$

$$K_n = \frac{E_a b_a}{t_a} \quad \text{(B3.27)}$$

with E_a, G_a, b_a and t_a being the modulus of elasticity, shear modulus, width and thickness of the adhesive layer respectively. $I_{trc,frp}$ is the second moment of area of the cracked plated section transformed into FRP, $x_{trc,frp}$ is the neutral axis depth of this transformed cracked section (distance from the compression face to the neutral axis), I_{frp} the second moment of area of the FRP plate alone, d_{frp} the distance from the compression face of the RC beam to the centroid of the FRP plate, and M_0 and V_0 are the bending moment and the shear force at the plate end respectively.

Substitution of Equations B3.22 and B3.23 into Equation B3.21 gives an expression for the shear force in the beam, at the plate-end, to cause plate-end interfacial debonding:

$$V_{db,end} = \frac{f_c'}{C_{R1}} \left[\frac{C}{\alpha_1 f_{ct} \left(1 + \alpha_2 C_{R2} \tan\phi\right)}\right]^{4/5} \quad \text{(B3.28)}$$

This relationship is subject to the constraint of $a/h < 3$ (Ziraba *et al.* 1994) where h is the depth of the RC beam. The following values for α_1, α_2 and ϕ were specified in Ziraba *et al.* (1994): $\alpha_1 = 35$, $\alpha_2 = 1.1$ and $\phi = 28°$. In Ziraba *et al.* (1994) two values of C were used for specific numerical examples, namely 2.68 MPa and 5.36 MPa, although the criteria used for the selection of these values were not made clear. Ziraba *et al.* (1995) suggested that C lies between 4.80 MPa and 9.50 MPa based on experimental and numerical findings. Smith and Teng (2001d) adopted a value of 7.15 MPa for C for this model in their work which is an average of the values given in Ziraba *et al.* (1995).

Model II: Concrete Cover Separation: In this model, Ziraba *et al.* (1994) assumed that the shear force resisted at debonding consists of the shear resistance of concrete V_c and a part of the shear resistance of steel stirrups V_s:

$$V_{db,end} = (V_c + kV_s) \tag{B3.29}$$

where V_c and V_s are given in their paper by

$$V_c = \frac{1}{6}\left(\sqrt{f_c'} + 100\rho_s\right) b_c d \tag{B3.30}$$

$$V_s = \left(A_{sv} f_{yv} d\right)/s \tag{B3.31}$$

In Equation B3.29, k is an empirically derived stirrup efficiency factor to adjust the contribution of the steel stirrups and is related to the peak interfacial normal stress at the plate end. Based on a regression analysis of the test results of 14 steel-plated RC beams which failed by concrete cover separation, Ziraba *et al.* (1994) proposed that

$$k = 2.4e^n \quad \text{and} \quad n = -0.08 C_{R1} C_{R2} \times 10^6 \tag{B3.32}$$

3.11.3.2 Varastehpour and Hamelin's Model

Varastehpour and Hamelin (1997) also developed a plate-end interfacial debonding strength model based on the Mohr–Coulomb failure criterion (Equation B3.21). In their model, the coefficient of cohesion C was calibrated from single lap shear tests and the angle of internal friction ϕ determined from small-scale FRP-plated beam tests that failed by debonding. Details of the particular mode of debonding exhibited by these test beams were not specified by them. An average value of 5.4 MPa was suggested for C, and a value of 33° was suggested for ϕ.

The shear stress required in the Mohr–Coulomb equation is given by

$$\tau = \frac{1}{2}\sqrt{\beta}\,(\lambda V_0)^{3/2} \tag{B3.33}$$

This equation was modified from the following expression for the shear stress between a steel soffit plate and an RC beam originally proposed by Jones *et al.* (1988):

$$\tau = \lambda V_0 \tag{B3.34}$$

where λ is the rigidity of the section defined by

$$\lambda = \frac{t_{frp} E_{frp}}{I_{trc,c} E_c}\left(d_{frp} - x_{trc,c}\right) \tag{B3.35}$$

where $x_{trc,c}$ is the neutral axis depth of a cracked plated section transformed to concrete. In Equation B3.33, the factor β was introduced to account for the effect of various variables that have a significant influence on the interfacial shear stress between the RC beam and the soffit plate, such as the rigidity and thickness of the plate, geometry of the section, and nature of loading based on a parametric study using a numerical model which considered slip between the plate and the RC beam. The following expression was developed by them for β:

$$\beta = \frac{1.26 \times 10^5 B}{h^{0.7} t_{frp} E_{frp}} \quad \text{(B3.36)}$$

The normal stress, σ_y as derived by Roberts (1989) is related to the shear stress as follows:

$$\sigma_y = C_{R2}\tau \quad \text{(B3.37)}$$

The shear force in the beam, at the plate end to cause debonding, can then be determined by

$$V_{db,end} = \frac{1.6\tau_{max}^{2/3}}{\lambda \beta^{1/3}} \quad \text{(B3.38)}$$

where

$$\tau_{max} = \frac{5.4}{1 + C_{R2}\tan 33^\circ} \quad \text{(B3.39)}$$

3.11.3.3 Saadatmanesh and Malek's Model

Saadatmanesh and Malek (1998) developed a debonding strength model to predict concrete cover separation in FRP-plated beams based on the assumption that concrete cover failure is related to high stresses at the plate end (Figure 3.15). Of the three stresses present at the end of the soffit plate, the shear stress τ and the normal stress σ_y are from a closed-form solution derived by Malek *et al.* (1998) while the longitudinal stress is from a section bending analysis, all based on an uncracked section. This closed-form solution was derived assuming the following quadratic distribution for the bending moment which can accommodate both point and uniformly distributed loads:

$$M = a_1(x + a)^2 + a_2(x + a) + a_3 \quad \text{(B3.40)}$$

where x is the distance along the soffit plate from its left end. For a simply supported beam subjected to three- or four-point bending, the bending moment at the plate end, M_0, for a soffit plate terminated in the shear span, is given by $M_0 = a_2 a + a_3$.

Under this bending moment distribution (Equation B3.40), the interfacial shear stress τ at the plate end is given as (Malek *et al.* 1998)

$$\tau = t_{frp}\left(b_3\sqrt{A} + b_2\right) \quad \text{(B3.41)}$$

where

$$A = \frac{G_a}{t_a t_{frp} E_{frp}} \tag{B3.42}$$

$$b_2 = \frac{E_{frp}}{I_{tru,c} E_c} \left(d_{frp} - x_{tru,c}\right) (2a_1 a + a_2) \tag{B3.43}$$

$$b_3 = E_{frp} \left[\frac{1}{I_{tru,c} E_c} \left(d_{frp} - x_{tru,c}\right) \left(a_1 a^2 + a_2 a + a_3\right) + 2 \frac{a_1 E_{frp}}{I_{tru,c} E_c} \left(d_{frp} - x_{tru,c}\right) \frac{t_a t_{frp}}{G_a} \right] \tag{B3.44}$$

where $I_{tru,c}$ is the uncracked second moment of area of the plated section transformed to concrete, and $x_{tru,c}$ is the neutral axis depth of this transformed section. The normal stress σ_y at the plate end is (Malek *et al.* 1998)

$$\sigma_y = \frac{K_n}{2b_a \beta^{*3}} \left(\frac{V_{frp}}{E_{frp} I_{frp}} - \frac{V_0^* + \beta^* M_0}{E_c I_c} \right) + \frac{q E_{frp} I_{frp}}{b_{frp} E_c I_c} \tag{B3.45}$$

where

$$V_0^* = V_0 - 0.5 h b_{frp} t_{frp} \left(b_3 \sqrt{A} + b_2 \right) \tag{B3.46}$$

$$V_{frp} = -0.5 b_{frp} t_{frp}^2 \left(b_3 \sqrt{A} + b_2 \right) \tag{B3.47}$$

and

$$\beta^* = \left(\frac{K_n b_{frp}}{4 b_a E_{frp} I_{frp}} \right)^{1/4} \tag{B3.48}$$

where I_c is the second moment of area of the beam and q a uniformly distributed load if such a load exists.

The longitudinal stress σ_x at the base of the RC beam, at the end of the soffit plate, due to a bending moment M_0 can be determined from a bending analysis of an uncracked section. The bending moment in the concrete beam at the plate end is increased by an amount M_m, defined as follows and attributed to the peak interfacial shear stress:

$$M_m = 0.5 h a b_{frp} \tau \tag{B3.49}$$

Once all three stress components at the plate end are determined, the maximum principal stress can be found as follows:

$$\sigma_1 = \left(\frac{\sigma_x + \sigma_y}{2} \right) + \sqrt{\left(\frac{\sigma_x - \sigma_y}{2} \right)^2 + \tau^2} \tag{B3.50}$$

At the plate end, the concrete is generally subjected to biaxial tension. Concrete cover separation is deemed to occur when the maximum principal stress (in MPa) reaches the concrete splitting tensile strength, that is when

$$\sigma_1 = f_{ct} = 0.295\left(f_c'\right)^{2/3} \tag{B3.51}$$

3.11.3.4 Tumialan *et al.*'s Model

Tumialan *et al.* (1999b) developed a concrete cover separation debonding strength model for FRP-plated beams. Their model is in principle similar to Saadatmanesh and Malek's model reviewed above, in that the stresses acting on a concrete element at the plate end (Figure 3.15) are determined and checked against a failure criterion for the concrete. The differences lie in how these stresses are evaluated. In Tumialan *et al.*'s model, the peak interfacial shear and normal stresses at the plate end are determined from the solution of Roberts (1989) as follows:

$$\tau = \overline{C}_{R1}\frac{E_{frp}}{E_c}V_0 \tag{B3.52}$$

$$\sigma_y = C_{R2}\tau \tag{B3.53}$$

where C_{R2} is defined by Equation B3.25 while $\overline{C}_{R1}$ is given by

$$\overline{C}_{R1} = \left[1 + \left(\frac{K_s}{E_{frp}b_{frp}t_{frp}}\right)^{1/2}\frac{M_0}{V_0}\right]\frac{b_{frp}t_{frp}}{I_{tru,c}b_a}\left(d_{frp} - x_{tru,c}\right) \tag{B3.54}$$

The factor $\overline{C}_{R1}$ here differs from C_{R1} in Roberts (1989) in that the neutral axis and corresponding second moment of area are based here on an uncracked section transformed to concrete as opposed to a cracked section transformed to the plate material. As a result, the modular ratio of E_{frp}/E_c is included in Equation B3.52. The longitudinal stress at the base of the RC beam, σ_x, at the plate end, is also determined from a bending analysis of an uncracked section. However, the additional component induced by the interfacial shear stress as considered by Saadatmanesh and Malek (1998) in Equation B3.49 is not included here.

Once the stresses at the plate end are determined, the maximum principal stress σ_1 can be easily determined using Equation B3.50. Failure by concrete cover separation is deemed to occur when the maximum principal stress σ_1 reaches the modulus of rupture of the concrete f_r which is taken as $0.689\sqrt{f_c'}$ (MPa) ($8.3\sqrt{f_c'}$ psi, Mirza *et al.* 1979).

References

ACI 318-95 (1999) *Building Code Requirements for Structural Concrete (318-95) and Commentary (318R-95)*, American Concrete Institute (ACI), Fifth Printing, Farmington Hills, Michigan, USA.

Ahmed, O. and van Gemert, D. (1999) 'Effect of longitudinal carbon fiber reinforced plastic laminates on shear capacity of reinforced concrete beams', *Proceedings of the Fourth International Symposium on Fiber Reinforced Polymer Reinforcement for Reinforced Concrete Structures*, Baltimore, Maryland, USA, edited by C. W. Dolan, S. H. Rizkalla and A. Nanni, pp. 933–943, American Concrete Institute, Farmington for Hills, Michigan, USA.

An, W., Saadatmanesh, H. and Ehsani, M. R. (1991) 'RC beams strengthened with FRP plates II: analysis and parametric study', *Journal of Structural Engineering, ASCE*, Vol. 117, No. 11, pp. 3434–3455.

Arduini, M. and Nanni, A. (1997) 'Behavior of precracked RC beams strengthened with carbon FRP sheets', *Journal of Composites for Construction, ASCE*, Vol. 1, No. 2, pp. 63–70.

AS 3600 (1988) *Concrete Structures, AS 3600*, Standards Australia, Sydney, Australia.

Ashrafuddin, M., Baluch, M. H., Sharif, A., Al-Sulaimani, G. J., Azad, A. K. and Khan, A. R. (1999) 'Peeling and diagonal tension failures in steel plated R/C beams', *Construction and Building Materials*, Vol. 13, pp. 459–467.

Basunbul, I. A., Gubati, A. A., Al-Sulaimani, G. J. and Baluch, M. H. (1990) 'Repaired reinforced concrete beams', *ACI Materials Journal*, Vol. 87, No. 4, pp. 348–354.

Bonacci, J. F. and Maalej, M. (2000) 'Externally bonded fiber-reinforced polymer for rehabilitation of corrosion damaged concrete beams', *ACI Structural Journal*, Vol. 97, No. 5, pp. 703–711.

BS 8110 (1997) *Structural Use of Concrete, Part 1. Code of Practice for Design and Construction*, British Standards Institution, London, UK.

Buyukozturk, O. and Hearing, B. (1998) 'Failure behavior of precracked concrete beams retrofitted with FRP', *Journal of Composites for Construction, ASCE*, Vol. 2, No. 3, pp. 138–144.

Chaallal, O., Nollet, M. -J. and Perraton, D. (1998) 'Strengthening of reinforced concrete beams with externally bonded fiber-reinforced-plastic plates: design guidelines for shear and flexure', *Canadian Journal of Civil Engineering*, Vol. 25, pp. 692–704.

Chajes, M. J., Thomson, T. A., Januszka, T. F. and Finch, W. W., Jr (1994) 'Flexural strengthening of concrete beams using externally bonded composite materials', *Construction and Building Materials*, Vol. 8, No. 3, pp. 191–201.

Chen, J. F. and Teng, J. G. (2001) 'Anchorage strength models for FRP and steel plates bonded to concrete', *Journal of Structural Engineering, ASCE*, Vol. 127, No. 7, pp. 784–791.

Chen, J. F., Yang, Z. J. and Holt, G. D. (2001) 'FRP or steel plate-to-concrete bonded joints: effect of test methods on experimental bond strength', *Steel and Composite Structures – An International Journal*, Vol. 1, No. 2, pp. 231–244.

CSA (1994) *Design of Concrete Structures*, CSA Standard A23.3-94, Canadian Standards Association, Ontario, Canada.

El-Mihilmy, M. T. and Tedesco, J. W. (2000) 'Analysis of reinforced concrete beams strengthened with FRP laminates', *Journal of Structural Engineering, ASCE*, Vol. 126, No. 6, pp. 684–691.

GangaRao, H. V. S. and Vijay, P. V. (1998) 'Bending behavior of concrete beams wrapped with carbon fabric', *Journal of Structural Engineering, ASCE*, Vol. 124, No. 1, pp. 3–10.

Garden, H. N. and Hollaway, L. C. (1998a) 'An experimental study of the influence of plate end anchorage of carbon fibre composite plates used to strengthen reinforced concrete beams', *Composite Structures*, Vol. 42, pp. 175–188.

Garden, H. N. and Hollaway, L. C. (1998b) 'An experimental study of the failure modes of reinforced concrete beams strengthened with prestressed carbon composite plates', *Composites Part B*, Vol. 29B, pp. 411–424.

Garden, H. N. and Mays, G. C. (1999) 'Structural strengthening of concrete beams using prestressed plates', *Strengthening of Reinforced Concrete Structures Using Externally-Bonded FRP Composites in Structural and Civil Engineering*, edited by L. C. Hollaway and M. B. Leeming, Woodhead Publishing, Cambridge, UK.

Garden, H. N., Hollaway, L. C. and Thorne, A. M. (1997) 'A preliminary evaluation of carbon fibre reinforced polymer plates for strengthening reinforced concrete members', *Proceedings of the Institution of Civil Engineers: Structures and Buildings*, Vol. 123, May, pp. 127–142.

Garden, H. N., Quantrill, R. J., Hollaway, L. C., Thorne, A. M. and Parke, G. A. R. (1998) 'An experimental study of the anchorage length of carbon fibre composite plates used to strengthen reinforced concrete beams', *Construction and Building Materials*, Vol. 12, pp. 203–219.

Grace, N. F., Soliman, A. K., Abdel-Sayed, G. and Saleh, K. R. (1998) 'Behavior and ductility of simple and continuous FRP reinforced beams', *Journal of Composites for Construction, ASCE*, Vol. 2, No. 4, pp. 186–194.

Grace, N. F., Sayed, G. A., Soliman, A. K. and Saleh, K. R. (1999) 'Strengthening reinforced concrete beams using fiber reinforced polymer (FRP) laminates', *ACI Structural Journal*, Vol. 96, No. 5, pp. 865–874.

Hau, K. M. (1999) *Experiments on Concrete Beams Strengthened by Bonding Fibre Reinforced Plastic Sheets*, Master of Science in Civil Engineering Thesis, The Hong Kong Polytechnic University.

Heffernan, P. J. and Erki, M. A. (1996) 'Equivalent capacity and efficiency of reinforced concrete beams strengthened with carbon fibre reinforced plastic sheets', *Canadian Journal of Civil Engineering*, Vol. 23, pp. 21–29.

Hollaway, L. C. and Leeming, M. B. (1999) 'Review of materials and techniques for plate bonding', *Strengthening of Reinforced Concrete Structures Using Externally-Bonded FRP Composites in Structural and Civil Engineering*, edited by L. C. Hollaway and M. B. Leeming, Woodhead Publishing, Cambridge, UK.

Hollaway, L. C. and Mays, G. C. (1999) 'Structural strengthening of concrete beams using unstressed composite plates', *Strengthening of Reinforced Concrete Structures Using Externally-Bonded FRP Composites in Structural and Civil Engineering*, edited by L. C. Hollaway and M. B. Leeming, Woodhead Publishing, Cambridge, UK.

Hussain, M., Sharif, A., Basunbul, I. A., Baluch, M. H. and Al-Sulaimani, G. J. (1995) 'Flexural behavior of precracked reinforced concrete beams strengthened externally by steel plates', *ACI Structural Journal*, Vol. 92, No. 1, pp. 14–22.

Hutchinson, A. R. and Quinn, J. (1999) 'Materials', *Strengthening of Reinforced Concrete Structures Using Externally-Bonded FRP Composites in Structural and Civil Engineering*, edited by L. C. Hollaway and M. B. Leeming, Woodhead Publishing, Cambridge, UK.

Jansze, W. (1997) *Strengthening of RC Members in Bending by Externally Bonded Steel Plates*, PhD Thesis, Delft University of Technology.

Jones, R., Swamy, R. N., Bloxham, J. and Bouderbalah, A. (1980) 'Composite behaviour of concrete beams with epoxy bonded external reinforcement', *The International Journal of Cement Composites* 2, No. 2, pp. 91–107.

Jones, R., Swamy, R. N. and Charif, A. (1988) 'Plate separation and anchorage of reinforced concrete beams strengthened with epoxy-bonded steel plates', *The Structural Engineer*, Vol. 66, No. 5, pp. 85–94.

Juvandes, L., Figueiras, J. A. and Marques, A. T. (1998) 'Performance of concrete beams strengthened with CFRP laminates', *Proceedings of the Second International Conference on Composites in Infrastructure (ICCI)*, Tucson, Arizona, USA, edited by H. Saadatmanesh and M. R. Ehsani.

Karbhari, V. M. and Zhao, L. (1998) 'Issue related to composite plating and environmental exposure effects on composite-concrete interface in external strengthening', *Composite Structures*, Vol. 40, Nos 3–4, pp. 293–304.

Kong, F. K and Evans, R. H. (1987) *Reinforced and Prestressed Concrete*, Third Edition, Chapman & Hall, London, UK.

Lam, L. and Teng, J. G. (2001) 'Strength of cantilever RC slabs bonded with GFRP strips', *Journal of Composites for construction, ASCE*, Vol. 5, No. 4.

Leeming, M. B. and Darby, J. J. (1999) 'Design and specifications for FRP plate bonding of beams', *Strengthening of Reinforced Concrete Structures Using Externally-Bonded FRP Composites in Structural and Civil Engineering*, edited by L. C. Hollaway and M. B. Leeming, Woodhead Publishing, Cambridge, UK.

Liu, Z. and Zhu, B. (1994) 'Analytical solutions for R/C beams strengthened by externally bonded steel plates', *Journal of Tongji University (in Chinese)*, Vol. 22, No. 1, pp. 21–26.

Lopez, M., Naaman, A. E. and Till, R. D. (1999) 'Bending behavior of reinforced concrete beams strengthened with carbon fiber reinforced polymer laminates and subjected to freeze-thaw cycles', *Proceedings of the Fourth International Symposium on Fiber Reinforced Polymer Reinforcement for Reinforced Concrete Structures*, Baltimore, Maryland, USA, edited by C. W. Dolan, S. H. Rizkalla and A. Nanni, pp. 559–576, American Concrete Institute, Farmington Hills, Michigan, USA.

Macdonald, M. D. and Calder, A. J. J. (1982) 'Bonded steel plating for strengthening concrete structures', *International Journal of Adhesion and Adhesives*, Vol. 2, No. 2, pp. 119–127.

Malek, A. M., Saadatmanesh, H. and Ehsani, M. R. (1998) 'Prediction of failure load of R/C beams strengthened with FRP plate due to stress concentration at the plate end', *ACI Structural Journal*, Vol. 95, No. 1, pp. 142–152.

Meier, U. (1995) 'Strengthening of structures using carbon fibre/epoxy composites', *Construction and Building Materials*, Vol. 9, No. 6, pp. 341–351.

Meier, U., Deuring, M., Meier, H. and Schwegler, G. (1993) 'CFRP bonded sheets', *Fibre-Reinforced-Plastic (FRP) Reinforcement for Concrete Structures: Properties and Applications*, edited by A. Nanni, Elsevier Science, Amsterdam, The Netherlands.

Mirza, S., Hatzinikolas, M. and MacGregor, J. (1979) 'Statistical descriptions of the strength of concrete', *Journal of Structural Engineering, ASCE*, Vol. 105, No. ST6, pp. 1021–1037.

Mukhopadhyaya, P. and Swamy, N. (2001) 'Interface shear stress: a new design criterion for plate debonding', *Journal of Composites for Construction, ASCE*, Vol. 5, No. 1, pp. 35–43.

Nanni, A., Focacci, F. and Cobb, C. A. (1998) 'Proposed procedure for the design of RC flexural members strengthened with FRP sheets', *Proceedings of the Second International Conference on Composites in Infrastructure (ICCI 1998)*, Tucson, Arizona, USA. edited by H. Saadatmanesh and M. R. Ehsani.

Nguyen, D. M., Chan, T. K. and Cheong, H. K. (2001) 'Brittle failure and bond development length of CFRP-concrete beams', *Journal of Composites for Construction, ASCE*, Vol. 5, No. 1, pp. 12–17.

Oehlers, D. J. (1992) 'Reinforced concrete beams with plates glued to their soffits', *Journal of Structural Engineering, ASCE*, Vol. 118, No. 8, pp. 2023–2038.

Oehlers, D. J. and Ali, M. (1998) 'Discussion on 'Predictions of the maximum plate end stresses of FRP strengthened beams: Part II – by Quantrill, R. J., Hollaway, L. C. and Thorne, A. M.", *Magazine of Concrete Research*, Vol. 50, No. 1, pp. 91–92.

Oehlers, D. J. and Moran, J. P. (1990) 'Premature failure of externally plated reinforced concrete beams', *Journal of Structural Engineering, ASCE*, Vol. 116, No. 4, pp. 978–995.

Quantrill, R. J., Hollaway, L. C. and Thorne, A. M. (1996a) 'Experimental and analytical investigation of FRP strengthened beam response: Part I', *Magazine of Concrete Research*, Vol. 48, No. 177, pp. 331–342.

Quantrill, R. J., Hollaway, L. C. and Thorne, A. M. (1996b) 'Predictions of the maximum plate end stresses of FRP strengthened beams: Part II', *Magazine of Concrete Research*, Vol. 48, No. 177, pp. 343–351.

Rabinovich, O. and Frostig, Y. (2000) 'Closed-form high-order analysis of RC beams strengthened with FRP strips', *Journal of Composites for Construction, ASCE*, Vol. 4, No. 2, pp. 65–74.

Rahimi, H. and Hutchinson, A. (2001) 'Concrete beams strengthened with externally bonded FRP plates', *Journal of Composites for Construction, ASCE*, Vol. 5, No. 1, pp. 44–56.

Ramana, V. P. V., Kant, T., Morton, S. E., Dutta, P. K., Mukherjee, A. and Desai, Y. M. (2000) 'Behavior of CFRPC strengthened reinforced concrete beams with varying degrees of strengthening', *Composites: Part B*, Vol. 31, pp. 461–470.

Raoof, M. and Hassanen, M. A. H. (2000) 'Peeling failure of reinforced concrete beams with fibre-reinforced plastic or steel plates glued to their soffits', *Proceedings of the Institution of Civil Engineers: Structures and Buildings*, Vol. 140, August, pp. 291–305.

Raoof, M. and Zhang, S. (1997) 'An insight into the structural behaviour of reinforced concrete beams with externally bonded plates', *Proceedings of the Institution of Civil Engineers: Structures and Buildings*, Vol. 122, November, pp. 477–492.

Ritchie, P. A., Thomas, D. A., Lu, L. W. and Connely, G. M. (1991) 'External reinforcement of concrete beams using fiber reinforced plastics', *ACI Structural Journal*, Vol. 88, No. 4, pp. 490–500.

Roberts, T. M. (1989) 'Approximate analysis of shear and normal stress concentrations in the adhesive layer of plated RC beams', *The Structural Engineer*, Vol. 67, No. 12, pp. 229–233.

Roberts, T. M. and Haji-Kazemi, H. (1989) 'Theoretical study of the behaviour of reinforced concrete beams strengthened by externally bonded steel plates', *Proceedings of the Institution of Civil Engineers*, Vol. 87, Part 2, pp. 39–55.

Ross, C. A., Jerome, D. M., Tedesco, J. W. and Hughes, M. L. (1999) 'Strengthening of reinforced concrete beams with externally bonded composite laminates', *ACI Structural Journal*, Vol. 96, No. 2, pp. 212–220.

Saadatmanesh, H. and Ehsani, M. R. (1991) 'RC beams strengthened with GFRP plates. I: experimental study', *Journal of Structural Engineering*, Vol. 117, No. 11, pp. 3417–3433.

Saadatmanesh, H. and Malek, A. M. (1998) 'Design guidelines for flexural strengthening of RC beams with FRP plates', *Journal of Composites for Construction, ASCE*, Vol. 2, No. 4, pp. 158–164.

Shahawy, M. A., Arockiasamy, M., Beitelman, T. and Sowrirajan, R. (1996) 'Reinforced concrete rectangular beams strengthened with CFRP laminates', *Composites: Part B*, Vol. 27B, pp. 225–233.

Sharif, A., Al-Sulaimani, G. J., Basunbul, I. A., Baluch, M. H. and Ghaleb, B. N. (1994) 'Strengthening of initially loaded reinforced concrete beams using FRP plates', *ACI Structural Journal*, Vol. 91, No. 2, pp. 160–168.

Shen, H. S., Teng, J. G. and Yang, J. (2001) 'Interfacial stresses in beams and slabs bonded with a thin plate', *Journal of Engineering Mechanics, ASCE*, Vol. 127, No. 4, pp. 399–406.

Smith, S. T. and Teng, J. G. (2001a) 'Plate end debonding failures in FRP-strengthened RC beams', *to be published*.

Smith, S. T. and Teng, J. G. (2001b) 'Interfacial stresses in plated RC beams', *Engineering Structures*, Vol. 23, pp. 857–871.

Smith, S. T. and Teng, J. G. (2001c) 'FRP strengthened RC beams–I: review of debonding strength models', *Engineering Structures, in press.*

Smith, S. T. and Teng, J. G. (2001d) 'FRP strengthened RC beams–II: assessment of debonding strength models', *Engineering Structures, in press.*

Smith, S. T., Teng, J. G. and Chen, J. F. (2001) 'Debonding in FRP plated RC beams induced by intermediate cracking', *ACUN-3 International Conference: Technology Convergence in Composite Applications*, Sydney, Australia, edited by S. Bandyopadhyay *et al.*, pp. 67–72, University of New South Wales Press, Sydney, Australia.

Spadea, G., Bencardino, F. and Swamy, R. N. (1998) 'Structural behavior of composite RC beams with externally bonded CFRP', *Journal of Composites for Construction, ASCE*, Vol. 2, No. 3, pp. 132–137.

Spadea, G., Bencardino, F. and Swamy, R. N. (2000) 'Optimizing the performance characteristics of beams strengthened with bonded CFRP laminates', *Materials and Structures*, Vol. 33, pp. 119–126.

Swamy, R. N. and Mukhopadhyaya, P. (1999) 'Debonding of carbon-fibre-reinforced polymer plate from concrete beams', *Proceedings of the Institution of Civil Engineers: Structures and Buildings*, Vol. 134, November, pp. 301–317.

Swamy, R. N., Jones, R. and Ang, T. H. (1982) 'Under- and over-reinforced concrete beams with glued steel plates', *The International Journal of Cement Composites and Lightweight Concrete 4*, Vol. 4, No. 1, pp. 19–32.

Swamy, R. N., Jones, R. and Bloxham, J. W. (1987) 'Structural behaviour of reinforced concrete beams strengthened by epoxy-bonded steel plates', *The Structural Engineer*, Vol. 65A, No. 2, pp. 59–68.

Swamy, R. N., Jones, R. and Charif, A. (1989) 'The effect of external plate reinforcement on the strengthening of structurally damaged RC beams', *The Structural Engineer*, Vol. 67, No. 3, pp. 45–56.

Takeda, K., Mitsui, Y., Murakami, K., Sakai, H. and Nakamura, M. (1996) 'Flexural behaviour of reinforced concrete beams strengthened with carbon fibre sheets', *Composites Part A*, Vol. 27A, pp. 981–987.

Täljsten, B. (1997) 'Strengthening of beams by plate bonding', *Journal of Materials in Civil Engineering, ASCE*, Vol. 9, No. 4, pp. 206–212.

Täljsten, B. (1999) 'Concrete beams strengthened for bending using CFRP-sheets', *Proceedings of the Eighth International Conference on Advanced Composites for Concrete Repair*, edited by M. C. Forde, Engineering Technics Press, Edinburgh, UK.

Tan, K. H. (1999) 'Towards a cost-effective application of FRP reinforcement in structural rehabilitation', *Proceedings of the Seventh East Asia-Pacific Conference on Structural Engineering and Construction*, Kochi, Japan, pp. 65–73.

Teng, J. G., Lam, L., Chan, W. and Wang, J. S. (2000a) 'Retrofitting of deficient RC cantilever slabs using GFRP strips', *Journal of Composites for Construction, ASCE*, Vol. 4, No. 2, pp. 75–84.

Teng, J. G., Chen, J. F., Smith, S. T. and Lam, L. (2000b) *RC Structures Strengthened with FRP Composites*, Research Centre for Advanced Technology in Structural Engineering, The Hong Kong Polytechnic University, Hong Kong, China, December.

Teng, J. G., Zhang, J. W. and Smith, S. T. (2001a) 'Interfacial stresses in RC beams bonded with a soffit plate: a finite element study', *Construction and Building Materials, submitted.*

Teng, J. G., Smith, S. T., Yao, J. and Chen, J. F. (2001b) 'Intermediate crack induced debonding in RC beams and slabs', *to be published.*

Triantafillou, T. C. and Plevris, N. (1992) 'Strengthening of RC beams with epoxy-bonded fibre-composite materials', *Materials and Structures*, Vol. 25, pp. 201–211.

Tumialan, G., Serra, P., Nanni, A. and Belarbi, A. (1999a) 'Concrete cover delamination in reinforced concrete beams strengthened with carbon fiber reinforced polymer sheets',

Proceedings of the Fourth International Symposium on Fiber Reinforced Polymer Reinforcement for Reinforced Concrete Structures, Baltimore, Maryland, USA, edited by C. W. Dolan, S. H. Rizkalla and A. Nanni, pp. 725–735, American Concrete Institute, Farmington Hills, Michigan, USA.

Tumialan, G., Belarbi, A. and Nanni, A. (1999b) *Reinforced Concrete Beams Strengthened with CFRP Composites: Failure Due to Concrete Cover Delamination*, Department of Civil Engineering, Center for Infrastructure Engineering Studies, Report No. CIES-99/01, University of Missouri-Rolla, USA.

van Gemert, D., Ahmed, O. and Brosens, K. (1999) 'Anchoring of externally bonded CFRP reinforcement', *Proceedings of the International Congress on Creating with Concrete*, Dundee, Scotland, edited by R. K. Dhir and M. J. McCarthy, pp. 81–92, Thomas Telford, London, UK.

Varastehpour, H. and Hamelin, P. (1997) 'Strengthening of concrete beams using fiber-reinforced plastics', *Materials and Structures*, Vol. 30, pp. 160–166.

Vecchio, F. J. and Bucci, F. (1999) 'Analysis of repaired reinforced concrete structures', *Journal of Structural Engineering, ASCE*, Vol. 125, No. 6, pp. 644–652.

Vilnay, O. (1988) 'The analysis of reinforced concrete beams strengthened by epoxy bonded steel plates', *The International Journal of Cement Composites and Lightweight Concrete*, Vol. 10, No. 2, pp. 73–78.

Wang, C. Y. and Ling, F. S. (1998) 'Prediction model for the debonding failure of cracked RC beams with externally bonded FRP sheets', *Proceedings of the Second International Conference of Composites in Infrastructure (ICCI 98)*, Tucson, Arizona, USA edited by H. Saadatmanesh and M. R. Ehsani, pp. 548–562.

Yao, J., Teng, J. G. and Lam, L. (2001) 'Debonding in FRP-strengthened RC cantilever slabs', *to be published*.

Zhang, S., Raoof, M. and Wood, L. A. (1995) 'Prediction of peeling failure of reinforced concrete beams with externally bonded steel plates', *Proceedings of the Institution of Civil Engineers: Structures and Buildings*, Vol. 110, August, pp. 257–268.

Ziraba, Y. N., Baluch, M. H., Basunbul, I. A., Sharif, A. M., Azad, A. K. and Al-Sulaimani, G. J. (1994) 'Guidelines towards the design of reinforced concrete beams with external plates', *ACI Structural Journal*, Vol. 91, No. 6, pp. 639–646.

Ziraba, Y. N., Baluch, M. H., Basunbul, I. A., Azad, A. K., Al-Sulaimani, G. J. and Sharif, A. M. (1995) 'Combined experimental-numerical approach to characterization of steel-concrete interface', *Materials and Structures*, Vol. 28, pp. 518–525.

4 Shear Strengthening of Beams

4.1 Introduction

Flexural and shear failures are the two main failure modes for normal (unstrengthened) RC beams. Flexural failure is generally preferred to shear failure as the strength-governing failure mode because the former is ductile, whilst the latter is brittle. A ductile failure allows stress redistribution and provides warning to occupants, whilst a brittle failure is sudden and thus catastrophic.

For a flexurally strengthened RC beam with an externally bonded FRP plate, its flexural failure mode shows much reduced ductility than a normal RC beam as was discussed in Chapter 3. However, it is still a more ductile mode than shear failure. Therefore, a strengthened beam must have a shear capacity which exceeds its flexural capacity.

When an RC beam is deficient in shear, or when its shear capacity is less than the flexural capacity after flexural strengthening, shear strengthening must be considered. It is critically important to examine the shear capacity of RC beams which are intended to be strengthened in flexure. Shear strengthening very often forms a key part of an effective strengthening strategy for RC structures.

Whilst a number of traditional methods are available for the shear strengthening of RC beams, the use of FRP plates/strips has attracted much recent attention (e.g. Al-Sulaimani *et al.* 1994, Chajes *et al.* 1995, Alexander 1996, Chaallal *et al.* 1998). Apart from the common advantages of corrosion resistance and high strength-to-weight ratio of FRPs, the versatility of FRPs in coping with different sectional shapes and corners is also a benefit for shear strengthening applications.

Studies on the shear strengthening of RC beams have been carried out since the early 1990s (e.g. Uji 1992, Al-Sulaimani *et al.* 1994, Arduini *et al.* 1994, Chajes *et al.* 1995, Alexander 1996, Sato *et al.* 1996, 1997a, Araki *et al.* 1997, Funakawa *et al.* 1997, Triantafillou 1997, 1998a, b, Chaallal *et al.* 1998, Malek and Saadatmanesh 1998, Mitsui *et al.* 1998, Fanning and Kelly 1999, Hutchinson and Rizkalla 1999, Kachlakev and Barnes 1999, Khalifa *et al.* 1999, Mutsuyoshi *et al.* 1999, Khalifa and Nanni 2000), but the work is much more limited compared with that on beams flexurally strengthened with FRP plates. Much more research is needed before the full potential of FRP shear strengthening can be realised.

This chapter first presents a systematic description of bonding schemes. The behaviour of strengthened beams is then summarised based on existing research. Existing design equations are reviewed, and a design procedure is outlined.

It may be noted that this chapter is concerned only with the shear strength of RC beams strengthened in shear. It has been observed that the FRP bonded to the soffit of an RC beam does not significantly change the shear behaviour from that of unstrengthened beams (Plevris *et al.* 1995, Vichit-Vadakan 1997, Buyukozturk and Hearing 1998). Therefore, the effect of FRP bonded to the soffit for flexural strengthening may be ignored in predicting the shear strength of the beam. It is also evident that the longitudinal FRP bonded to the sides of a beam does not provide significant help to the shear strength of the beam (e.g. Chajes *et al.* 1995, Khalifa *et al.* 1999). These statements have, however, been based on very limited experimental evidence. It should be emphasised that, although fibres running longitudinally along a beam do not make a significant contribution to the shear strength, fibres running at other angles on the sides of a beam do make contributions. However, fibres at all directions, except those exactly parallel to the shear crack, are helpful in limiting the shear crack width.

4.2 Methods of Strengthening

4.2.1 Strengthening Schemes

Various FRP bonding schemes have been used to increase the shear resistance of RC beams. These include bonding FRP to the sides of a beam only, bonding FRP U jackets to both the sides and the tension face, and wrapping FRP around the whole cross-section of a beam. Both FRP strips and continuous sheets have been used.

Because FRPs are strong only in the directions of fibres, the fibres may be orientated in such directions as to control shear cracks best. Furthermore, shear forces in a beam may be reversed under reversed cyclic loading and earthquake attacks. Fibres may thus be arranged at two different directions to satisfy the requirement of shear strengthening in both directions. The use of fibres in two directions can obviously be beneficial to shear resistance even if strengthening for reversed cyclic loading is not required, except for the unlikely case in which one of the fibre directions is exactly parallel to the shear cracks. In this sense, FRPs with fibres in three directions (e.g. 0°/60°/120°) may also be used.

The combination of different bonding configurations, fibre orientations and fibre distributions can result in many different strengthening schemes. Symbolic representations are introduced here so that each shear strengthening scheme is identified by a set of clearly defined symbols. Tables 4.1 to 4.3 present symbolic

Table 4.1 FRP bonding configurations

Scheme	Side bonding	U jacketing	Wrapping
Symbol	S	U	W
	FRP bonded to sides only	The use of three separate plates is unacceptable, as the U jacket must be an integral U section. Normal U jackets (sides + tension face) are preferred but reversed U jackets (sides + compression face) may be used if unavoidable	The FRP must be wrapped around the cross-section with sufficient overlapping over the compression face. Overlapping over the compression face is preferred to minimise the relative strain between the FRP and the concrete in the direction of the main fibres. Overlapping over the tension face may be used if unavoidable
Notes	The free ends/edges of the strips/sheets should be properly anchored if possible		
		To avoid FRP failure due to high bending stresses at corners and debonding of FRP on the compression/tension faces, the corners of the beam should be properly rounded	

Table 4.2 Fibre distributions

	Strips	Plates/sheets
Symbol	S	P

Table 4.3 Fibre orientations

	First fibre/strip orientation, β	Second fibre/strip orientation, ϕ
Symbol	$0^\circ \leqslant \beta < 180^\circ$	$0^\circ \leqslant \phi < 180^\circ$

representations of each of the three aspects mentioned above. Each shear strengthening scheme can then be denoted by one symbol representing the bonding configuration (Table 4.1), followed by a second symbol representing the fibre distribution (Table 4.2) and followed by two sets of numbers representing the first and second fibre orientations (Table 4.3). For example, US45/135 represents U jacketing with FRP strips at 45° and 135°. Figure 4.1 shows additional examples. A photograph of a U-jacketed RC test beam is shown in Figure 4.2.

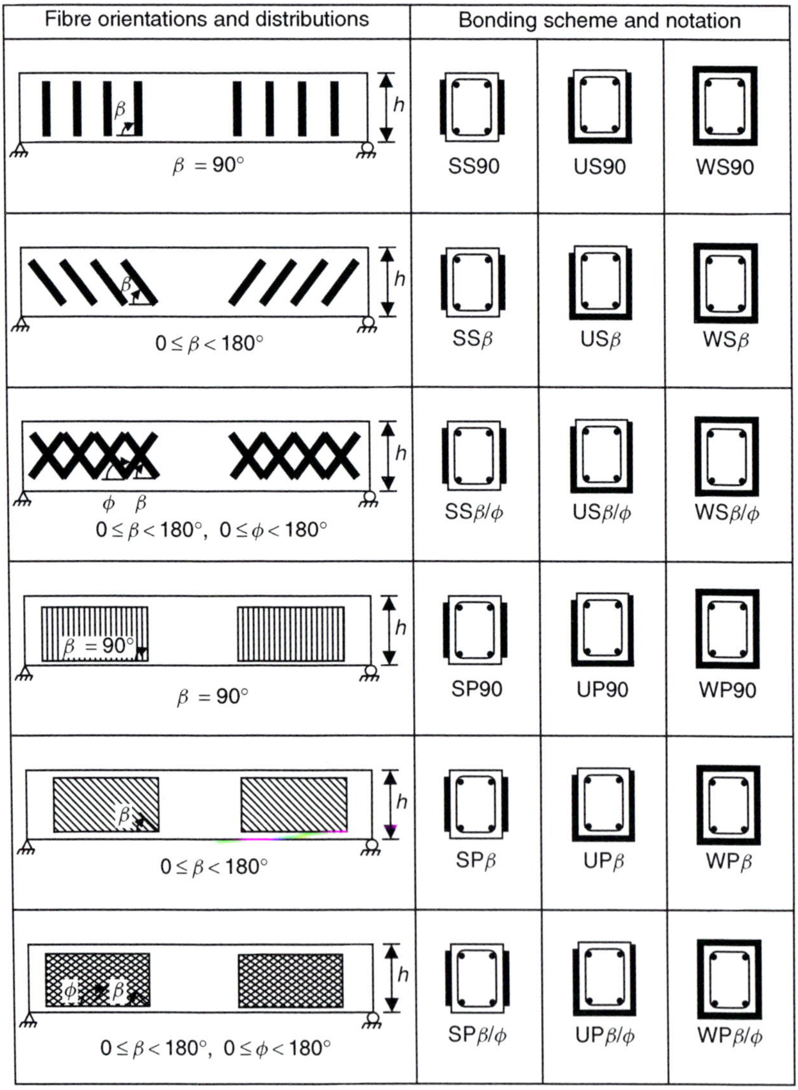

Figure 4.1 FRP shear strengthening schemes

Figure 4.2 A U-jacketed RC beam

For both side bonding and U jacketing, mechanical anchors may be required at the free ends of FRP strips/sheets.

4.2.2 Selection of Strengthening Schemes

A suitable strengthening scheme needs to be selected for a given situation based on a number of factors. These may include:

- accessibility of the site such as whether the whole perimeter of a beam can be accessed so that it may be strengthened by wrapping;
- strengthening requirement such as whether it only needs to be strengthened for monotonic loading or needs to be strengthened for reversed cyclic loading as well;
- amount of increase required in the shear capacity;
- availability of FRP materials; and
- economic considerations.

Tables 4.4 to 4.6 respectively list some advantages and disadvantages of different bonding configurations, fibre distributions and fibre/strip orientations, which may be used as a reference in the selection of a suitable strengthening scheme.

Based on available experimental data, side-bonded FRPs are most vulnerable to debonding (or premature failure/peeling off), whilst FRPs wrapped around the cross-section are most effective, with FRP U jackets lying in between. Therefore, FRP wrapping should be used whenever possible, followed by U jacketing.

When both flexural strengthening by soffit plating and shear strengthening by either U jacketing or wrapping are required, it has been suggested that the FRP for shear strengthening should be applied first (Tan 1999). In this way, the tensile

Table 4.4 Advantages and disadvantages of different FRP bonding configurations

Configuration	Advantages	Disadvantages
Side bonding	Easiest to apply Needs least amount of FRP for a small increase in shear capacity	Most vulnerable to debonding Least effective
U jacketing	Bottom of U well anchored Less vulnerable to debonding Moderately effective Acting as mechanical anchors for flexural strengthening FRP	May need mechanical anchors at the free ends of U
Wrapping	Least vulnerable to debonding Most effective Acting as mechanical anchors for flexural strengthening FRP	Not possible or very difficult if at least one side of the beam is not accessible, as in the case of a slab-supporting beam

Table 4.5 Advantages and disadvantages of different fibre distributions

Fibre distribution	Advantages	Disadvantages
Strips	Flexibility in controlling the amount of FRP Possible savings in material Uniform adhesive layer more easily achieved	More laborious
Plates/sheets	Ease in site application Beam protected from further environmental degradation if wholly covered	Reduced flexibility in controlling the amount of FRP Uniform adhesive layer more difficult to achieve

Table 4.6 Advantages and disadvantages of different fibre orientations

Fibre orientation	Advantages	Disadvantages
Vertical ($\beta = 90^\circ$)	Easiest to apply Effective for strengthening for reversed shear	Less effective than inclined fibres/strips for shear crack control
Inclined (mostly $\beta = 45^\circ$)	More effective for shear crack control	Impossible for U jacketing and wrapping using unidirectional sheets and wide strips
Bidirectional sheets/strips (mostly $0^\circ/90^\circ$ or $45^\circ/135^\circ$)	Most effective for shear crack control Effective for strengthening for reversed shear	More fibres required

force in the soffit plate is partly transferred to the FRP on the beam sides, thus reducing the risk of premature debonding failure of the soffit plate from the beam.

4.3 Failure Modes

4.3.1 General

A number of failure modes have been observed in experiments on RC beams shear strengthened with bonded FRPs. These include shear failure with FRP rupture, shear failure without FRP rupture, shear failure due to FRP debonding and local failures.

As for normal RC beams, shear failure modes include shear tension, flexural shear, shear compression and deep beam failure. Shear tension and flexural shear failures occur when shear span-to-depth ratios are large ($6 > B/d > 2.5$ for normal RC beams; shear failure is rare when $B/d > 6$). Shear compression failures usually occur in beams with smaller shear span-to-depth ratios ($2.5 > B/d > 1$ for normal RC beams). The deep beam failure mode usually occurs in beams with very small shear span-to-depth ratios ($B/d < 1$ for normal RC beams). Further details of these failure modes can be found in many textbooks on RC structures (e.g. Kong and Evans 1987).

For RC beams shear strengthened with externally bonded FRPs, it is possible that the boundaries between these failure modes are different. The behaviour and contribution of FRP in each of these failure modes may also differ. However, these questions cannot yet be answered with certainty and much further research is required. A general description of the failure modes, based on the failure behaviour of the FRP as observed in experimental studies, is presented as follows.

4.3.2 Shear Failure with FRP Rupture

This type of failure occurs most often with a diagonal tension shear crack (Figure 4.3). Vertical flexural cracks originating from the tensile face occur first. A crack near the support may propagate towards the loading point and become

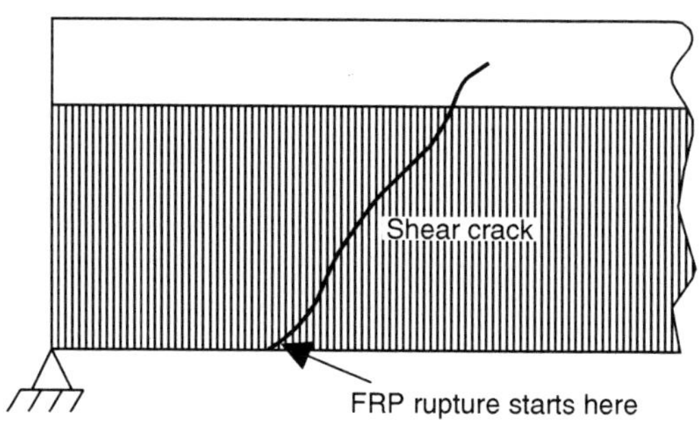

Figure 4.3 Shear tension with FRP rupture

an inclined crack. In some cases, a diagonal crack may appear abruptly. As the width of the diagonal crack increases, the maximum strain in the FRP eventually reaches its ultimate strain, which often occurs at the lower end of the crack. The FRP is immediately torn when its ultimate strength is reached. Rupture of the FRP then propagates along the diagonal shear crack in the concrete, leading to total failure of the beam in a brittle manner. Partial debonding of the FRP from the beam sides often occurs prior to failure in most cases, but eventual failure is due to FRP rupture. Therefore, FRP debonding does not affect the ultimate shear capacity of this mode. Limited existing experimental data indicate that almost all beams with wrapped FRPs and some with U jackets (e.g. Chajes *et al.* 1995, Araki *et al.* 1997, Funakawa *et al.* 1997, Kage *et al.* 1997, Sato *et al.* 1997b, Chaallal *et al.* 1998) failed in this mode.

4.3.3 Shear Failure without FRP Rupture

The failure process of this mode is essentially the same as that of shear failure with FRP rupture, except that the FRP is not torn at failure in this mode of failure. The FRP can still carry a significant load after the concrete fails. Examples of this failure mode include two beams strengthened with an aramid FRP which had an ultimate strain of 2.25 % (Chajes *et al.* 1995). The AFRP allowed the failed beams to continue to carry a significant load.

4.3.4 Shear Failure due to FRP Debonding

A shear-strengthened RC beam may fail owing to the debonding of the FRP from the beam. Figures 4.4 and 4.5 show the possible debonding zones for U jackets and side plates respectively. Once the FRP starts to peel off, the beam will fail very quickly. The ductility of beams failing in this mode is usually very limited. Available test data indicate that almost all beams strengthened with side-bonded FRP only, and many strengthened with U jackets, failed in this mode. Figures 4.6(a–c) show the FRP debonding failure of the U-jacketed RC beam of Figure 4.2.

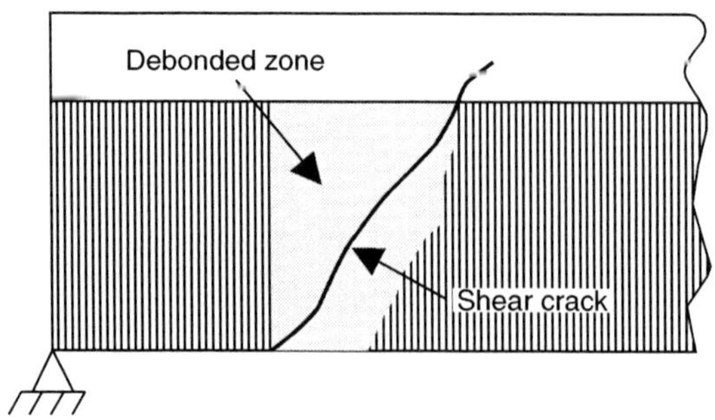

Figure 4.4 Shear failure due to FRP debonding: U jacket

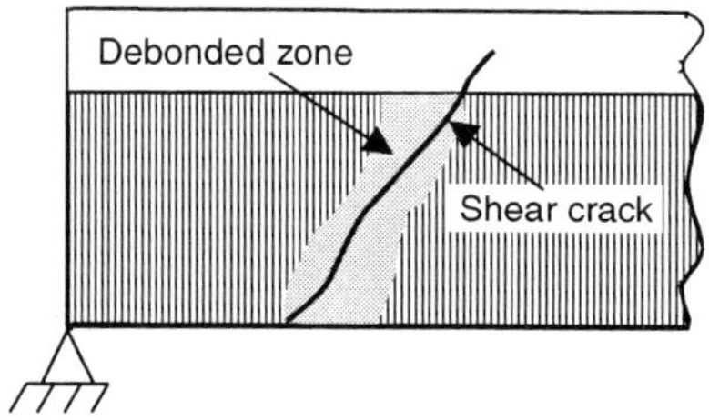

Figure 4.5 Shear failure due to FRP debonding: side-bonded FRP

(a)
(b)
(c)

Figure 4.6 FRP debonding failure of a U-jacketed RC beam: (a) overall view; (b) close-up view of one side; (c) close-up view of the other side

4.3.5 Failures Near Mechanical Anchors

If mechanical anchors are used to anchor the FRP on the sides of an RC beam, it is possible for failures to occur adjacent to the anchors. For example, mechanical anchors used to anchor U jackets beneath the flange of a T-beam may cause delamination at the flange/web junction. Around the mechanical anchors, either FRP or concrete may fail locally. These failure modes have been observed in experiments by Sato *et al.* (1997b). Extra care must be taken in design before a better understanding of these aspects is achieved.

4.4 Existing Research on Shear Strength Models

RC beams strengthened in shear using externally bonded FRPs show complex behaviour. Both the shear strength and the failure mode are influenced by many factors such as the size and geometry of the beam, the strength of the concrete, internal shear and flexural reinforcements, loading conditions, the method of strengthening, and the properties of the bonded FRP. Shear strengthening using bonded FRPs is also much less investigated compared with flexural strengthening (Swamy and Mukhopadhyaya 1995). The complexity of the problem, combined with the limited research available, makes it difficult to develop robust predictive strength models suitable for practical design. Nevertheless, several attempts have been made in developing such models as detailed below.

All existing models use the following expression to calculate the shear strength of a strengthened RC beam, V_n:

$$V_n = V_c + V_s + V_{frp} \tag{4.1}$$

where V_c is the shear capacity of the concrete which consists of shear contributions of concrete in compression, aggregate interlock and dowel action of steel flexural reinforcement, V_s is the contribution of the steel stirrups and bent-up bars and V_{frp} is the contribution of the FRP. V_c and V_s may be calculated according to provisions in existing design codes, so the main differences between available models lie in the evaluation of the FRP contribution V_{frp}. They are briefly discussed as follows. Note that all models presented in this chapter apply to both strips and continuous plates, with continuous plates being a special case of strips, unless specified otherwise.

Chaallal *et al.* (1998) proposed the following equation to calculate the shear contribution of bonded FRPs in the ACI format (ACI 318-95 1999):

$$V_{frp} = \phi_{frp} A_{frp} f_{frp} \frac{(\sin\beta + \cos\beta)d}{s_{frp}} \tag{4.2}$$

where f_{frp} is the tensile strength of FRP, $\phi_{frp} = 0.8$ is the material reduction factor for the FRP, A_{frp} is the cross-sectional area of a pair of FRP strips, β is the angle of the fibre orientation measured clockwise from the horizontal direction for the left side of the beam, d is the effective depth of the beam, and s_{frp} is the spacing of FRP strips measured along the longitudinal axis. In addition, the design average shear stress between the FRP and the concrete is required not to exceed half the maximum debonding stress (see Section 2.3).

The basic assumption behind this design proposal is that the bonded FRP contributes to the shear resistance in the same way as that of internal steel shear reinforcement and that all the FRP intersected by the shear crack reaches its full tensile strength at beam failure unless otherwise limited by insufficient bond lengths.

This model suffers from two major drawbacks. First, the assumption that all the FRP intersected by the shear crack reaches its full tensile strength at beam failure is over-optimistic (Chen and Teng 2001a). Second, Chaallal *et al.*'s bond strength model used for limiting the stress level in the FRP likely to debond does not fit well with experimental data (Chen and Teng 2001b, see also Section 2.3). Furthermore, their bond strength model is given in terms of the average shear stress between the FRP and the concrete. This can be misleading, because it implies that the tensile strength of the FRP can always be fully utilised if there is a sufficiently long bond length (e.g. the beam depth is large enough), while in reality the stress level in the FRP at debonding does not increase with bond length if the bond length is already sufficiently long (i.e. longer than the effective bond length as discussed in Chapter 2).

It may be noted that similar equations using the full tensile capacity of FRP have been previously used in the design for strengthening actual engineering structures (e.g. Gergely *et al.* 1998). Although a large 'safety' reserve might have been used to achieve safe design in these particular cases, the use of the concept could lead to misunderstanding and unconservative design.

Triantafillou (1998a) argued that to predict accurately the contribution of FRP to shear resistance is impossible, because it depends on the failure mechanism which in turn depends on various factors. He thus used a semi-quantitative approach, leading to a design equation for continuous sheets/plates in the Eurocode format (ENV 1992-1-1 1991):

$$V_{frp} = \frac{0.9}{\gamma_{frp}} \rho_{frp} E_{frp} \varepsilon_{frp,e} b_w d(1 + \cot\beta)\sin\beta \tag{4.3}$$

where E_{frp} is the modulus of elasticity of FRP, $\rho_{frp} = 2t_{frp}/b_w$ is the FRP area fraction, t_{frp} is the thickness of the FRP, b_w is the width of the RC beam, γ_{frp} is the partial safety factor for the tensile strength of FRP which was proposed to be about 1.15, 1.20 and 1.25 for CFRP, AFRP and GFRP respectively (Triantafillou and Fardis 1997), and $\varepsilon_{frp,e}$ is the effective strain of FRP at failure which was obtained from the regression of experimental data of beam tests:

$$\varepsilon_{frp,e} = 0.0119 - 0.0205\rho_{frp}E_{frp} + 0.0104\left(\rho_{frp}E_{frp}\right)^2$$
$$\text{for } 0 \leqslant \rho_{frp}E_{frp} < 1 \tag{4.4a}$$

$$\varepsilon_{frp,e} = -0.000\,65\rho_{frp}E_{frp} + 0.002\,45 \text{ for } 0 \leqslant \rho_{frp}E_{frp} < 1 \tag{4.4b}$$

Equation 4.2 is equally applicable to beams strengthened with strips if appropriate ρ_{frp} values are used. Triantafillou (1998a) actually used it to calculate the contribution of FRP strips, but it is not clear how he evaluated ρ_{frp}. The chief disadvantage of this model is that no distinction is made between the different strengthening schemes or failure modes. As a result, the predicted FRP contribution is the same, for example, for wrapping and side bonding, which is clearly incorrect.

More recently, Triantafillou and Antonopoulos (2000) published an extension of the above work in which different effective strain expressions were proposed for different types of FRP based on a regression analysis of a larger database. Different coefficients were proposed for CFRP wrapping and other strengthening schemes. However, no distinction was made between side bonding and U jacketing. A close examination of the data presented in the paper by Chen and Teng (2001a) showed that the average ratio of the observed to the predicted FRP strain by their model is 1.07 with a standard deviation of 0.4. This is statistically not satisfactory for safe practical design.

Khalifa *et al.* (1998) proposed a modification to Triantafillou's effective strain model. They used the ratio of effective stress (or strain) in the FRP to its ultimate strength (or strain), R, which is given by $R = f_{frp,e}/f_{frp} = \varepsilon_{frp,e}/\varepsilon_{frp,rup}$. Regression of experimental data led to the following expression for the ratio of effective stress R:

$$R = 0.5622\left(\rho_{frp}E_{frp}\right)^2 - 1.2188\rho_{frp}E_{frp} + 0.778 \leqslant 0.50 \tag{4.5}$$

Recognising that the effective stress (or strain) model may only be applicable when failure is governed by FRP rupture, Khalifa *et al.* (1998) further proposed a bond mechanism design approach based on the bond strength model of Maeda *et al.* (1997). This model gives a stress (or strain) reduction factor of

$$R = \frac{f_{frp,e}}{f_{frp}} = \frac{0.0042\left(f_c'\right)^{2/3} w_{frp}}{\left(E_{frp}t_{frp}\right)^{0.58} \varepsilon_{frp,rup}d_{frp}} \leqslant 0.5 \tag{4.6}$$

where f_c' is the cylinder compressive strength of concrete, w_{frp} is the width of strips and d_{frp} is the effective depth of FRP reinforcement (equal to the depth d from the compression face to the centroid of the steel tension reinforcement for rectangular sections). Khalifa *et al.* (1998) suggested the use of a strength reduction factor of 0.7 for the calculated V_{frp}. However, this was based on the fitting of the shear experimental results to the above equations rather than the actual behaviour of FRPs.

Khalifa *et al.*'s model also features two important deficiencies. First, Equation 4.5 means that on average no more than half of the tensile capacity of the FRP can be made use of at the shear failure of the beam. No satisfactory explanation for this small effective strain ratio R (which has an upper limit of 0.5) obtained from the regression of test data was offered by Khalifa *et al.* (1998). This empirical effective strain approach thus lacks a rational theoretical basis. Second, the bond strength model of Maeda *et al.* (1997) adopted in Khalifa *et al.*'s design proposal is an invalid model because it cannot correctly predict the effective bond length (Chen and Teng 2001b; see also Chapter 2).

Recognising the deficiencies of the previous models, Chen and Teng developed rational shear strength models for both the FRP rupture failure mode (Chen and Teng 2001a) and the debonding failure mode (Chen and Teng 2001c) which can be used for practical design. The separate treatment of these two failure modes is essential because of the distinctive difference between them. These two models are presented in detail in the next section.

4.5 Chen and Teng's Shear Strength Model

4.5.1 General Expressions

Chen and Teng's (2001a, c) shear strength models were derived by assuming that discrete FRP strips can be modelled as an equivalent continuous FRP sheet/plate. As a result, the models are applicable to both FRP strips and continuous sheets/plates. As this approach involves some simplification for beams strengthened with strips, some strip spacing limitations will apply when these strength models are used in practical design, as discussed in detail in Section 4.6. As continuous sheets are a special case of strips, the equations of Chen and Teng's models are in terms of strips only.

For a general strengthening scheme, if the shear crack is inclined to the longitudinal axis of the beam by an angle θ and the FRP strips are of the same width and bonded on both sides of the beam (Figure 4.7), the contribution of the FRP to the shear strength is given by

$$V_{frp} = 2 f_{frp,e} t_{frp} w_{frp} \frac{h_{frp,e}(\cot\theta + \cot\beta)\sin\beta}{s_{frp}} \tag{4.7}$$

where $f_{frp,e}$ is the average stress of the FRP intersected by the shear crack at ultimate limit state, t_{frp} is the thickness of the FRP, w_{frp} is the width of each individual FRP strip (perpendicular to the fibre orientation), s_{frp} is the horizontal

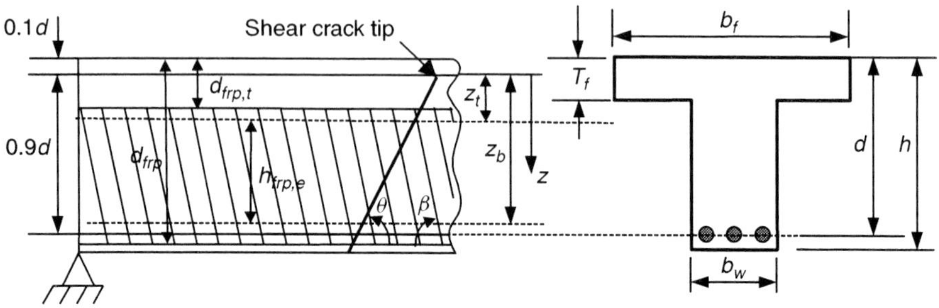

Figure 4.7 Notation for a general shear strengthening scheme

spacing of FRP strips (i.e. the centre-to-centre distance of FRP strips along the longitudinal axis of the beam), β is the angle of the inclination of fibres in the FRP to the longitudinal axis of the beam (measured clockwise for the left side of the beam as shown in Figure 4.1), and $h_{frp,e}$ is the effective height of the FRP bonded on the web.

Chen and Teng (2001a, c) assumed that the shear crack ends at a distance of $0.1d$ below the compression face of the beam, where d is the effective depth of the beam measured from the compression face to the centre of steel tension reinforcement. Employing a downward-oriented coordinate originating from the upper tip of the shear crack (Figure 4.7), the effective height of the FRP $h_{frp,e}$ can be expressed as

$$h_{frp,e} = z_b - z_t \tag{4.8}$$

where z_t and z_b are the coordinates of the top and bottom ends of the effective FRP respectively. They can be found from

$$z_t = \left(0.1d + d_{frp,t}\right) - 0.1d = d_{frp,t} \tag{4.9a}$$

$$z_b = \left[d - \left(h - d_{frp}\right)\right] - 0.1d \tag{4.9b}$$

in which $d_{frp,t}$ is the distance from the compression face to the top edge of the FRP (thus $d_{frp,t} = 0$ for wrapping), h is the height of the beam, and d_{frp} is the distance from the compression face to the lower edge of the FRP.

Equation 4.9b means that the lower end of the effective FRP is always taken to be $(h - d_{frp})$ above the actual lower end. It is thus taken to be at the centre of the steel tension reinforcement for wrapping and U jacketing ($d_{frp} = h$). For side plates, the lower end of the effective FRP is taken to be at the centre of the steel tension reinforcement if the FRP terminates at the base of the RC beam, but at a place above the centre of the tension reinforcement if the FRP is terminated above the base. Figure 4.7 shows an example where the side plates terminate slightly above the base of the RC beam.

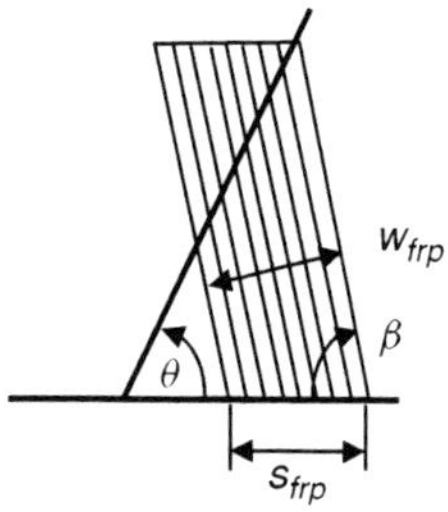

Figure 4.8 Relationship between w_{frp} and s_{frp} for continuous sheets/plates

It may be noted that for continuous FRP plates/sheets (Figure 4.8),

$$s_{frp} = \frac{w_{frp}}{\sin \beta} \tag{4.10}$$

Therefore, $s_{frp} = w_{frp}$ only if $\beta = 90^\circ$ (fibres vertically oriented). This is different from steel plates and the difference arises from the fact that FRPs can only bear a significant load in their main fibre direction. This is worth noting because some researchers (e.g. Chaallal *et al.* 1998) have used $s_{frp} = w_{frp}$ without giving due consideration to the fibre orientation.

Because the shear crack angle θ can be assumed to be 45° for practical design, the shear resistance contributed by the FRP (Equation 4.7) can be obtained if the average stress in the FRP $f_{frp,e}$ is known.

Chen and Teng (2001a, c) pointed out that the stress distribution in the FRP along a shear crack is nonuniform at the ultimate limit state for both FRP rupture and FRP debonding failures, for reasons detailed in the next two sections. The average stress in the FRP intersected by the shear crack at the ultimate limit state $f_{frp,e}$ can thus be written as

$$f_{frp,e} = D_{frp}\sigma_{frp,max} \tag{4.11}$$

in which $\sigma_{frp,max}$ is the maximum stress that can be reached in the FRP intersected by the shear crack and D_{frp} is the stress distribution factor defined by

$$D_{frp} = \frac{f_{frp,e}}{\sigma_{frp,max}} \tag{4.12}$$

The values of $\sigma_{frp,max}$ and D_{frp} depend on whether the shear failure is controlled by FRP rupture or FRP debonding. They are determined in the next two sections.

4.5.2 Shear Failure Controlled by FRP Rupture

This section is concerned with shear-strengthened RC beams whose failure is controlled by FRP rupture. For FRP-wrapped RC beams failing without FRP rupture, their shear strength may be evaluated using the model presented in this section by setting a suitable limit for the maximum strain in the FRP. The model of this section is thus applicable to all RC beams wrapped with FRPs around the whole perimeter. Beams strengthened with U jackets may fail by either FRP rupture or FRP debonding. Therefore, their shear strength should be evaluated using both the strength model presented in this section for failure controlled by FRP rupture and that controlled by FRP debonding presented in the next section. If proper mechanical anchors are provided at the free ends of FRP U jackets and side plates to ensure that failure is controlled by FRP rupture, the shear strength can also be evaluated using this model.

Chen and Teng (2001a) noted that the strain distribution in the FRP along a shear crack is nonuniform and FRPs are brittle materials. Therefore, for RC beams whose shear strength is controlled by FRP rupture, the failure process starts when the most highly stressed point in the FRP intersected by the shear crack reaches the ultimate tensile strength of the FRP. Rupture of adjacent parts of the FRP then follows quickly, as the intact parts of the FRP need to take over the forces released by the ruptured parts. Therefore, rupture of the FRP, once started, will propagate rapidly along the shear crack, leading to catastrophic failure of the beam. It is evident that in this process of failure, the FRP intersected by the shear crack is not stressed to the same ultimate tensile strength at any instant of the failure process.

To develop an accurate shear strength model, a rational stress distribution in the FRP needs to be assumed and considered in conjunction with the brittleness of the FRP. Chen and Teng (2001a) proposed a general parabolic stress distribution along a shear crack, with a linear distribution as a simple degenerated case. For a linear stress distribution in the FRP along the shear crack, the stress in the FRP increases linearly from zero at the crack tip to the ultimate tensile strength at the lower end of the shear crack.

Experimental studies have shown that most FRP rupture failures occurred in beams strengthened by FRP wrapping. It is also most likely that the FRP has debonded over the full height of the FRP over the major part of the shear crack before FRP rupture occurs. The stress (strain) in the FRP is thus approximately proportional to the width of the shear crack, which may be closely represented by a linear variation increasing from zero at the crack tip for most of such shear failures. Therefore, the linear model is believed to be a good and yet simple approximation of the actual stress distribution.

As a conservative and simple model, the linear stress distribution model was recommended by Chen and Teng (2001a) for practical use. For a general strengthening scheme as shown in Figure 4.7, the stress distribution factor is expressed as

$$D_{frp} = \frac{1+\zeta}{2} \tag{4.13}$$

where

$$\zeta = \frac{z_t}{z_b} \tag{4.14}$$

For strengthening schemes in which FRPs are bonded over the full height of the sides (e.g. wrapping), $\zeta = 0$ so $D_{frp} = 0.5$.

The FRP starts to rupture if the maximum stress in the FRP, which occurs at the lower end of the FRP intersected by the shear crack, reaches its ultimate tensile strength. The maximum stress in the FRP is thus

$$\sigma_{frp,max} = f_{frp} \tag{4.15}$$

It should be noted that Equation 4.15 cannot be used in cases where the ultimate shear strength of the beam is reached before FRP rupture. Such situations are encountered in beams strengthened with FRPs with a large ultimate strain. A crude explanation of this phenomenon is given as follows.

Take for example an RC beam without any steel shear reinforcement and strengthened by bonding FRPs on its sides. The total shear force in its critical cross-section is carried by the concrete and the external FRP reinforcement. The force carried by the FRP continuously increases as the width of the shear crack increases, whilst the force carried by the concrete (including dowel action, aggregate interlock and that carried by the uncracked concrete section) may also increase or remain stable if the width of the crack is small. When the crack width further increases, the force carried by the concrete may reduce but the total force carried by the concrete and the FRP still increases if the shear crack is not very wide. In this case, the shear strength of the beam reaches its peak value when the maximum stress in the FRP reaches its ultimate tensile strength, leading to FRP rupture (Figure 4.9a).

However, if the width of the crack becomes very large before FRP rupture, the shear force carried by the concrete (especially aggregate interlock) can decrease rapidly. When this decrease equals the corresponding increase in the shear force carried by the FRP, the total shear force of the beam peaks. The beam thus reaches its ultimate shear strength before the rupture of the FRP (Figure 4.9b). In this case, it is necessary to limit the maximum usable strain of the FRP as discussed in Section 4.6.

4.5.3 Shear Failure Controlled by FRP Debonding

As discussed earlier in this chapter, debonding between the FRP and the concrete has been the predominant failure mode for FRPs bonded on the sides of beams. This mode also controls the strength for most beams strengthened with U jackets. The shear strength of such beams is controlled by the bond strength between the FRP and the concrete. This section deals with the determination of the maximum stress in the FRP, $\sigma_{frp,max}$, and the stress distribution factor D_{frp} for such beams.

The key issue here is the ultimate bond strength between the FRP and the concrete at failure. As discussed in Chapter 2, substantial experimental and theoretical work exists on the bond strength of FRP-to-concrete joints subject to

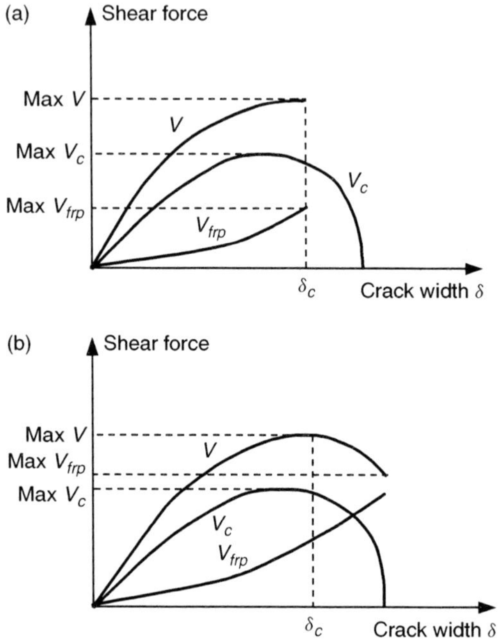

Figure 4.9 Shear forces carried by the concrete and the FRP: (a) failure with FRP rupture; (b) failure without FRP rupture

tension (e.g. van Gemert 1980, Swamy *et al.* 1986, Triantafillou and Plevris 1992, Kobatake *et al.* 1993, Autocon 1994, Holzenkämpfer 1994, Täljsten 1994, 1997, Chajes *et al.* 1995, 1996, Ziraba *et al.* 1995, Bizindavyi and Neale 1997, 1999, Brosens and van Gemert 1997, Fukuzawa *et al.* 1997, Hiroyuki and Wu 1997, Maeda *et al.* 1997, Neubauer and Rostásy 1997, Chaallal *et al.* 1998, Khalifa *et al.* 1998, Yuan and Wu 1999, Yuan *et al.* 2001). A very important aspect of this bond behaviour is that there exists an effective bond length beyond which an extension of the bond length cannot increase the bond strength. This is a fundamental difference between externally bonded reinforcement and internal reinforcement. For the latter, a sufficiently long bond length can always be found so that the full tensile strength of the reinforcement can be achieved, provided there exists a sufficient concrete cover.

Based on simple shear tests, Chen and Teng (2001b) developed a simple, rational and accurate model for predicting the bond strength and the effective bond length of FRP-to-concrete joints. Chen and Teng (2001c) thus adopted this bond strength model in predicting the shear strength of FRP-strengthened RC beams controlled by debonding. A detailed description of this model was presented in Chapter 2.

For the debonding failure of FRP strips/sheets crossing a shear crack, the maximum stress in the FRP occurs at the location where the FRP has the longest bond length. The maximum stress in the FRP ($\sigma_{frp,max}$ in Equation 4.11) is thus limited by either the ultimate bond strength of Chen and Teng's (2001b) model or the ultimate tensile strength of the FRP:

$$\sigma_{frp,max} = \min \begin{cases} f_{frp} \\ 0.427\beta_w \beta_L \sqrt{\dfrac{E_{frp}\sqrt{f'_c}}{t_{frp}}} \end{cases} \tag{4.16}$$

where β_L reflects the effect of bond length and β_w the effect of FRP-to-concrete width ratio. For shear strengthening considered here, they can be expressed as

$$\beta_L = \begin{cases} 1 & \text{if } \lambda \geqslant 1 \\ \sin(\pi\lambda/2) & \text{if } \lambda < 1 \end{cases} \tag{4.17a}$$

$$\beta_w = \sqrt{\frac{2 - w_{frp}/\left(s_{frp}\sin\beta\right)}{1 + w_{frp}/\left(s_{frp}\sin\beta\right)}},$$

$$\beta_w = \frac{\sqrt{2}}{2} \quad \text{for continuous sheets/plates} \tag{4.17b}$$

in which the normalised maximum bond length λ, the maximum bond length L_{max} and the effective bond length L_e are respectively given by

$$\lambda = \frac{L_{max}}{L_e} \tag{4.18a}$$

$$L_{max} = \begin{cases} h_{frp,e}/\sin\beta & \text{for U jackets} \\ h_{frp,e}/(2\sin\beta) & \text{for side plates} \end{cases} \tag{4.18b}$$

$$L_e = \sqrt{\frac{E_{frp}t_{frp}}{\sqrt{f'_c}}} \tag{4.18c}$$

Among the fibres crossing a shear crack, the one at the lower end has the maximum bond length for U jacketing. For side plates, the fibre at the middle of the shear crack has the maximum bond length. This is the reason that the number 2 appears in the denominator for side plates in Equation 4.18b.

Because an FRP-to-concrete bonded joint generally has a certain amount of slip after reaching the ultimate bond strength (i.e. it shows pseudo-plastic behaviour), Chen and Teng (2001c) assumed that all the FRP intersected by the shear crack can develop its bond strength fully. Note, however, that the bond strength of a particular part or strip depends on where the shear crack is located relative to the ends of the part or strip. Under this assumption, Chen and Teng (2001c) derived the stress distribution factor for debonding failure D_{frp}:

$$D_{frp} = \begin{cases} \dfrac{2}{\pi\lambda}\dfrac{1-\cos(\pi/2)\lambda}{\sin(\pi/2)\lambda} & \text{if } \lambda \leqslant 1 \\ 1-\dfrac{\pi-2}{\pi\lambda} & \text{if } \lambda > 1 \end{cases} \tag{4.19}$$

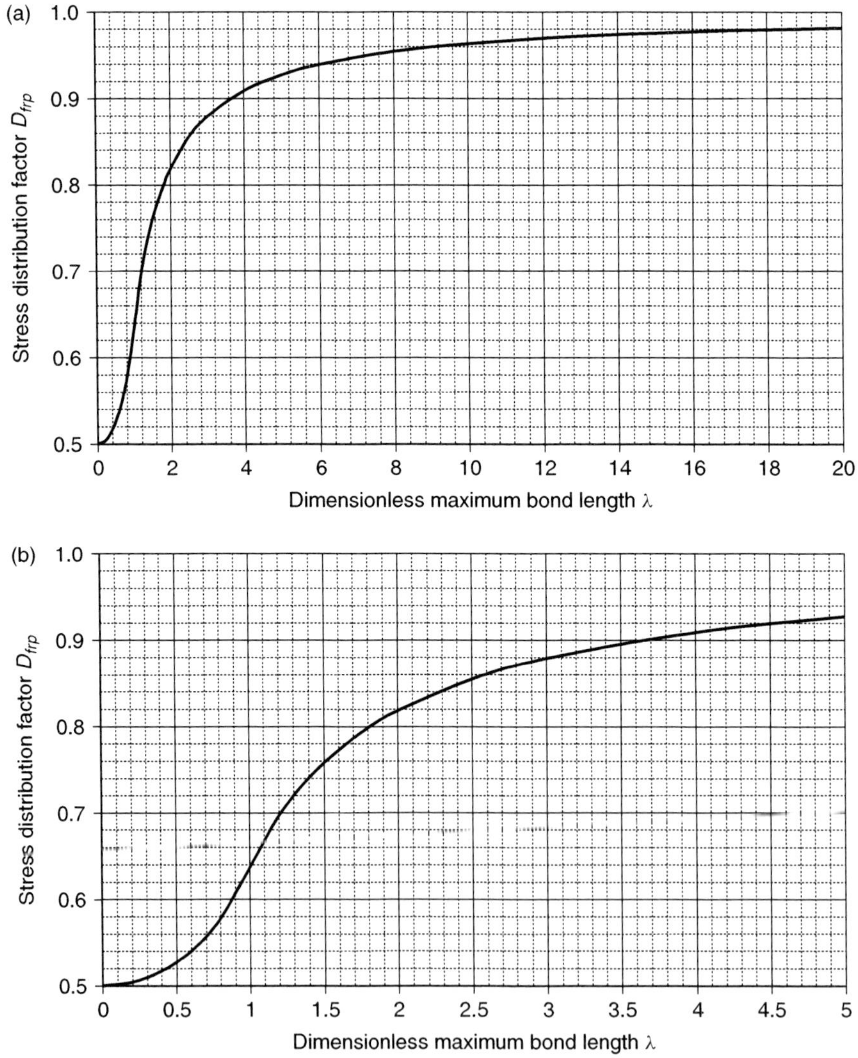

Figure 4.10 Stress distribution factor for U jacketing and side bonding: (a) D_{frp}, $0 < \lambda < 20$; (b) D_{frp}, $0 < \lambda < 5$

The average stress in the FRP along the shear crack is calculated using $f_{frp,e} = D_{frp}\sigma_{frp,max}$ (Equation 4.11).

Equation 4.19 is applicable to both U jackets and side strips/plates. However, the actual calculated values are different for theses two cases even if the bond geometry is the same on the beam sides. This is because the maximum bond length L_{max} for U jackets in Equation 4.18b is twice that for side strips/plates (Equation 4.18b).

Figure 4.10(a) shows the variation of D_{frp} with λ. A more detailed picture of this variation for $0 < \lambda < 5$ is shown in Figure 4.10(b). The figure shows that D_{frp} increases from 0.5 as λ increases from 0 and approaches unity when λ approaches infinity. For the same bonding geometry on the sides of a beam, D_{frp} is larger for U jacketing than for side bonding, because of the difference in the values of λ. This reflects the fact that U jacketing is more effective in shear strengthening than side bonding.

4.6 Design Recommendation

The design proposal given by Chen and Teng (2001a, c), based on the shear strength models described above, is recommended here for use in practical design. This design proposal is presented below.

4.6.1 General Equations

The total shear capacity of a shear-strengthened RC beam can be calculated by assuming that it is the summation of the concrete contribution V_c, the contribution by steel shear reinforcement V_s, and the FRP contribution $V_{frp,}$ as given by Equation 4.1. V_c and V_s may be obtained directly from formulae in existing codes of practice for RC structures (e.g. BS 8110 1997). However, it must emphasised that the authors have some reservation about the accuracy of V_s calculated in such a manner because not all the steel shear reinforcement may reach its yield strength in shear-strengthened RC beams with externally bonded FRPs. Therefore, care must be exercised in practical design until a better understanding becomes available from future research.

Based on the strength models presented in Section 4.5, and assuming that $\theta = 45°$ for practical design, the FRP contribution to the design shear strength (applicable to all bonding schemes) is given by (Chen and Teng 2001a, c)

$$V_{frp} = 2 f_{frp,e} t_{frp} w_{frp} \frac{h_{frp,e}(\sin\beta + \cos\beta)}{s_{frp}} \tag{4.20}$$

where

$$h_{frp,e} = z_b - z_t \tag{4.21a}$$

$$z_t = d_{frp,t} \tag{4.21b}$$

$$z_b = d_{frp} - h + 0.9d \tag{4.21c}$$

$$f_{frp,e} = D_{frp}\sigma_{frp,max} \tag{4.21d}$$

4.6.2 Shear Failure Controlled by FRP Rupture

For wrapping, the shear strength should be evaluated for the FRP rupture failure mode. For U jacketing, the smaller predictions of the FRP rupture mode and the debonding mode should be used. It is generally unnecessary to check the shear strength for the FRP rupture mode for beams strengthened by side bonding, except for very large beams or for beams with FRP strips soundly anchored at strip ends using mechanical anchors.

For the shear failure mode controlled by FRP rupture, the stress distribution factor is

$$D_{frp} = \frac{1+\zeta}{2} \tag{4.22a}$$

where

$$\zeta = \frac{z_t}{z_b} \tag{4.22b}$$

The maximum stress in the FRP can be obtained from

$$\sigma_{frp,max} = \begin{cases} 0.8 f_{frp}/\gamma_{frp} & \text{if } f_{frp}/E_{frp} \leqslant \varepsilon_{max} \\ 0.8\,\varepsilon_{max} E_{frp}/\gamma_{frp} & \text{if } f_{frp}/E_{frp} > \varepsilon_{max} \end{cases} \tag{4.23}$$

in which γ_{frp} is the partial safety factor for the FRP tensile strength as specified in Section 1.6.

The factor of 0.8 was introduced by Chen and Teng (2001a) to account for factors not considered in their strength models. One such factor is the detrimental effect of the corners of beams on the tensile strength of the FRP, if the corners are not properly rounded.

Equation 4.23 also includes a limit on the maximum usable strain of the FRP to control the width of shear cracks for design use. There is, however, insufficient information in the literature to determine a suitable value for this maximum usable strain quantitatively so further research is needed to clarify this. A value of $\varepsilon_{max} = 1.5\,\%$ may be used until a soundly based proposal is available.

4.6.3 Shear Failure Controlled by FRP Debonding

For shear strengthening using side strips/plates, the shear strength will be evaluated for the failure mode of debonding. U-jacketed beams are also likely to fail by debonding, so it is essential that this mode is checked for them.

For shear failure controlled by FRP debonding, the 95 percentile char acteristic bond strength given in Section 2.5 (Equation 2.20) can be used to determine the design ultimate stress in the FRP:

$$\sigma_{frp,max} = \min \begin{cases} 0.8 f_{frp}/\gamma_{frp} \\ \dfrac{0.3}{\gamma_b}\beta_w \beta_L \sqrt{\dfrac{E_{frp}}{t_{frp}}\sqrt{f_{cu}}} \end{cases} \tag{4.24}$$

where f_{cu} is the cube compressive strength of concrete, γ_b is the partial safety factor for bond strength as given in Section 1.6, and

$$\beta_L = \begin{cases} 1 & \text{if } \lambda \geqslant 1 \\ \sin(\pi\lambda/2) & \text{if } \lambda < 1 \end{cases} \tag{4.25a}$$

$$\beta_w = \sqrt{\frac{2 - w_{frp}/\left(s_{frp}\sin\beta\right)}{1 + w_{frp}/\left(s_{frp}\sin\beta\right)}},$$

$$\beta_w = \frac{\sqrt{2}}{2} \text{ for continuous sheets/plates} \tag{4.25b}$$

$$\lambda = \frac{L_{max}}{L_e} \tag{4.25c}$$

$$L_{max} = \begin{cases} h_{frp,e}/\sin\beta & \text{for U jackets} \\ h_{frp,e}/(2\sin\beta) & \text{for side plates/wings} \end{cases} \tag{4.25d}$$

$$L_e = \sqrt{\frac{E_{frp}t_{frp}}{\sqrt{0.8 f_{cu}}}} \tag{4.25e}$$

If Equation 4.24 gives a higher stress prediction than that of Equation 4.23, the result of Equation 4.23 should be used. MPa and mm should be used in Equation 4.25e.

The stress distribution factor can be obtained using either Figure 4.10, or the following equation:

$$D_{frp} = \begin{cases} \dfrac{2}{\pi\lambda}\dfrac{1-\cos(\pi/2)\lambda}{\sin(\pi/2)\lambda} & \text{if } \lambda \leqslant 1 \\ 1 - \dfrac{\pi - 2}{\pi\lambda} & \text{if } \lambda > 1 \end{cases} \tag{4.26}$$

In the design of a shear strengthening scheme by U jacketing or side bonding with strips, an iterative procedure is required because the coefficient β_w in Equation 4.25 is related to the ratio of strip width to strip spacing w_{frp}/s_{frp} ($\beta_w = 0.707$ for continuous sheets/plates). An initial value of $\beta_w = 1$ may be used. The iterative process will converge very quickly, with three iterations usually being sufficient.

4.6.4 Spacing Requirement for FRP Strips

It may be noted that the preceding design equations were derived by treating strips as equivalent continuous sheets/plates. For this treatment to be accurate, the number of strips intersected by the shear crack should be sufficient (e.g. two or more). Otherwise, the treatment can lead to either conservative or unconservative predictions, depending on the locations of the strips. Some extreme cases as shown in Figure 4.11 are discussed as follows.

Take for example the diagonal shear failure where the shear crack is widest at the lower end: a strip is most effective if it is at the lower end of the shear crack (Figure 4.11a) for wrapping because the ultimate tensile strength of the FRP can be reached while the shear crack is not so large so that the concrete shear strength and aggregate interlock are maintained. The most effective position is also the lower end for U jacketing (Figure 4.11a), but at the middle for side bonding (Figure 4.11b), because the bond length is largest in both cases.

A strip near the upper end of the crack contributes little to the shear strength (Figure 4.11c). The reason is that the strip needs to be sufficiently deformed for it to be effective for wrapping, but this means that the crack at the lower edge has to become very wide, leading to loss of aggregate interlock. Therefore, if only a single strip is intersected by the shear crack near its upper end, the beam will fail before the FRP strip becomes effective. A strip located at the upper end for U jacketing (Figure 4.11c) and either end of the shear crack for

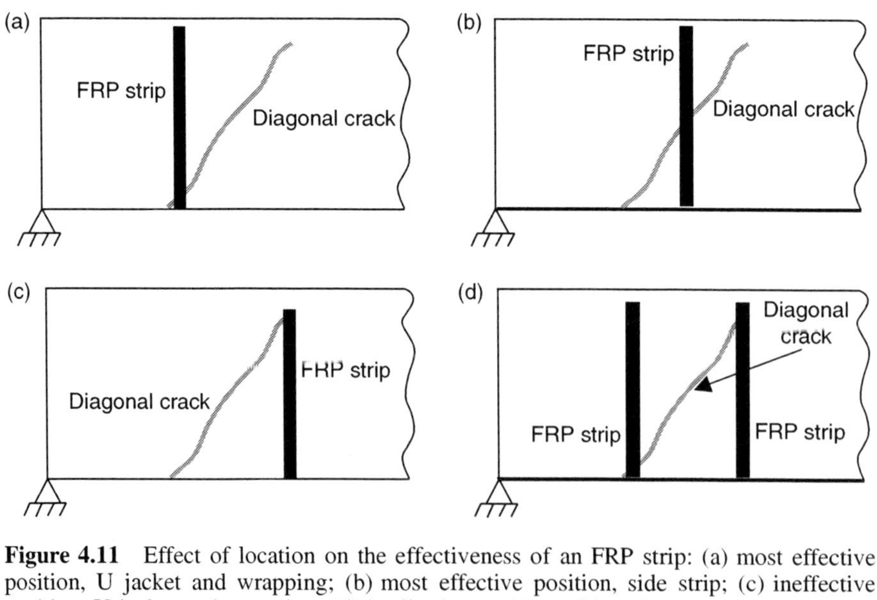

Figure 4.11 Effect of location on the effectiveness of an FRP strip: (a) most effective position, U jacket and wrapping; (b) most effective position, side strip; (c) ineffective position, U jacket and wrapping; (d) ineffective position, side strip

side bonding (Figure 4.11d) is completely ineffective owing to a lack of bond length. Therefore, a strengthening scheme can be ineffective if only one strip is intersected by the shear crack.

For a shear strengthening scheme to be effective, it should therefore be ensured that the strip spacing does not exceed half the horizontal distance spanned by the shear crack, so that in all situations at least two strips will cross the assumed diagonal crack with at least one being effective, i.e.

$$s_{frp} \leqslant s_{frp,max} = \frac{h_{frp,e}(\sin\beta + \cos\beta)}{2} = \frac{h_{frp,e}}{2} \quad \text{if } \beta = 90^\circ \tag{4.27}$$

4.6.5 Strengthening Design for Indeterminate Beams

The above recommended design equations are equally applicable in the design of shear strengthening of statically indeterminate beams. However, extra care must be taken when dealing with negative moment regions, where the upper face is in tension but the lower face is in compression. Because shear cracks are expected to start from the upper face of the beam in these regions, the dimensional quantities shown in Figure 4.7 should be measured from the lower face of the beam instead. Similarly, reversed U jackets must be used to bond FRPs on the sides and the upper face of the beam (instead of the sides and the lower face in positive moment regions), otherwise U jackets are only as effective as side plates/strips.

4.7 Comparison with Experimental Observations

The shear strength models presented in Section 4.5 explicitly include the effect of nonuniform stress distribution in the FRP along the shear crack. These models satisfactorily explain the phenomenon that the contribution of the FRP to the shear strength of a beam is less than its ultimate tensile strength.

Based on a comprehensive survey of experimental data in the published literature, Chen and Teng (2001a, c) compared their strength models with the data gathered by them. Figure 4.12 shows that their models have a good agreement with experimental data for all the strengthening schemes. Details of the statistical performance are given in Table 4.7. In these comparisons, different angles estimated from the relevant literature were used for the shear crack. More details about these test data and the comparisons can be found in Chen and Teng (2001a, c).

Figure 4.13 compares experimental observations with Chen and Teng's design proposal (2001a, c) as described in Section 4.6. All the material partial safety factors (γ_s, γ_c and γ_{frp}) were set to one for the purpose of this comparison. Only very few data points have predicted values greater than the experimental observations. The design proposal is thus suitable for practical use.

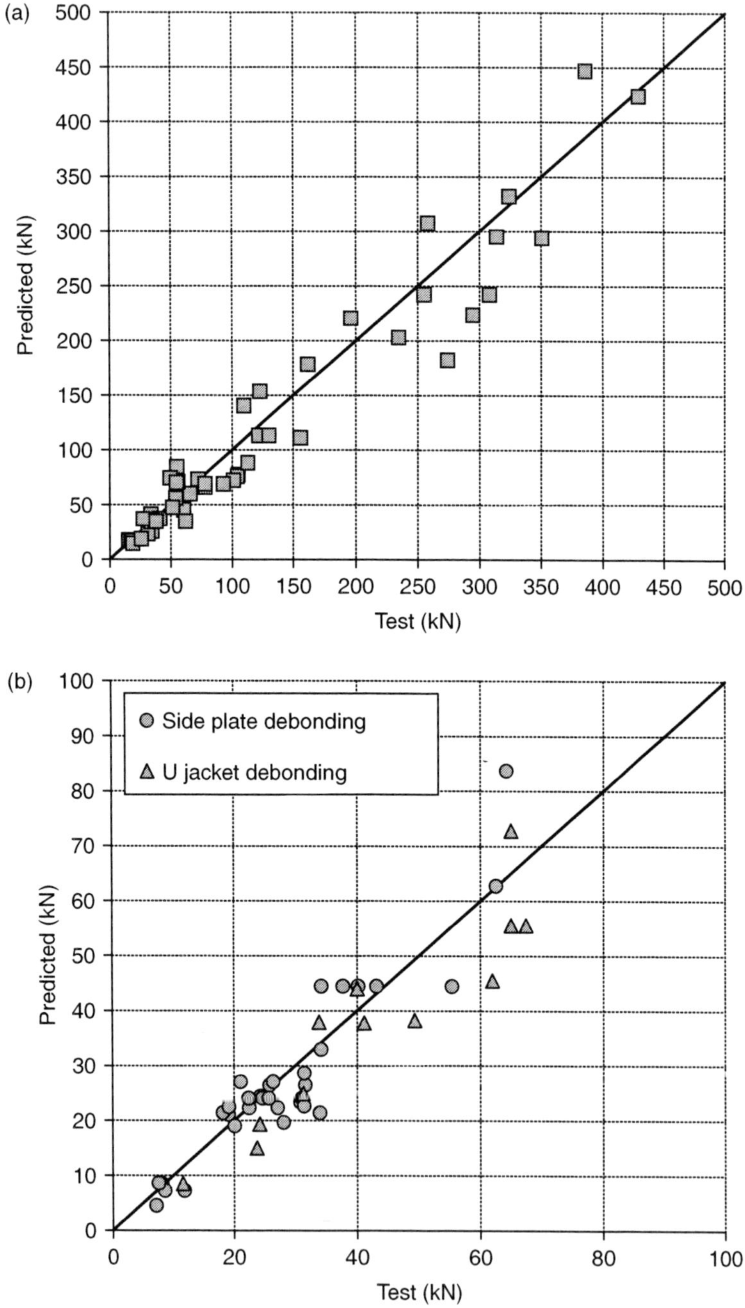

Figure 4.12 Chen and Teng's predictions versus test results: (a) FRP rupture; (b) FRP debonding

Table 4.7 Statistical performance of Chen and Teng's shear strength models

Failure mode	FRP rupture failure	Debonding of side plates	Debonding of U jackets
No. of valid experimental data	50	32	13
Average of test-to-predicted strength ratio	1.06	1.07	1.20
Standard deviation	0.229	0.226	0.226
Coefficient of variation	21.5 %	21.1 %	18.9 %

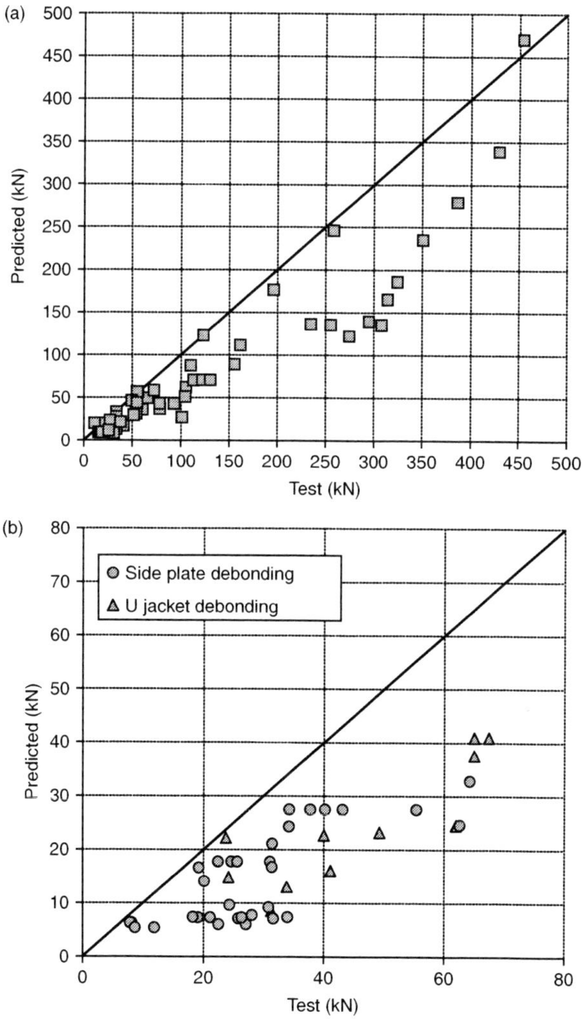

Figure 4.13 Chen and Teng's design proposal versus test results: (a) FRP rupture; (b) FRP debonding

4.8 Conclusions

This chapter has presented a detailed treatment of shear strengthening of RC beams using FRP composites. Various strengthening schemes have been classified, and their advantages and disadvantages discussed. Some guidelines for the selection of a suitable shear strengthening scheme for a given situation were also presented. The main focus of the chapter, however, has been the presentation of shear strength models for FRP-strengthened RC beams, based on the recent work of Chen and Teng (2001a, c). Among the existing models, the shear strength models of Chen and Teng are the most rational and accurate. A design proposal developed by Chen and Teng (2001a, c) based on these models is therefore recommended for design use.

It must be emphasised that the technique of shear strengthening of RC beams using FRP composites is still at its early stage of development. Current understanding of the behaviour of the strengthened beam is limited and much further research is needed. While the design method recommended here is believed to be conservative as far as the contribution of bonded FRPs to shear resistance is concerned, it should be kept in mind that the steel shear reinforcement in a strengthened beam may contribute less than that predicted by existing codes of practice for RC structures. There is thus still considerable uncertainty concerning the total shear capacity of an RC beam with externally bonded FRP shear reinforcement. Therefore, care must be exercised in practical design and expert advice or experimental evidence should be sought wherever necessary.

References

ACI 318-95 (1999) *Building Code Requirements for Structural Concrete (318-95) and Commentary (318R-95)*, American Concrete Institute (ACI), Fifth Printing, Farmington Hills, Michigan, USA.

Alexander, J. G. S. (1996) 'Shear strengthening of small scale concrete beams with carbon fibre reinforced plastic sheets', *Proceedings of Annual Conference of Canadian Society for Civil Engineering*, Vol. 2, No. A, pp. 167–178.

Al-Sulaimani, G. J., Sharif, A. M., Basunbul, I. A., Baluch, M. H. and Ghaleb, B. N. (1994) 'Shear repair for reinforced concrete by fibreglass plate bonding', *ACI Structural Journal*, Vol. 91, No. 3, pp. 458–464.

Araki, N., Matsuzaki, Y., Nakano, K., Kataka, T. and Fukuyama, H. (1997) 'Shear capacity of retrofitted RC members with continuous fibre sheets', *Non-Metallic (FRP) Reinforcement for Concrete Structures, Proceedings of the Third International Symposium*, Sapporo, Japan, pp. 512–522.

Arduini, M., D'Ambrisi, A. and Di Tommaso, A. (1994) 'Shear failure of concrete beams reinforced with FRP plates', *Infrastructure: New Materials and Methods of Repair, Proceedings of the Materials Engineering Conference 804, ASCE*, New York, USA, pp. 123–130.

Autocon (1994) *FORCA Tow Sheets Technical Notes*, Autocon Composites Incorporated, Ontario, Canada.

Bizindavyi, L. and Neale, K. W. (1997) 'Experimental and theoretical investigation of transfer lengths for composite laminates bonded to concrete', *Structures–Composites Materials, Structural Systems, Telecommunication Towers, Proceedings of Annual Conference of Canadian Society for Civil Engineering*, Quebec, Canada, Vol. 6, pp. 51–60.

Bizindavyi, L. and Neale, K. W. (1999) 'Transfer lengths and bond strengths for composites bonded to concrete', *Journal of Composites for Constructions, ASCE*, Vol. 3, No. 4, pp. 153–160.

Brosens, K. and van Gemert, D. (1997) 'Anchoring stresses between concrete and carbon fibre reinforced laminates', *Non-Metallic (FRP) Reinforcement for Concrete Structures, Proceedings of the Third International Symposium*, Sapporo, Japan, Vol. 1, pp. 271–278.

BS 8110 (1997) *Structural Use of Concrete, Part 1. Code of Practice for Design and Construction*, British Standards Institution, London, UK.

Buyukozturk, O. and Hearing, B. (1998) 'Failure behavior of precracked concrete beams retrofitted with FRP', *Journal of Composites for Construction, ASCE*, Vol. 2, No. 3, pp. 138–144.

Chaallal, O., Nollet, M. J. and Perraton, D. (1998) 'Strengthening of reinforced concrete beams with externally bonded fibre-reinforced-plastic plates: design guidelines for shear and flexure', *Canadian Journal of Civil Engineering*, Vol. 25, No. 4, pp. 692–704.

Chajes, M. J., Januszka, T. F., Mertz, D. R., Thomson, T. A., Jr, and Finch, W. W., Jr (1995) 'Shear strengthening of reinforced concrete beams using externally applied composite fabrics', *ACI Structural Journal*, Vol. 92, No. 3, pp. 295–303.

Chajes, M. J., Finch, W. W., Jr, Januszka, T. F. and Thomson, T. A., Jr (1996) 'Bond and force transfer of composite material plates bonded to concrete', *ACI Structural Journal*, Vol. 93, No. 2, pp. 295–303.

Chen, J. F. and Teng, J. G. (2001a) 'Shear capacity of FRP strengthened RC beams: FRP rupture', *to be published.*

Chen, J. F. and Teng, J. G. (2001b) 'Anchorage strength models for FRP and steel plates attached to concrete', *Journal of Structural Engineering, ASCE*, Vol. 127, No. 7, pp. 784–791.

Chen, J. F. and Teng, J. G. (2001c) 'Shear capacity of FRP strengthened RC beams: FRP debonding', *to be published.*

ENV 1992-1-1 (1991) *Eurocode 2: Design of Concrete Structures, Part 1: General Rules and Rules for Buildings*, CEN, Brussels, Belgium.

Fanning, P. and Kelly, O. (1999) 'Shear strengthening of reinforced concrete beams: an experimental study using CFRP plates', *Proceedings of the Eighth International Conference on Structural Faults and Repair*, edited by M. C. Forde, Engineering Technics Press, Edinburgh, UK.

Fukuzawa, K., Numao, T., Wu, Z., Yoshizawa, H. and Mitsui, M. (1997) 'Critical strain energy release rate of interface debonding between carbon fibre sheet and mortar', *Non-Metallic (FRP) Reinforcement for Concrete Structures, Proceedings of the Third International Symposium*, Sapporo, Japan, pp. 295–302.

Funakawa, I., Shimono, K., Watanabe, T., Asada, S. and Ushijima, S. (1997) 'Experimental study on shear strengthening with continuous fibre reinforcement sheet and methyl methacrylate resin', *Non-Metallic (FRP) Reinforcement for Concrete Structures, Proceedings of the Third International Symposium*, Sapporo, Japan, pp. 475–482.

Gergely, I., Pantelides, C. P., Nuismer, R. J. and Reaveley, L. D. (1998) 'Bridge pier retrofit using fibre-reinforced plastic composites', *Journal of Composites for Construction, ASCE*, Vol. 2, No. 4, pp. 165–174.

Hiroyuki, Y. and Wu, Z. (1997) 'Analysis of debonding fracture properties of CFS strengthened member subject to tension', *Non-Metallic (FRP) Reinforcement for*

Concrete Structures, Proceedings of the Third International Symposium, Sapporo, Japan, pp. 287–294.

Holzenkämpfer, O. (1994) *Ingenieurmodelle des Verbundes geklebter Bewehrung für Betonbauteile*, Dissertation, TU Braunschweig.

Hutchinson, R. L. and Rizkalla, S. H. (1999) 'Shear strengthening of AASHTO bridge girders using carbon fiber reinforced polymer sheets', *Proceedings of the Fourth International Symposium on Fiber Reinforced Polymer Reinforcement for Reinforced Concrete Structures*, Baltimore, Maryland, USA, edited by C. W. Dolan, S. H. Rizkalla and A. Nanni, pp. 945–958, American Concrete Institute, Farmington Hills, Michigan, USA.

Kachlakev, D. I. and Barnes, W. A. (1999) 'Flexural and shear performance of concrete beams strengthened with fiber reinforced polymer laminates', *Proceedings of the Fourth International Symposium on Fiber Reinforced Polymer Reinforcement for Reinforced Concrete Structures*, Baltimore, Maryland, USA, edited by C. W. Dolan, S. H. Rizkalla and A. Nanni, pp. 959–971, American Concrete Institute, Farmington Hills, Michigan, USA.

Kage, T., Abe, M., Lee, H. S. and Tomosawa, F. (1997) 'Effect of CFRP sheets on shear strengthening of RC beams damaged by corrosion of stirrup', *Non-Metallic (FRP) Reinforcement for Concrete Structures, Proceedings of the Third International Symposium*, Sapporo, Japan, pp. 443–450.

Khalifa, A. and Nanni, A. (2000) 'Improving shear capacity of existing RC T-section beams using CFRP composites', *Cement and Concrete Composites*, Vol. 22, pp. 165–174.

Khalifa, A., Gold, W. J., Nanni, A. and Aziz, A. (1998) 'Contribution of externally bonded FRP to shear capacity of RC flexural members', *Journal of Composites for Construction, ASCE*, Vol. 2, No. 4, pp. 195–203.

Khalifa, A., Tumialan, G., Nanni, A. and Belarbi, A. (1999) 'Shear strengthening of continuous reinforced concrete beams using externally bonded carbon fiber reinforced polymer sheets', *Proceedings of the Fourth International Symposium on Fiber Reinforced Polymer Reinforcement for Reinforced Concrete Structures*, Baltimore, Maryland, USA, edited by C. W. Dolan, S. H. Rizkalla and A. Nanni, pp. 995–1008, American Institute, Farmington Hills, Michigan, USA.

Kobatake, Y., Kimura, K. and Katsumata, H. (1993) 'A retrofitting method for reinforced concrete structures using carbon fibre', *Fibre-Reinforced-Plastic (FRP) Reinforcement for Concrete Structures: Properties and Applications*, edited by A. Nanni, Elsevier Science, Amsterdam, The Netherlands, pp. 435–450.

Kong, F. K. and Evans, R. H. (1987) *Reinforced and Prestressed Concrete*, Third Edition, Chapman & Hall, London, UK.

Maeda, T., Asano, Y., Sato, Y., Ueda, T. and Kakuta, Y. (1997) 'A study on bond mechanism of carbon fiber sheet', *Non-Metallic (FRP) Reinforcement for Concrete Structures, Proceedings of the Third International Symposium*, Sapporo, Japan, pp. 279–285.

Malek, A. M. and Saadatmanesh, H. (1998) 'Analytical study of reinforced concrete beams strengthened with web-bonded fiber reinforced plates or fabrics', *ACI Structural Journal*, Vol. 95, No. 3, pp. 343–351.

Mitsui, Y., Murakami, K., Takeda, K. and Sakai, H. (1998) 'A study on shear reinforcement of reinforced concrete beams externally bonded with carbon fiber sheets', *Composite Interfaces*, Vol. 5, No. 4, pp. 285–295.

Mutsuyoshi, H., Ishibashi, T., Okano, M. and Katsuki, F. (1999) 'New design method for seismic retrofit of bridge columns with continuous fibre sheet – performance-based design', *Proceedings of the Fourth International Symposium on Fiber Reinforced Polymer Reinforcement for Reinforced Concrete Structures*, Baltimore, Maryland, USA, edited by C. W. Dolan, S. H. Rizkalla and A. Nanni, pp. 229–241, American Concrete Institute, Farmington Hills, Michigan, USA.

Neubauer, U. and Rostásy, F. S. (1997) 'Design aspects of concrete structures strengthened with externally bonded CFRP plates', *Proceedings of the Seventh International Conference on Structural Faults and Repairs*, edited by M. C. Forde, Engineering Technics Press, Edinburgh, UK, pp. 109–118.

Plevris, N., Triantafillou, T. C. and Veneziano, D. (1995) 'Reliability of RC members strengthened with CFRP laminates', *Journal of Structural Engineering, ASCE*, Vol. 121, No. 7, pp. 1037–1044.

Sato, Y., Ueda, T., Kakuta, Y. and Tanaka, T. (1996) 'Shear reinforcing effect of carbon fibre sheet attached to side of reinforced concrete beams', *Advanced Composite Materials in Bridges and Structures, Proceedings of the Second International Conference*, Quebec, Canada, edited by M. M. El-Badry, pp. 621–627, Canadian Society for Civil Engineering, Montreal, Canada.

Sato, Y., Ueda, T., Kakuta, Y. and Ono, S. (1997a) 'Ultimate shear capacity of reinforced concrete beams with carbon fibre sheet', *Non-Metallic (FRP) Reinforcement for Concrete Structures, Proceedings of the Third International Symposium*, Sapporo, Japan, pp. 499–505.

Sato, Y., Katsumata, H., and Kobatake, Y. (1997b) 'Shear strengthening of existing reinforced concrete beams by CFRP sheet', *Non-Metallic (FRP) Reinforcement for Concrete Structures, Proceedings of the Third International Symposium*, Sapporo, Japan, pp. 507–513.

Swamy, R. N. and Mukhopadhyaya, P. (1995) 'Role and effectiveness of non-metallic plates in strengthening concrete structures', *Non-Metallic (FRP) Reinforcement for Concrete Structures, Proceedings of the Second International RILEM Symposium*, Ghent, Belgium, edited by L. Taerwe, pp. 473–482, E & FN Spon, London, UK.

Swamy, R. N., Jones, R. and Charif, A. (1986) 'Shear adhesion properties of epoxy resin adhesives', *Adhesion between Polymers and Concrete, Proceedings of an International Symposium*, Paris, France, edited by H. R. Sasse, pp. 741–755, Chapman & Hall, London, UK.

Täljsten, B. (1994) *Plate Bonding. Strengthening of Existing Concrete Structures with Epoxy Bonded Plates of Steel or Fibre Reinforced Plastics*, Doctoral Thesis, Luleå University of Technology.

Täljsten, B. (1997) 'Defining anchor lengths of steel and CFRP plates bonded to concrete', *International Journal of Adhesion and Adhesives*, Vol. 17, No. 4, pp. 319–327.

Tan, K. H. (1999) 'Towards a cost-effective application of FRP reinforcement in structural rehabilitation', *Proceedings of the Seventh East Asia-Pacific Conference on Structural Engineering & Construction*, Kochi, Japan, pp. 65–73.

Triantafillou, T. C. (1997) 'Shear strengthening of concrete members using composites', *Non-Metallic (FRP) Reinforcement for Concrete Structures, Proceedings of the Third International Symposium*, Sapporo, Japan, pp. 523–530.

Triantafillou, T. C. (1998a) 'Shear strengthening of reinforced concrete beams using epoxy-bonded FRP composites', *ACI Structural Journal*, Vol. 95, No. 2, pp. 107–115.

Triantafillou, T. C. (1998b) 'Composites: a new possibility for the shear strengthening of concrete, masonry and wood', *Composites Science and Technology*, Vol. 58, No. 8, pp. 1285–1295.

Triantafillou, T. C. and Antonopoulos, C. P. (2000) 'Design of concrete flexural members strengthened in shear with FRP', *Journal of Composites for Construction ASCE*, Vol. 4, No. 4, pp. 198–205.

Triantafillou, T. C. and Fardis, M. N. (1997) 'Strengthening of historic masonry structures with composite materials', *Materials and Structures*, Vol. 30, pp. 486–496.

Triantafillou, T. C. and Plevris, N. (1992) 'Strengthening of RC beams with epoxy-bonded fibre-composite materials', *Materials and Structures*, Vol. 25, pp. 201–211.

Uji, K. (1992) 'Improving shear capacity of existing reinforced concrete members by applying carbon fibre sheets', *Transactions of the Japan Concrete Institute*, Vol. 14, pp. 253–266.

van Gemert, D. (1980) 'Force transfer in epoxy-bonded steel-concrete joints', *International Journal of Adhesion and Adhesives*, No. 1, pp. 67–72.

Vichit-Vadakan, W. (1997) *Shear Behaviour of Pre-Cracked, Reinforced Concrete Beams Retrofitted with Glass Fiber Reinforced Plastic Sheet*, Master's Thesis, Massachusetts Institute of Technology.

Yuan, H. and Wu, Z. (1999) 'Interfacial fracture theory in structures strengthened with composite of continuous fiber', *Symposium of China and Japan, Science and Technology of 21st Century*, Tokyo, Japan, pp. 142–155.

Yuan, H., Wu, Z. and Yoshizawa, H. (2001) 'Theoretical solutions on interfacial stress transfer of externally bonded steel/composite laminates', *Journal of Structural Mechanics and Earthquake Engineering, JSCE*, No. 675/1–55, pp. 27–39.

Ziraba, Y. N., Baluch, M. H., Basunbul, A. M., Azad, A. K., Al-Sulaimani, G. J. and Sharif, I. A. (1995) 'Combined experimental – numerical approach to characterisation of steel-glue-concrete interface', *Materials and Structures*, Vol. 28, pp. 518–525.

5 Flexural Strengthening of Slabs

5.1 Introduction

Compared with the flexural strengthening of beams, much less work has been done on the flexural strengthening of slabs. This is believed to be due to two reasons. First, unlike beams, tests on slabs, particularly two-way slabs, are rather more involved. Second, one-way slabs are expected to behave similarly to beams, and the design of two-way slabs using the maximum moment in each direction (Kong and Evans 1987) may be described as one-way slab design in two directions.

Whilst the existing research on the flexural strengthening of RC beams is indeed applicable to the flexural strengthening of slabs, there are also some important differences. One such difference is that slabs are more prone to intermediate crack-induced debonding rather than plate-end debonding, whilst in RC beams the latter is more likely.

5.2 Methods of Strengthening

5.2.1 General

The basic procedure of flexural strengthening of slabs using FRP composites is to bond FRP strips or sheets to the tension face of the slab. Research has been conducted on slabs bonded with either FRP strips covering part of the surface (Ichimasu *et al.* 1993, Erki and Heffernan 1995, Hisabe *et al.* 1997, Karbhari *et al.* 1999, Karbhari and Seible 2000), or FRP sheets covering the entire surface (Arockiasamy *et al.* 1995, Erki and Heffernan 1995, Arockiasamy *et al.* 1996,

Shahawy *et al.* 1996, Karbhari *et al.* 1999, Karbhari and Seible 2000). FRP sheets covering the entire soffit of the slab make it difficult to check the quality of the bond (Ichimasu *et al.* 1993) and can stop the free movement of moisture out of the slab (Karbhari *et al.* 1999), which increases the risk of degradation of the bond. Nevertheless, for the same amount of FRP to be bonded to the slab, wide strips are believed to be more desirable than narrow strips as the former offer a larger contact area between the slab and the FRP and thus reduce the risk of debonding failures. In some cases, the amount of FRP required, particularly when CFRP is used, may not be sufficient to cover the entire surface by even one layer. In such cases, the use of FRP strips is necessary to avoid wastage of material.

5.2.2 Simply Supported One-way or Two-way Slabs

Simply supported one-way slabs are strengthened by bonding FRP strips/sheets to the soffit along the longitudinal direction (Figure 5.1). For two-way slabs, strengthening needs to be carried out for both directions, by bonding FRP strips in both directions (Figure 5.2), or using FRPs with fibres in two directions. The bonded FRP contributes to the flexural resistance of the slab section as additional reinforcement.

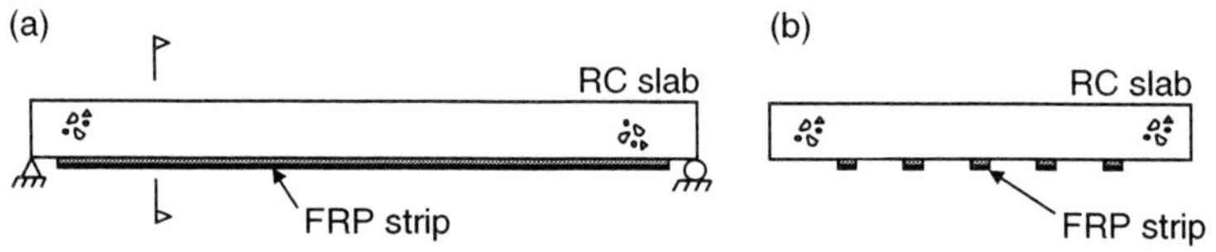

Figure 5.1 One-way slab bonded with longitudinal FRP strips: (a) elevation; (b) cross-section

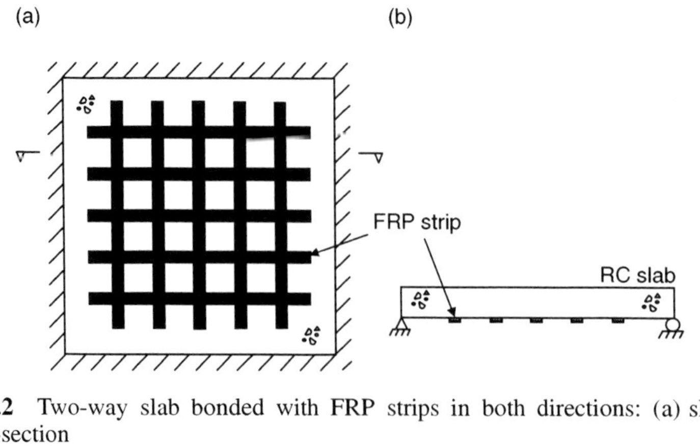

Figure 5.2 Two-way slab bonded with FRP strips in both directions: (a) slab soffit; (b) cross-section

In a recent study by Zhang *et al.* (2001), two-way RC slabs subject to a central concentrated load bonded with a steel plate in the central region were found to fail by formation of yield lines (Kong and Evans 1987) around the perimeter of the bonded plate and along the diagonal lines of the unstrengthened part of the slab (Figure 5.3). Although these slabs were bonded with a steel plate, slabs bonded with an equivalent amount of FRP in the central region are expected to fail in the same manner. This leads to the interesting idea of slab strengthening by modification of the yield line pattern (i.e. the collapse mechanism), in which the designer has the freedom to choose to strengthen part of the slab to push any yield lines away from the strengthened region. The modified yield line pattern leads to an increase in the ultimate strength of the slab. Therefore, strengthening of a two-way RC slab can be concentrated in the central region (Figure 5.4a) so that FRP strips/sheets can be terminated far away from the edges or in the region with the greatest moments (Figure 5.4b). The strengthened region needs to be strong

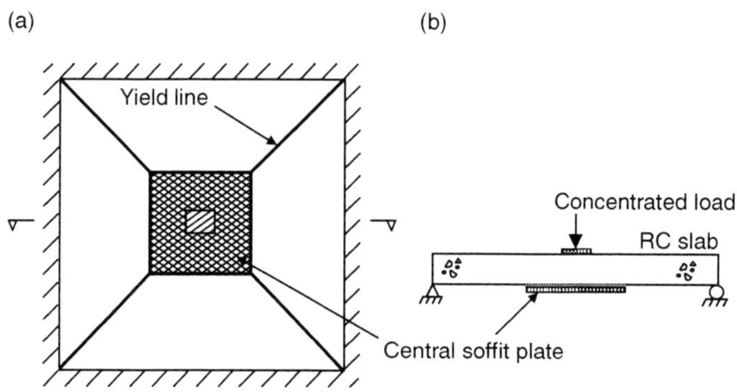

Figure 5.3 Yield lines in a two-way slab bonded with a steel plate: (a) slab soffit; (b) cross-section

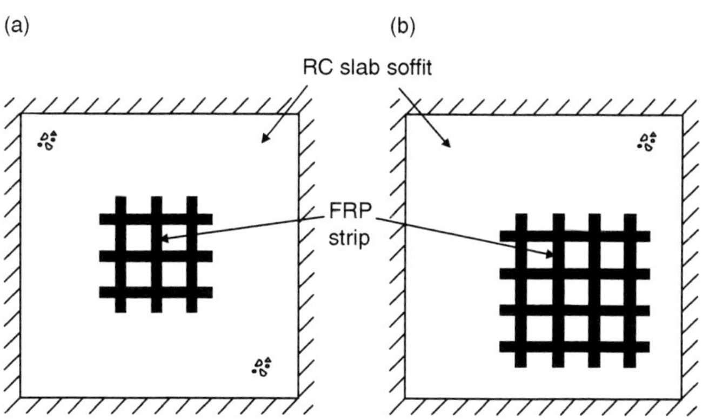

Figure 5.4 Two-way slabs strengthened in a small region: (a) in the central region; (b) in the greatest moment region

enough to avoid any failure, so that failure will occur only by the formation of yield lines in the unstrengthened region. The strength of such strengthened slabs can be predicted by a yield line analysis, as the part of the RC slab without bonded FRP possesses sufficient ductility for the formation of yield lines.

5.2.3 Cantilever Slabs

Cantilever slabs differ from one-way or two-way simply supported slabs in that the fixed end support in a cantilever slab is subject to the greatest moment in the slab. Therefore, FRP strips/sheets cannot be terminated before the fixed end. Instead, they need to be properly anchored at the fixed end. A number of different anchorage schemes are possible at the fixed end, depending on the particular configuration of the slab.

For slabs cantilevered from a wall or a large beam, a simple option is to bend the FRP strips/sheets onto the wall surface (Figure 5.5) if a wet lay-up process is adopted. This was found to be ineffective by Teng *et al.* (2000) in a test during which the slab failed by debonding of the FRP strips from the wall when the tensile stresses in the FRP strips were still low. To achieve sound anchorage, FRP strips need to be inserted into holes predrilled in the wall (Figure 5.6). Owing to

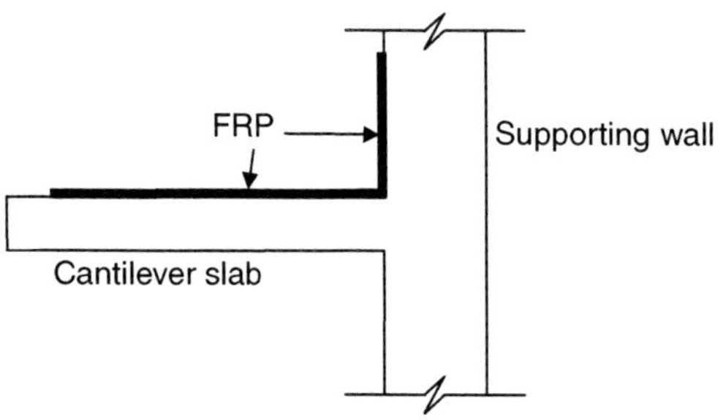

Figure 5.5 Fixed end anchorage for wall-supported cantilever slabs: simple bonding of FRP on the wall

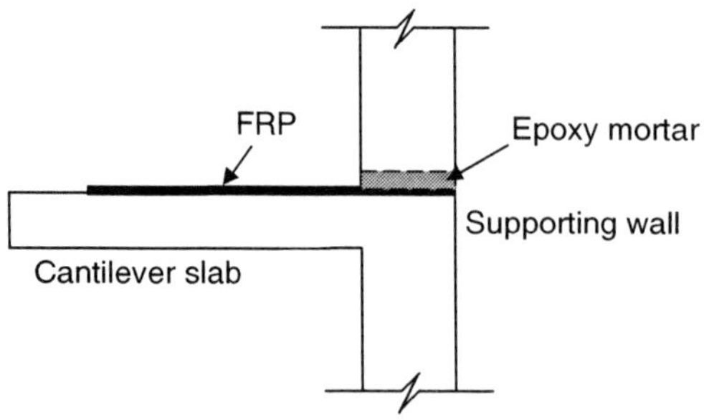

Figure 5.6 Fixed end anchorage for wall-supported cantilever slabs: inserting FRP strips in slots in the wall

the presence of internal steel reinforcement, this type of anchorage can only be used for sufficiently narrow strips so that they can be inserted into holes drilled between steel reinforcing bars. Most of the available tests have been conducted on model wall-supported slabs adopting this anchorage method, with either one or two FRP strips bonded to the slab. This method of anchorage was found to perform well in all tests conducted by Teng *et al.* (2000, 2001), Lam and Teng (2001) and Yao *et al.* (2001). The holes should be refilled with epoxy mortar as the excessive shrinkage of pure cement mortar may compromise the anchorage strength. The test set-up used by Yao *et al.* (2001) for slabs bonded with a single FRP strip is shown in Figure 5.7.

For continuous cantilever slabs, the anchorage of the FRP strips/sheets may be achieved by extending the FRP strips/sheets to the inside slab for a sufficient distance as shown in Figure 5.8. No research has been conducted on slabs of this type, but it is believed to be safe to use an anchorage length on the inside span which is no less than that proposed for soffit plates bonded to beams (Equation 3.29).

Figure 5.7 Laboratory test set-up for model wall-supported slabs

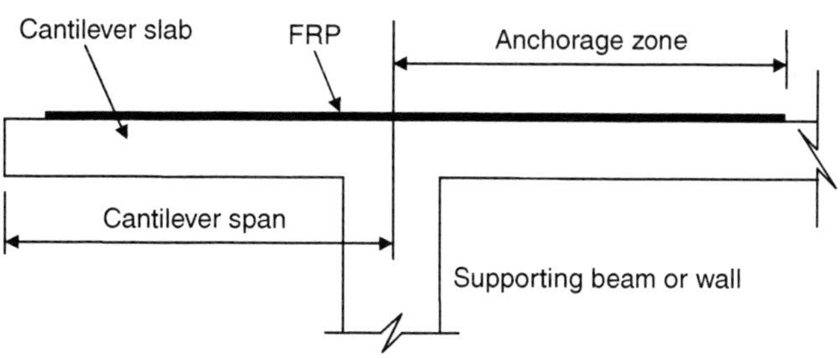

Figure 5.8 Fixed end anchorage for continuous cantilever slabs

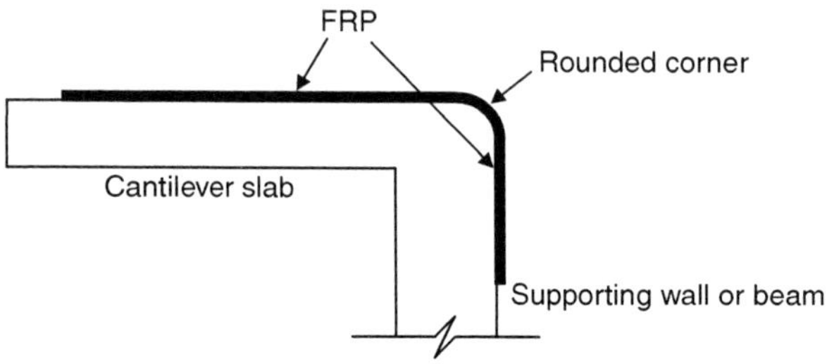

Figure 5.9 Fixed end anchorage for beam- or wall-top cantilever slabs

For slabs cantilevering from the top of an RC beam or wall, the FRP strips/sheets can be wrapped around the edge (Figure 5.9). Such wrapping can be achieved only in a wet lay-up process. Provided the length of the FRP strips/sheets on the wall is sufficient, this should provide strong anchorage. The sharp corner of the beam or wall needs to be rounded to a suitable radius to avoid premature FRP rupture here due to stress concentration. It should be noted that the radius of the rounded corner is limited by the presence of internal steel reinforcement. Again no research has been carried out on the FRP strengthening of such cantilever slabs.

5.3 Failure Modes and Typical Behaviour

5.3.1 Cantilever Slabs and One-way Slabs

Cantilever slabs bonded with FRP strips have been found to fail in one of two modes: (a) debonding of FRP strips which starts at a major flexural crack near the fixed end and then propagates towards the free end (Figure 5.10); and (b) flexural failure by tensile rupture of the FRP strip or crushing of concrete. The first mode was referred to as intermediate crack-induced debonding in Chapter 3, and this name is still used here, although the crack is in fact near the fixed end.

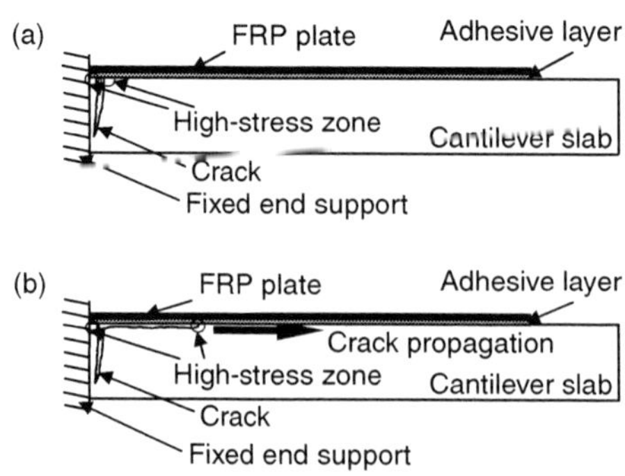

Figure 5.10 Intermediate crack-induced debonding in an RC cantilever slab: (a) before debonding; (b) propagation of debonding

The differentiation between flexural crack-induced debonding and flexural shear crack-induced debonding, as done for beams in Chapter 3, is unimportant for slabs, as the latter was shown to occur at a higher load in Section 3.7 and shear cracking is much less likely in thin slabs.

Figure 5.11 shows a GFRP strip in the process of debonding propagation (Yao *et al.* 2001). The debonded zone in reality can be easily identified as the part of the FRP that changed from its initial colour of yellow to white. Figure 5.12 shows a CFRP strip completely debonded from a cantilever slab. A thin layer of concrete was pulled off the slab by the CFRP strip (Figure 5.12), which is typical of such debonding failures, although in some of Yao *et al.*'s tests which

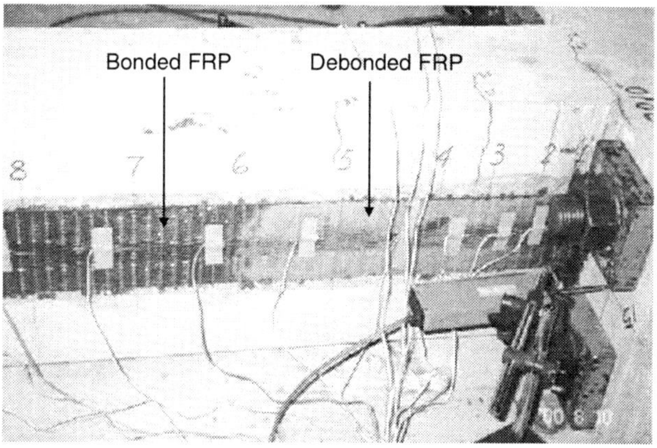

Figure 5.11 FRP-strengthened RC cantilever slab in the process of debonding propagation

Figure 5.12 Complete debonding of an FRP strip from concrete in an RC cantilever slab test

had a high concrete strength, the amount of concrete attached to the debonded plate was found to be very small (Yao *et al.* 2001).

Debonding at the end of the FRP strip (i.e. plate-end debonding) by concrete cover separation, which occurs often in FRP-plated beams, has not been observed so far in tests on cantilever slabs. As shown in Section 3.6, the plate-end debonding load of an RC beam is higher than the shear strength of the concrete in an RC beam without the contribution of the shear reinforcement. For cantilever slabs, since there is normally no shear reinforcement, the plate-end debonding load is generally higher than the shear strength of the slab. This means that, in any such slabs, plate-end debonding cannot occur as shear failure always occurs before plate-end debonding failure. Furthermore, shear failure is generally not critical in such slabs. For thick slabs with steel shear reinforcement, no research exists which can prove the impossibility of concrete cover separation or otherwise.

For slabs bonded with FRP strips, there is another factor which reduces the likeliness of concrete cover separation: the width of the FRP strip is generally much smaller than the centre-to-centre spacings of the FRP strips. In a beam where the FRP plate has about the same width as that of the beam section, the total force transmitted from the FRP plate is resisted only by concrete along the failure plane which is perpendicular to the sides of the beam. In slabs, this force is resisted by this failure plane plus the concrete on the sides of the strip. As a result, concrete cover separation is unlikely, and if plate-end debonding is going to occur, it is likely to be an interfacial debonding failure in the concrete adjacent to the adhesive-to-concrete interface.

To alleviate concerns with premature debonding, Teng *et al.* (2000) adopted the innovative Fibre Anchor® system marketed by L & M Specialist Construction Ltd to hold the FRP strips down to the slab. The fibre anchors used in Teng *et al.* (2000) and Lam and Teng (2001) were made of the same woven fabric and epoxy resin as those used for constructing the GFRP composite. They were installed near the fixed end of the slab to arrest propagating debonding. Provided the fibre anchors do not fail (either in the concrete or in the anchors themselves) and a sufficient number of them are used, the propagation of debonding can be arrested.

Cantilever slabs can be viewed as half of a simply supported one-way slab. Simply supported slabs are thus expected to behave in a similar manner. Hisabe *et al.* (1997) and Karbhari *et al.* (1999) have observed flexural failure by tensile rupture of the FRP, whilst Shahawy *et al.* (1996) observed flexural failure by concrete crushing in simply supported slabs.

5.3.2 Two-way Slabs

Two-way slabs are indeterminate structures, so failure does not occur immediately when the most highly stressed zone in the slab reaches the flexural capacity of the slab section. If punching shear is excluded from consideration, then failure occurs by the formation of yield lines. As discussed earlier, owing to the brittleness of the bonded FRPs, it is not clear whether there is sufficient ductility in the

FRP-strengthened section to allow a collapse mechanism to be fully developed so that a yield line analysis is valid.

If, however, strengthening is done for a limited region so that any yield lines are pushed outside this strengthened region, then failure occurs by the formation of a collapse mechanism consisting of yield lines in the unstrengthened region. For such slabs, a yield line analysis is valid.

Debonding failures in two-way slabs do not appear to have been studied at all. The tests by Zhang *et al.* (2001) on two-way slabs bonded with steel plates did not suffer from any debonding failures, even when a rather thick steel plate was bonded to a small central region (i.e. the plate was terminated far away from the support, a condition favourable for the occurrence of debonding in RC beams). This indicates that in two-way slabs, debonding failures are less likely. Again, intermediate crack-induced debonding is believed to be the more likely debonding mode rather than plate-end debonding, but the two-way action in these slabs means that it is harder for the former to occur than in one-way slabs where a major crack can run across the whole width and can open up rapidly with further loading.

5.4 Design Recommendation for One-way Slabs

In design, the strength of one-way slabs needs to be checked for the following modes of failure:

(a) flexural failure;
(b) intermediate crack-induced debonding;
(c) plate-end interfacial debonding;
(d) fibre anchor failure;
(e) fixed end anchorage failure (cantilever slabs only).

For the first two modes, the design recommendations presented in Chapter 3 for RC beams can be used. Suitable design rules are not available for failure modes (c), (d) and (e). Recourse should be made to mock-up tests in practical applications to ensure that these failure modes do not occur or to determine the strength of the slab failing in one of these modes.

In addition, if steel shear reinforcement is present in the slab, the possibility of concrete cover separation needs to be investigated. Again no design rule is available and mock-up tests should be relied on.

The same end anchorage requirement for beams should be used for one-way slabs (Equation 3.29).

5.5 Design Recommendation for Two-way Slabs

5.5.1 *Flexural Strength by the Moment Coefficient Method*

The flexural strength of a slab can be evaluated using the moment coefficient approach of BS 8110 (1997). In this approach, maximum bending moments in both directions are found first. The reinforcement requirement is then determined for each direction. The moment capacity of the slab in each direction can be evaluated using the design equations already presented in Section 3.4.

5.5.2 Flexural Strength by Yield Line Analysis

The ultimate strength of two-way slabs can be determined by a yield line analysis which is permitted by BS 8110 (1997). For slabs with bonded FRP, if flexural failures occur within the strengthened region, there is uncertainty whether a yield line analysis is valid. This is because the brittle FRP limits the rotational capacity of a slab section, so a plastic failure mechanism cannot be formed properly before the ultimate strength of the slab is achieved. The use of the yield line analysis is thus not recommended for such slabs. Instead, they should be designed using the moment coefficient method discussed above.

However, for two-way slabs with the bonded FRP concentrated in a small region either at the centre or elsewhere (Figure 5.4), the FRP can be so designed using the above moment coefficient method that failure occurs outside the strengthened region, then the yield line analysis can be used as RC slabs are known to possess sufficient ductility for a yield line analysis to be valid. In yield line analysis, the correct collapse mechanism with the lowest failure load needs to be identified by either trial and error and/or a minimisation process. Figure 5.3 shows the yield line pattern (Zhang *et al.* 2001) for an RC slab bonded with a steel plate in the central region which is believed to have similar behaviour to that of an RC slab heavily strengthened with FRP in the central region.

5.5.3 Debonding Strengths

Two modes of debonding failures can occur as discussed previously:

(a) intermediate crack-induced debonding;
(b) plate-end interfacial debonding.

If the moment coefficient approach is adopted in design, the stress limit of the FRP, before the application of the partial safety factor, should be reduced to the limiting stress value of intermediate crack-induced debonding as recommended in Section 3.8 if this is smaller than the ultimate tensile strength of the FRP.

Alternatively, if a yield line analysis is carried out, the maximum moment in the strengthened zone of the slab in each direction needs to be found under the design load. The stress value in the FRP from a conventional cracked section analysis should then be checked against the smaller of the FRP tensile strength and the stress level at intermediate crack-induced debonding based on the recommendation of Section 3.8.

No suitable rule can be recommended for the plate-end interfacial debonding mode, and it is believed to be an unlikely mode of failure.

5.5.4 Anchorage of FRP Strips/Sheets

The FRP strips/sheets should extend beyond the region to be strengthened on all sides by no less than the anchorage length recommended in Section 3.8 for beams (Equation 3.29).

5.6 Conclusions

Existing research suggests that one-way slabs are likely to fail by either section flexural failure or intermediate crack-induced debonding. For both modes, the design recommendations made in Chapter 3 are applicable. Concrete cover separation is an unlikely mode, at least for slabs without steel shear reinforcement, because its strength is higher than the shear strength of the slab. In addition, for slabs bonded with FRP strips, the difference between strip widths and strip centre-to-centre spacings also reduces the likelihood of concrete cover separation. Plate-end interfacial debonding is a potential failure mode, and there is some uncertainty concerning the applicability of the debonding strength model developed for RC beams to this mode in slabs.

Much less is known about two-way slabs. No study appears to have examined debonding failures in FRP-strengthened two-way slabs.

One point worth noting is that the propagation of intermediate crack-induced debonding can be arrested using fibre anchors so that eventual failure is by FRP rupture or concrete crushing. For such slabs, the strength of the slab can be predicted by the flexural strength equations presented in Chapter 3. Potentially, fibre anchors can be used to prevent intermediate crack-induced debonding in both slabs and beams, although existing studies have been limited to cantilever slabs. There is at present insufficient information to evaluate the capacity of fibre anchors in resisting the propagation of debonding.

References

Arockiasamy, M., Sowrirajan, R., Shahawy, M. and Beitelman, T. E. (1995) 'Repair of damaged pretensioned solid slab using CFRP laminates', *Non-Metallic (FRP) Reinforcement for Concrete Structures, Proceedings of the Second International RILEM Symposium*, Ghent, Belgium, edited by L. Taerwe, pp. 492–500, E & FN Spon, London, UK.

Arockiasamy, M., Amer, A. and Shahawy, M. (1996) 'Concrete beams and slabs retrofitted with CFRP laminates', *Proceedings of the Eleventh Conference on Engineering Mechanics, ASCE*, edited by Y. K. Lin and T. C. Su, New York, pp. 776–779.

BS 8110 (1997) *Structural Use of Concrete, Part 1. Code of Practice for Design and Construction*, British Standards Institution, London, UK.

Erki, M. A. and Heffernan, P. J. (1995) 'Reinforced concrete slabs externally strengthened with fibre-reinforced plastic materials', *Non-Metallic (FRP) Reinforcement for Concrete*

Structures, Proceedings of the Second International RILEM Symposium, Ghent, Belgium, edited by L. Taerwe, pp. 509–516, E & FN Spon, London, UK.

Hisabe, N., Sakai, H., Otaguro, H. and Yagi, K. (1997) 'Flexural reinforcing effect of reinforced concrete slab strengthened with carbon fiber sheet under oscillation loading', *Non-Metallic (FRP) Reinforcement for Concrete Structures, Proceedings of the Third International Symposium*, Sapporo, Japan, pp. 335–342.

Ichimasu, H., Maruyama, M., Watanabe, H. and Hirose, T. (1993) 'RC slabs strengthened by bonded carbon FRP plates: Part 1 – laboratory study', *Fiber-Reinforced-Plastic Reinforcement for Concrete Structures, International Symposium*, edited by A. Nanni and C. W. Dolan, pp. 933–955, American Concrete Institute, Farmington Hills, Michigan, USA.

Karbhari, V. M. and Seible, F. (2000) 'Fiber reinforced composites – advanced materials for the renewal of civil infrastructure', *Applied Composite Materials*, Vol. 7, pp. 95–124.

Karbhari, V. M., Seible, F., Seim, W. and Vasquez, A. (1999) 'Post-strengthening of concrete slabs', *Proceedings of the Fourth International Symposium on Fiber Reinforced Polymer Reinforcement for Reinforced Concrete Structures*, Baltimore, Maryland, USA, edited by C. W. Dolan, S. H. Rizkalla and A. Nanni, pp. 1163–1173, American Concrete Institute, Farmington Hills, Michigan, USA.

Kong, F. K. and Evans, R. H. (1987) *Reinforced and Prestressed Concrete*, Third Edition, Chapman & Hall, London, UK.

Lam, L. and Teng, J. G. (2001) 'Strength of cantilever slabs bonded with GFRP strips', *Journal of Composites for Construction, ASCE*, Vol. 5, No. 4.

Shahawy, M. A., Beitelman, T., Arockiasamy, M. and Sowrirajan, R. (1996) 'Experimental investigation on structural repair and strengthening of damaged prestressed concrete slabs utilising externally bonded carbon laminates', *Composites: Part B*, Vol. 27B, pp. 217–224.

Teng, J. G., Lam, L., Chan, W. and Wang, J. (2000) 'Retrofitting of deficient RC cantilever slabs using GFRP strips', *Journal of Composites for Construction, ASCE*, Vol. 4, No. 2, pp. 75–84.

Teng, J. G., Cao, S. Y. and Lam, L. (2001) 'Behaviour of GFRP-strengthened RC cantilever slabs', *Construction and Building Materials*, Vol. 15, No. 7, pp. 339–349.

Yao, J., Teng, J. G. and Lam, L. (2001) 'Debonding in FRP-strengthened RC cantilever slabs', *to be published*.

Zhang, J. W., Teng, J. G., Wong, Y. L. and Lu, Z. T. (2001) 'Behaviour of two-way RC slabs externally bonded with a steel plate', *Journal of Structural Engineering, ASCE*, Vol. 127, No. 4, pp. 390–397.

6 Strengthening of Axially and Eccentrically Loaded Columns

6.1 Introduction

Until the early 1990s, constructing an additional reinforced concrete cage and installing grout-injected steel jackets were the two common methods for strengthening a deficient RC column (Ballinger *et al.* 1993). Steel jacketing is more effective than caging, because the latter results in a substantial increase in the cross-sectional area and self-weight of the structure. Both methods are, however, labour intensive and sometimes difficult to implement on site. In addition to being heavy, steel jackets are also poor in resisting weather attacks (Ballinger *et al.* 1993, Demers and Neale 1999). In recent years, the technique of strengthening RC columns using FRP composites has been used increasingly widely to replace steel jacketing (Ballinger *et al.* 1993, Saadatmanesh *et al.* 1996, Seible *et al.* 1997). The most common form of FRP column strengthening involves the external wrapping of FRP sheets/straps.

The strengthening of existing RC columns using steel or FRP jacketing is based on a well-established fact that lateral confinement of concrete can substantially enhance its axial compressive strength and ductility (Richart *et al.* 1928, 1929, Ahmad and Shah 1982, Mander *et al.* 1988). Many studies have been conducted on the compressive strength and stress–strain behaviour of FRP-confined concrete (e.g. Fardis and Khalili 1981, Ahmad *et al.* 1991, Harmon and Slattery 1992, Demers and Neale 1994, Howie and Karbhari 1994, Nanni and Bradford 1995, Karbhari and Gao 1997, Mirmiran *et al.* 1998a, Demers and Neale 1999, Miyauchi *et al.* 1999, Purba and Mufti 1999, Saafi *et al.* 1999, Toutanji

1999, Rochette and Labossiere 2000, Xiao and Wu 2000, Zhang *et al.* 2000). These studies have shown that FRP-confined concrete behaves differently from steel-confined concrete (Mirmiran and Shahawy 1997, Samaan *et al.* 1998, Saafi *et al.* 1999, Spoelstra and Monti 1999), so design recommendations developed for steel-confined concrete columns cannot be applied to FRP-confined columns despite the apparent similarity between these two types of columns.

This chapter is concerned with the behaviour and strength of RC columns strengthened by FRP jacketing subject to axial or eccentric loads. The chapter starts with a summary of methods of strengthening, followed by a detailed treatment of the behaviour of FRP-confined concrete in compression. The behaviour of FRP-confined rectangular and elliptical concrete columns is next addressed. Finally, design recommendations are presented for FRP strengthening of columns subject to either axial or eccentric loads.

6.2 Methods of Strengthening

6.2.1 General

A number of different techniques have been developed for strengthening existing RC columns using FRP composites. The aspects described in this section are common to both strengthening under static loads and seismic retrofit, but specific issues in seismic retrofit are addressed in Section 7.3. The methods of strengthening can be classified into the following three categories in terms of the method adopted for constructing the FRP composite: (a) wrapping; (b) filament winding; and (c) prefabricated shell jacketing.

6.2.2 Wrapping

In situ FRP wrapping has been the most commonly used technique for column strengthening using FRP composites. In this method, unidirectional fibre sheets or woven fabric sheets are impregnated with polymer resins and wrapped around columns in a wet lay-up process, with the main fibres orientated in the hoop direction. A column can be fully wrapped with FRP sheets in single or multiple layers (Figure 6.1a). It can also be partially wrapped using FRP straps in a continuous spiral (Figure 6.1b) or discrete rings (Figure 6.1c). The compressive strength enhancement of concrete due to the external wrapping of FRP was first demonstrated by Fardis and Khalili (1981, 1982). This concept was first applied to the strengthening of real RC columns in Japan in the mid 1980s (ACI 1996). There have been many reports of the application of this technique in the retrofitting of bridge and building columns (e.g. ACI 1996, Neale and Labossiere 1997, Tan 1997) since then.

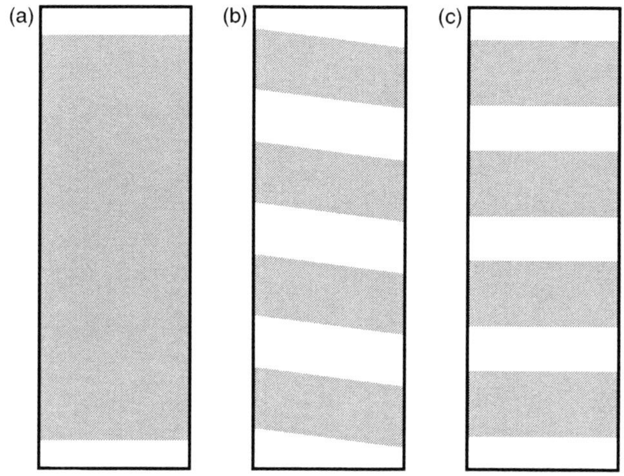

Figure 6.1 Typical FRP wrapping methods for RC columns: (a) full wrapping using FRP sheets; (b) partial wrapping using FRP straps in a continuous spiral; and (c) partial wrapping using FRP straps in discrete rings

6.2.3 Filament Winding

The principle of filament winding is similar to that of wrapping, except that the filament winding technique uses continuous fibre strands instead of sheets/straps so that winding can be processed automatically by means of a computer-controlled winding machine. An FRP jacket with controlled thickness, fibre orientation and volume fraction can be obtained in this process. The idea of confining concrete by winding continuous resin-impregnated fibre strands was first mentioned by Fardis and Khalili (1981). The first winding machine for column retrofitting was developed in Japan in the mid 1980s (ACI 1996).

6.2.4 Prefabricated Shell Jacketing

Existing RC columns can also be strengthened using prefabricated FRP shells. The shells are fabricated under controlled conditions using fibre sheets or strands with the impregnation of resins effected prior to field installation. They can be fabricated in half circles or half rectangles (Nanni and Norris 1995, Ohno *et al.* 1997) and circles with a slit or in continuous rolls (Xiao and Ma 1997), so that they can be opened up and placed around columns. For effective FRP confinement to be achieved, a full contact between the column and the FRP shell is essential. This can be ensured by either bonding the shell to the column using adhesives (Xiao and Ma 1997), or injecting shrinkage-compensated cement grout or mortar into the space between the shell and the column (Nanni and Norris 1995, Ohno *et al.* 1997). An interesting application of prefabricated shells is to use them as formwork in modifying column shapes as part of a strengthening measure. This has recently been discussed by Teng and Lam (2001) who suggested that square or rectangular columns (hereafter referred to as rectangular columns only unless the differentiation between square and rectangular sections is important) can be strengthened by reshaping them into circular or elliptical columns in which a prefabricated FRP shell is used to provide both the permanent formwork for casting the additional concrete and the required confinement following the curing of concrete.

6.2.5 Comparison of Strengthening Methods

Each of the three methods discussed above has its advantages and disadvantages. These are listed in Table 6.1. Overall, external wrapping appears to be the most popular method as its advantages (flexibility and ease in site handling) appear to be far more important than its disadvantages. Filament winding bears much similarity to wrapping as both involve a wet lay-up process, so in the rest of the chapter, the term 'wrapping' is used to cover both methods except when a specific distinction has to be made between the two.

6.2.6 Constructional Aspects

Prior to the application of FRPs using any one of the three techniques described above, the surface of the column to be strengthened should be properly prepared to provide a hard, dry and clean surface. Any damaged or deteriorated parts should be removed and patched up with good concrete, cement mortar or epoxy putty as appropriate. The repaired surface should be trowelled.

Table 6.1 A comparison of different methods of column strengthening

Method	Advantages	Disadvantages
Wrapping	Flexibility in coping with different column shapes	Least quality control
	Ease in site handling, without the need for special equipment	Most labour intensive
Filament winding	Improved quality control	Reduced flexibility in coping with different column shapes
	Reduced on-site labour	Special equipment required
Prefabricated shells	Best quality control	Limited flexibility in coping with different column shapes
	Least on-site labour	
	Useful for column shape modification	Prefabrication cost

For rectangular columns to be strengthened by wrapping, their corners should be rounded. This rounding is necessary to reduce any detrimental effect of the sharp corners on the tensile strength of the FRP, and to enhance the effectiveness of confinement. If rectangular prefabricated shells are used instead, the shells are generally slightly oversized and their corners rounded, with the small gap between the shell and the column filled with expansive or nonexpansive cement grout (Nanni and Norris 1995, Ohno *et al.* 1997). The same rounding treatment should be applied to other column shapes featuring shape corners.

Whether wrapping or prefabricated shell jacketing is used, one or more vertical joints generally exist in the FRP. Such joints should be made strong enough so that FRP joint failure does not become the strength-controlling failure mode, as otherwise the strength of the FRP is not fully utilised. When the FRP is wrapped continuously, a sound joint can be achieved by extending the end of the final layer of FRP to form a sufficient overlap. When an FRP shell with a vertical slit in each layer is used, either an additional FRP strip should be bonded over the vertical seam (for a shell consisting of a single or a small number of FRP layers), or the slits should be staggered (for a shell consisting of a large number of FRP layers) to avoid the concentration of seam weakness. In the latter case, the effective number of layers should be taken as the total number minus one, to account for the weakening effect of slits (Xiao and Ma 1997).

In most cases, the FRP confinement obtained is passive in nature, with hoop tensile stresses in the FRP developing as the concrete expands. Active confinement methods with FRP jackets have also been explored (Saadatmanesh *et al.* 1996). To achieve active confinement, the FRP jacket is slightly oversized. The space between the FRP shell and the original column is then filled with expansive cement grout or pressure injected with epoxy resin (Priestley and Seible 1995, Saadatmanesh *et al.* 1996, 1997). Alternatively, the fibres may be prestressed during wrapping. A confining pressure is therefore developed to the column before any subsequent expansion of concrete occurs.

6.2.7 Shape Modification

FRP jacketing provides very effective confinement to circular columns, but is much less effective for rectangular columns. For these columns, more effective confinement can be achieved by modifying the column section into a circular or elliptical section before FRP jacketing.

This shape modification idea appears to have been first explored by Priestley and Seible (1995). In their approach, precast concrete bolsters are added to obtain a circular or elliptical section before wrapping FRP sheets. Shape modification before wrapping has also been achieved using fast-cured cement (Saadatmanesh *et al.* 1997). Rectangular columns are generally modified into elliptical columns instead of circular ones to avoid excessive sectional enlargement, although circular sections are the natural choice for square columns. The general idea of shape modification followed by FRP confinement can be extended to many other sectional shapes.

Prefabricated circular or elliptical FRP shells can also be used for strengthening rectangular columns (Teng and Lam 2001). For strengthening rectangular columns, an FRP shell of elliptical section is placed around the column (Figure 6.2a). The space between the FRP shell and the original column can then be filled with concrete. The FRP shell serves both as the formwork for the concrete in-fill and as confinement for strength enhancement for the modified column of elliptical section. To reduce further the size of the strengthened column, the corners of the original column may be rounded, leading to a scheme as shown in Figure 6.2(b).

The FRP shell can be fabricated as a single shell with a slit as shown in Figure 6.3. Owing to the thinness of such shells, they can be easily opened up for placing around the column as shown in Figure 6.3(b), followed by bonding a vertical FRP strip to seal the joint. The fibres in the vertical FRP strip should, however, be orientated horizontally. When a large number of such shells (layers) are placed on one another to achieve increased strength enhancement, the approach of Xiao and Ma (1997) as described in Section 6.2.4 can be followed. Other alternatives include fabricating a single shell with a vertical overlap at the joint or as two halves with an overlap at both joints. The elliptical carbon-FRP (CFRP) shell shown in Figure 6.3 was formed in the laboratory by wrapping

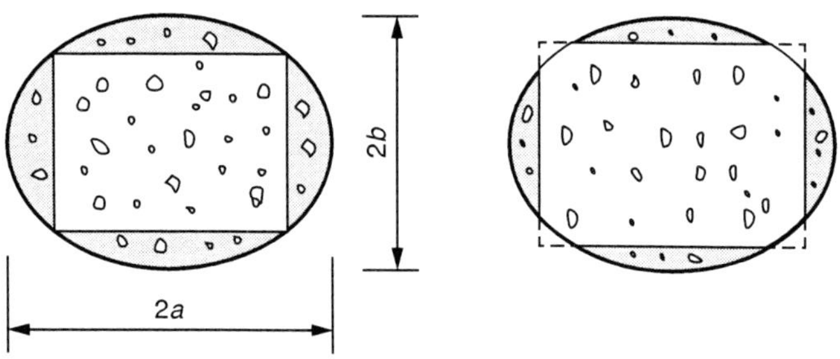

Figure 6.2 Strengthening of rectangular columns by shape modification and FRP confinement, without (a) and with (b) corner rounding

Figure 6.3 Prefabricated FRP shell: (a) mandrel (left) and shell (right); (b) opening up the shell

a fibre sheet around a mandrel (Teng and Lam 2001). In practice, such shells can be fabricated by filament winding. The shell of Figure 6.3 formed from two layers of CFRP is already easy to open up for placement, and a single-layer shell should be even more flexible.

6.3 Failure Modes and Typical Behaviour of FRP-confined Concrete Columns

6.3.1 Circular Concrete Columns

When an FRP-confined column is subject to axial compression, the concrete expands laterally and this expansion is confined by the FRP. The confining action of FRP composite for circular concrete columns is shown in Figure 6.4. For circular columns, the concrete is subject to uniform confinement, and the maximum confining pressure provided by FRP composite is related to the amount and strength of FRP and the diameter of the confined concrete core. This confining pressure is given by

$$f_l = \frac{2 f_{frp} t_{frp}}{d} = \frac{\rho_{frp} f_{frp}}{2} \tag{6.1}$$

where f_l is the lateral confining pressure, f_{frp} is the tensile strength of FRP in the hoop direction, t_{frp} is the total thickness of FRP (or the total equivalent thickness in the case of wrapped FRP strands/straps with small gaps), d is the diameter of the confined concrete, and ρ_{frp} is the FRP volumetric ratio given by the following equation for full wrapping (i.e. no gaps):

$$\rho_{frp} = \frac{\pi d t_{frp}}{\pi d^2/4} = \frac{4t_{frp}}{d} \tag{6.2}$$

As the FRP is subjected to tension in the hoop direction (Figure 6.4), eventual failure occurs when its hoop tensile strength is reached (Figure 6.5). This mode of failure has been reported for the vast majority of existing tests on FRP-confined

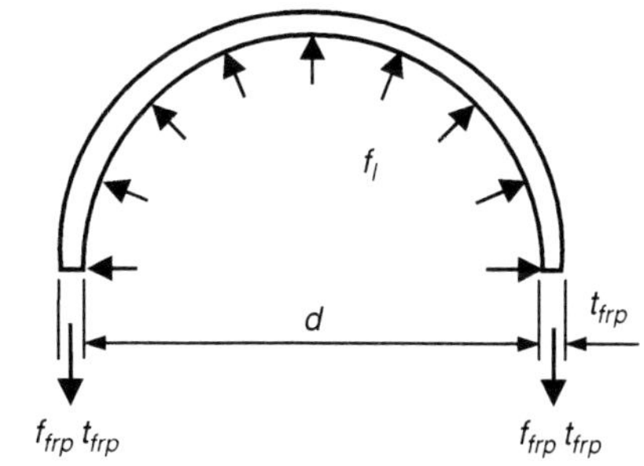

Figure 6.4 Confining action of FRP composite

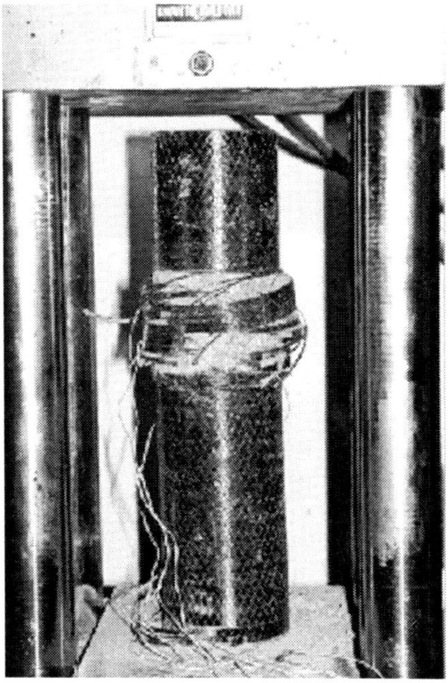

Figure 6.5 Failure of an FRP-confined circular concrete specimen by FRP rupture

circular concrete specimens (Lam and Teng 2001a), although premature failure due to the separation of the FRP at the lap joint has also been reported in some studies (e.g. Demers and Neale 1994, Nanni and Bradford 1995) for specimens with an insufficient lap length.

Existing tests also showed that the stress–strain curve of FRP-confined concrete in circular columns has a characteristic bilinear shape with sharp softening in a transition zone around the stress level of the unconfined concrete strength (Samaan *et al.* 1998, Toutanji 1999) (Figure 6.6), unless the amount of FRP provided is so small that no strength enhancement may be expected as shown by some of the tests conducted by Xiao and Wu (2000). In the first linear stage, the slope of the stress–strain curve is similar to that of the unconfined concrete, indicating that the behaviour is little affected by the FRP. In the second linear stage, the concrete is cracked and FRP confinement is activated (Toutanji 1999). The stress of the confined concrete increases linearly with increasing FRP strain and reaches its maximum at FRP rupture. In such cases, the ultimate (i.e. peak) stress of the concrete and the ultimate strain of the FRP are reached simultaneously (Mirmiran *et al.* 1998a).

However, when a concrete column is provided with little confinement, the bilinear stress–strain behaviour described above may not exist. For circular columns, such a situation arises when the amount of FRP provided is very small.

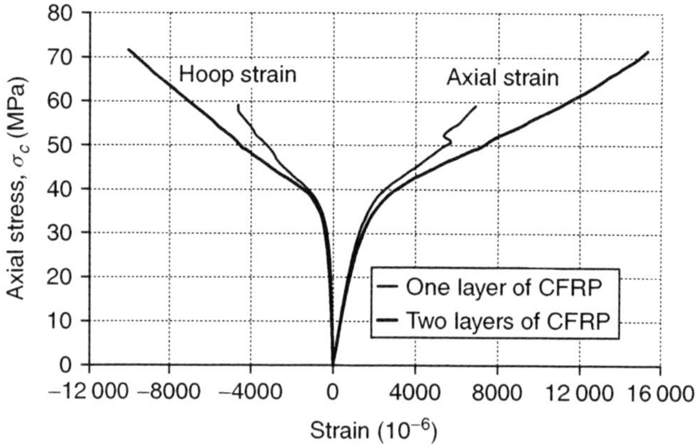

Figure 6.6 Typical stress–strain curves of FRP-confined circular concrete specimens

6.3.2 Rectangular Concrete Columns

It has been well established that FRP confinement is much less effective for rectangular columns than for circular columns even with the rounding of corners. This is because, in the former, the confining pressure is nonuniformly distributed and only part of the concrete core is effectively confined.

Failure generally occurs at the corners by FRP tensile rupture (Figure 6.7). The stress–strain curves are more likely to feature a descending branch, but

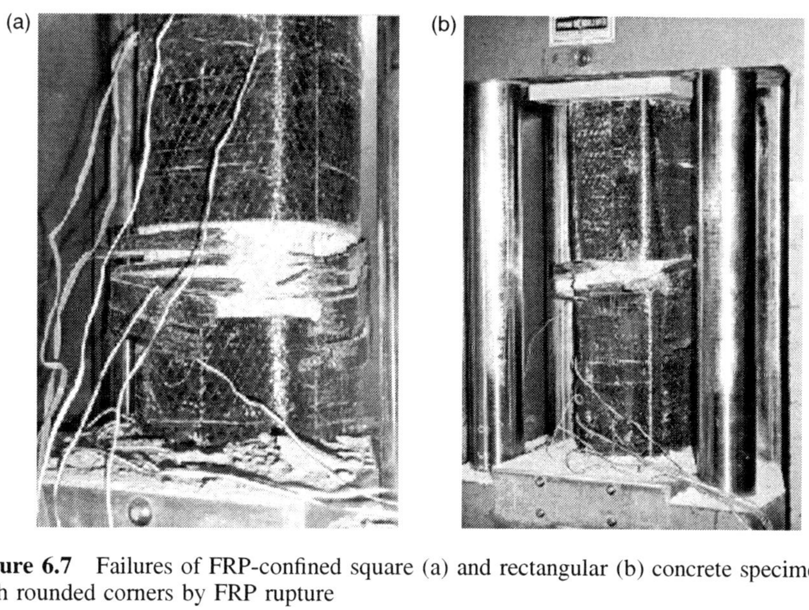

Figure 6.7 Failures of FRP-confined square (a) and rectangular (b) concrete specimens with rounded corners by FRP rupture

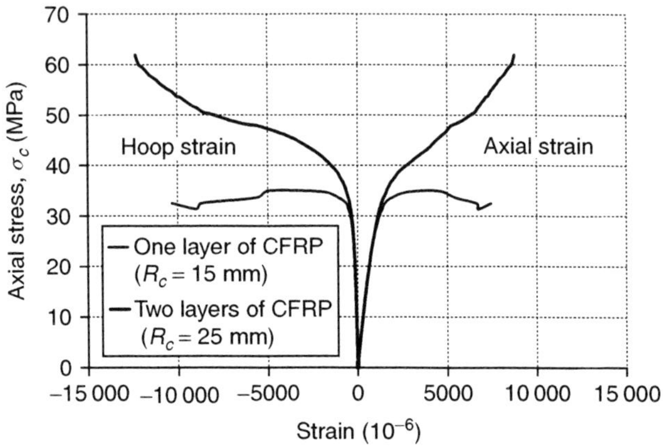

Figure 6.8 Typical stress–strain curves of FRP-confined square concrete specimens

in such cases FRP confinement normally provides little strength enhancement. Figure 6.8 shows two typical stress–strain curves for square specimens with different amounts of confinement as determined by the corner radius and the amount of CFRP. Of the two curves, the one for a smaller corner radius and a single layer of CFRP features a descending branch. In such cases, the ultimate strength of the FRP-confined concrete is reached before the ultimate strain of the FRP is reached, and the stress–strain curve ends at the ultimate strain of FRP with a substantial stress carried by the FRP-confined concrete. This stress may or may not exceed the strength of the unconfined concrete (Mirmiran *et al.* 1998a, Rochette and Labossiere 2000, Xiao and Wu 2000).

6.3.3 Elliptical Concrete Columns

Figure 6.9 shows an ellipse with a major axis (A–A' axis) length of $2a$ and a minor axis (B–B' axis) length of $2b$. A and A' are referred to as the major vertices while B and B' are referred to as minor vertices of the ellipse. In the mathematical literature, the shape of an ellipse is determined by its eccentricity $e = \sqrt{a^2 - b^2}/a$. In the present chapter, the simple ratio a/b of the major axis to minor axis length (or simply the aspect ratio) is used as the key shape parameter instead.

For the same FRP volumetric ratio, the amount of confinement provided by the FRP jacket to an elliptical column is between that of circular and rectangular columns. In more specific terms, if a rectangular column is confined by an FRP jacket with shape modification allowed, an elliptical jacket is less effective than a circular one but more effective than a rectangular one. The behaviour depends strongly on the aspect ratio. If this ratio is small and close to one, the behaviour is similar to that of circular columns. As the aspect ratio increases, the effectiveness of FRP confinement reduces.

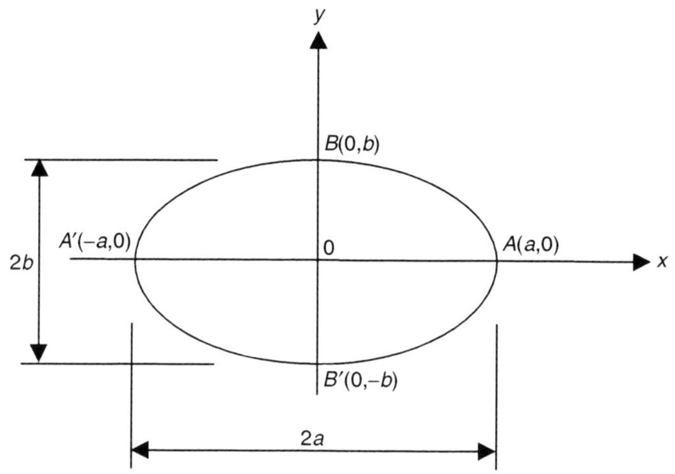

Figure 6.9 Definition of an ellipse

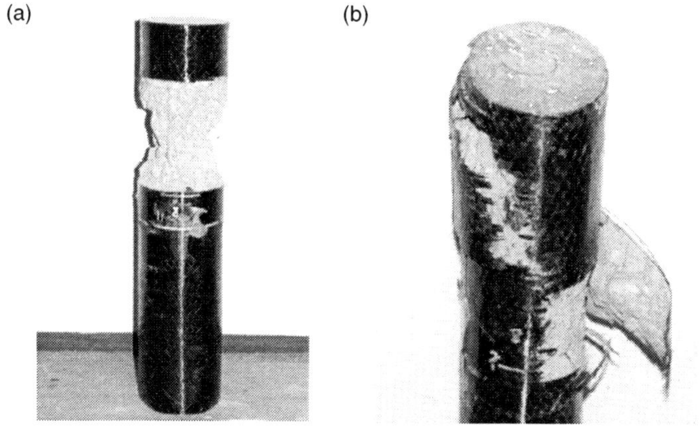

Figure 6.10 Failures of FRP-confined elliptical concrete specimens: (a) $a/b = 5/4$; (b) $a/b = 5/3$

The location of FRP rupture is generally at a corner when the aspect ratio is large (Figure 6.10b), but when this ratio is small, this location was found to vary in tests conducted by Teng and Lam (2001). Figure 6.11 shows the stress–strain curves of FRP-confined elliptical specimens of three different aspect ratios. The bilinear stress–strain behaviour is clearly seen for the two aspect ratios of 5/4 and 5/3, but the other curve for an aspect ratio of 5/2 features a descending branch. These two curves illustrate clearly the effect of aspect ratio on the behaviour of FRP-confined elliptical specimens.

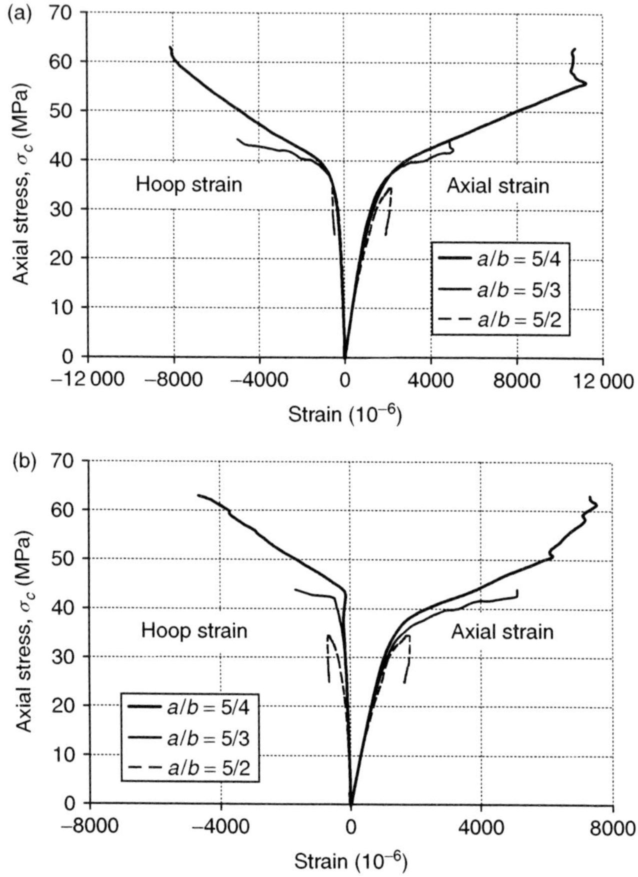

Figure 6.11 Typical stress–strain curves of FRP-confined elliptical concrete specimens: (a) at minor vertices; (b) at major vertices

6.4 Compressive Strength of FRP-confined Concrete

6.4.1 General

For the design of strengthening measures for columns, it is necessary to be able to calculate the strength gain of concrete due to FRP confinement. Although a great deal of information exists on steel-confined concrete (e.g. Richart *et al.* 1928, 1929, Mander *et al.* 1988, Cusson and Paultre 1995), recent studies have revealed that existing models for the axial compressive strength of steel-confined concrete are unconservative and cannot be used for FRP-confined concrete (Mirmiran and Shahawy 1997, Samaan *et al.* 1988, Saafi *et al.* 1999). Consequently, strength models specifically developed for FRP-confined concrete have been presented by a number of researchers (Karbhari and Gao 1997, Mirmiran *et al.* 1998a,

Samaan *et al.* 1998, Miyauchi *et al.* 1999, Saafi *et al.* 1999, Toutanji 1999, Lam and Teng 2001a) based on tests on FRP-confined circular concrete specimens in which the concrete is uniformly confined. These models, except the recent model by Lam and Teng (2001a), are first reviewed and assessed using the test data gathered by Lam and Teng (2001a) in this section. The Lam and Teng model, which is recommended for design use later, is next discussed in some detail.

6.4.2 Review of Strength Models

Several of the existing strength models for FRP-confined concrete take the following form (Karbhari and Gao 1997, Samaan *et al.* 1998, Miyauchi *et al.* 1999, Saafi *et al.* 1999, Toutanji 1999):

$$\frac{f'_{cc}}{f'_{co}} = 1 + k_1 \frac{f_l}{f'_{co}} \tag{6.3}$$

where f'_{cc} and f'_{co} are the compressive strengths of the confined and the unconfined concrete respectively, f_l is the lateral confining pressure defined by Equation 6.1 and k_1 is the confinement effectiveness coefficient. This form was first proposed by Richart *et al.* (1928) for actively confined concrete with a value of 4.1 for k_1. Richart *et al.* (1929) subsequently showed that their model is also suitable for steel-confined concrete. Fardis and Khalili (1982) suggested that Richart *et al.* 's (1928) model could be directly used for FRP-confined concrete.

However, recent studies revealed that existing models for the axial compressive strength of steel-confined concrete are unconservative and cannot be used for FRP-confined concrete (Mirmiran and Shahawy 1997, Samaan *et al.* 1998, Saafi *et al.* 1999). A number of strength models have recently been proposed for FRP-confined concrete which employ Equation 6.3 with modified expressions for k_1 (Karbhari and Gao 1997, Samaan *et al.* 1998, Miyauchi *et al.* 1999, Saafi *et al.* 1999, Toutanji 1999). Miyauchi *et al.* (1999) proposed $k_1 = 2.98$ based on their experimental results on CFRP-confined concrete specimens with f'_{co} varying from 20 to 100 MPa. In other models (Karbhari and Gao 1997, Samaan *et al.* 1998, Saafi *et al.*, 1999, Toutanji 1999), k_1 is not a constant, and instead is dependent on either f_l/f'_{co} or f_l as detailed in Table 6.2. Among these models, the models proposed by Samaan *et al.* (1998) and Saafi *et al.* (1999) were based on the test results of concrete-filled FRP tubes, while other models were based mainly on FRP-wrapped specimens. It should be noted that in Table 6.2, the model proposed by Karbhari and Gao (1997) is referred to as Karbhari and Gao's (1997) model I, as these authors also presented another model in the same paper.

Furthermore, Mirmiran *et al.* (1998a) took into account the effect of the length-to-diameter ratio (L/d) of test specimens and proposed the following equation to relate the confined concrete strength to that of a specimen with $L/d = 2$ denoted

Table 6.2 Various strength models for FRP-confined concrete

Source	Karbhari and Gao (1997):I	Samaan *et al.* (1998)	Miyauchi *et al.* (1999)	Saafi. *et al.* (1999)	Toutanji (1999)
Equation			$\frac{f'_{cc}}{f'_{co}} = 1 + k_1 \frac{f_l}{f'_{co}}$		
k_1	$2.1\left(\frac{f_l}{f'_{co}}\right)^{-0.13}$	$6.0 f_l^{-0.3}$	2.98	$2.2\left(\frac{f_l}{f'_{co}}\right)^{-0.16}$	$3.5\left(\frac{f_l}{f'_{co}}\right)^{-0.15}$

below by $f'_{cc2:1}$:

$$f'_{cc} = f'_{cc2:1}\left[0.0288\left(\frac{L}{d}\right)^2 - 0.263\left(\frac{L}{d}\right) + 1.418\right] \tag{6.4}$$

Apart from that described in Table 6.2 which is in the form of Equation 6.3, Karbhari and Gao (1997) also proposed an alternative model which is given by

$$f'_{cc} = f'_{co} + 3.1 f'_{co} \nu_c \frac{2t_{frp}}{d}\frac{E_{frp}}{E_c} + \frac{2 f_{frp} t_{frp}}{d} \tag{6.5}$$

where ν_c and E_c are Poisson's ratio and the modulus of elasticity of concrete, and E_{frp} is the modulus of elasticity of FRP. This alternative model is referred to as Karbhari and Gao's (1997) model II to distinguish it from the other model which is referred to as Karbhari and Gao's (1997) model I.

On the other hand, the model of Mander *et al.* (1988) which was developed from a multi-axial failure surface for concrete has also been used to predict the strength of FRP-confined concrete (Saadatmanesh *et al.* 1994, Purba and Mufti 1999, Spoelstra and Monti 1999):

$$\frac{f'_{cc}}{f'_{co}} = -1.254 + 2.254\sqrt{1 + \frac{7.94 f_l}{f'_{co}}} - 2\frac{f_l}{f'_{co}} \tag{6.6}$$

For partially wrapped columns with significant gaps between the FRP straps, Saadatmanesh *et al.* (1994) also proposed the use of Equation 6.6, with f_l replaced by f'_l, which can be written as

$$f'_l = k_g f_l \tag{6.7}$$

where f_l' is the effective lateral confining pressure, and k_g is referred to as the gap factor herein and is given by

$$k_g = \frac{\left(1 - s_{frp}/(2d)\right)^2}{1 - \rho_{sc}} \tag{6.8}$$

in which s_{frp} is the clear vertical spacing between straps, and ρ_{sc} is the ratio of the cross-sectional area of the longitudinal steel reinforcement to the gross area of the column section.

In addition, existing stress–strain models of FRP-confined concrete (e.g. Xiao and Wu 2000) include a formula for the ultimate condition of FRP rupture which may be used to predict the peak strength when the effect of confinement is substantial. However, when the effect of confinement is small, the load at this ultimate condition may not coincide with the peak load achieved. For this reason, these models are not further discussed in this section.

6.4.3 Experimental Data

Over the last few years, a large number of tests on FRP-confined concrete have been carried out by Lam and Teng (2001a). An extensive survey of these tests has recently been conducted (Lam and Teng 2001a), based on which a test database on the compressive strength of FRP-confined concrete has been established. This database contains over 200 test results on FRP-confined circular concrete specimens, with some of them representing the average of up to five nominally identical specimens. The database covers plain concrete specimens with the unconfined compressive strength ranging from 18 MPa to 109.5 MPa, the diameter from 51 mm to 190.6 mm and the length-to-diameter ratio L/d from 2 to 5. All three types of FRPs, namely AFRP, CFRP and GFRP, have been used in these tests. When necessary, this database is referred to as the strength database to differentiate it from the strain database introduced in Section 6.5 which contains values of key parameters of stress–strain curves.

Lam and Teng (2001a) employed this strength database to study the effectiveness of FRP confinement and to assess existing strength models. For the assessment to be more reliable, some of the test results were excluded from their comparisons. These excluded cases were specimens with L/d ratios different from those of control specimens, partially wrapped specimens with gaps between the FRP straps or strands, and specimens failing in a mode other than FRP rupture. As the vast majority of the specimens had unconfined concrete strengths less than 65 MPa, the remaining specimens which had unconfined compressive strengths exceeding 100 MPa were also excluded. With these exclusions, a total of 199 test results were available for the assessment of strength models, with many of them representing the average of up to five specimens. These data are all for fully confined plain concrete specimens with unconfined concrete strengths not greater than 65 MPa, which were obtained from control specimens of the same dimensions. Further details of the data can be found in Lam and Teng (2001a). In the following two sections, experimental trends as well as the performance of existing compressive strength models as revealed by these data and presented in Lam and Teng (2001a) are summarised.

6.4.4 Trends of Experimental Data

Of the 199 test results in Lam and Teng's (2001a) strength database retained for comparison with strength models, about two-thirds were reported with FRP material properties determined by the researchers themselves from either tensile tests of flat coupons (e.g. ASTM D3039/D3039M-95a 1995) or splitting tests of ring specimens (e.g. ASTM D2290-92 1992). For the rest of the tests, the FRP material properties supplied by manufacturers were reported, or the source of the reported FRP material properties was not clearly stated. Thus, Lam and Teng (2001a) divided the test results into three sets: set 1 for those with FRP properties from tensile tests of flat coupons by researchers, set 2 for those with FRP properties from splitting tests of ring specimens by researchers, and set 3 for the rest.

The three data sets are shown separately in Figure 6.12 to compare the relationships exhibited by them between compressive strength gains and lateral confining pressures. The statistical characteristic of these three sets of data are given in Table 6.3. It is seen from both Figure 6.12 and Table 6.3 that the relationship between the strengthening ratio f'_{cc}/f'_{co} and the confinement ratio f_l/f'_{co} is linear for all three sets of data, although for data set 2, the 10 data points with a high degree of confinement (i.e. $f_l/f'_{co} > 1$) are seen to fall below the linear trend. Table 6.3 also shows that the set 3 data lead to a much greater average confinement effectiveness coefficient k_1. This can be attributed to the fact that the values of FRP material properties supplied by manufacturers are generally on the conservative side. In addition, data set 1 has a lower average value of k_1 than data set 2 (Table 6.3). This difference may be due to the different methods of tensile tests employed for the two sets of data, as the rupture strain of the confining FRP in a circular specimen has been found to be lower than the ultimate tensile strain from flat-coupon tests (Shahawy *et al.* 2000, Xiao and Wu 2000), while the ring-splitting test is believed to provide a closer estimate of the hoop rupture strain of FRP in a circular specimen (Shahawy *et al.* 2000).

Most of the test results in the database are for circular concrete specimens confined by externally wrapped sheets or externally wound strands, but a significant number of them are for FRP tubes filled with concrete. The latter tests were conducted in the context of concrete-filled FRP tubular columns for new construction. The difference between these two types of specimens was also discussed in Lam and Teng (2001a). The main difference between them appears to be that the FRP and the concrete are bonded with an adhesive in an FRP-wrapped (or FRP-wound) specimen but not in a concrete-filled FRP tube. Mirmiran *et al.* (1998a)

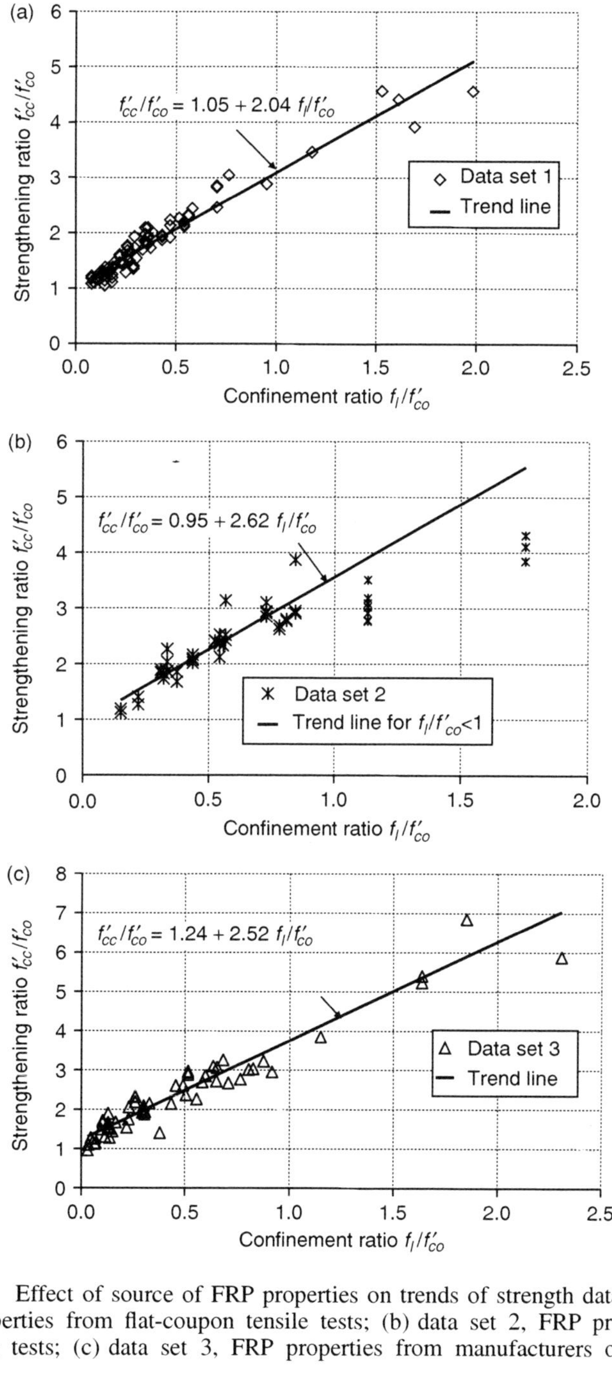

Figure 6.12 Effect of source of FRP properties on trends of strength data: (a) data set 1, FRP properties from flat-coupon tensile tests; (b) data set 2, FRP properties from ring-splitting tests; (c) data set 3, FRP properties from manufacturers or unspecified sources

Table 6.3 Compressive strength of FRP-confined concrete: statistical characteristics of the test data

Data	No. of data points	Confinement effectiveness coefficient, $k_1 = (f'_{cc} - f'_{co})/f_l$		Best-fit trend line	
		Average	Standard deviation	Equation	Correlation coefficient, R^2
Set 1	83	2.10	0.59	$\frac{f'_{cc}}{f'_{co}} = 1.05 + 2.04\frac{f_l}{f'_{co}}$	0.94
Set 2	47	2.31	0.60	$\frac{f'_{cc}}{f'_{co}} = 0.95 + 2.62\frac{f_l}{f'_{co}}$ (for $f_l/f'_{co} < 1$)	0.83
Set 3	69	3.45	1.46	$\frac{f'_{cc}}{f'_{co}} = 1.24 + 2.52\frac{f_l}{f'_{co}}$	0.93
All	199	2.62	1.15	$\frac{f'_{cc}}{f'_{co}} = 1.15 + 2.16\frac{f_l}{f'_{co}}$	0.87

suggested that there was no difference between these two types of specimens, but Saafi *et al.* (1999) believed that there was a significant difference in confinement effectiveness between them.

Lam and Teng (2001a) compared the results of these two types of specimens using their database. Figure 6.13 shows that the trend line for wrapped specimens is similar to that for concrete-filled FRP tubes with $f_l/f'_{co} < 1$. This appears to suggest that there is no obvious difference between these two types of specimens, but the reality may be more complicated because the effect of the two different methods of tensile tests is hidden in this comparison.

In addition, it should be noted that the 10 data points falling far below the linear trend lines in both Figures 6.12(b) and 6.13(b) were all reported by Tegola and Manni (1999) for concrete-filled GFRP tubes produced by filament winding mostly with a winding angle of ±55.83°. The significant longitudinal stiffness of the GFRP tube may have affected the results of these specimens.

Furthermore, Lam and Teng (2001a) studied the effects of unconfined concrete strength, column diameter, length-to-diameter ratio, and type of FRP (CFRP, AFRP and GFRP). They found that within the ranges examined, these factors had little effect on the effectiveness of FRP confinement in terms of strength enhancement.

6.4.5 Performance of Strength Models

As flat-coupon tensile tests (e.g. ASTM D3039/D3039M-95a 1995) are most commonly used to obtain FRP tensile properties, Lam and Teng (2001a) provided

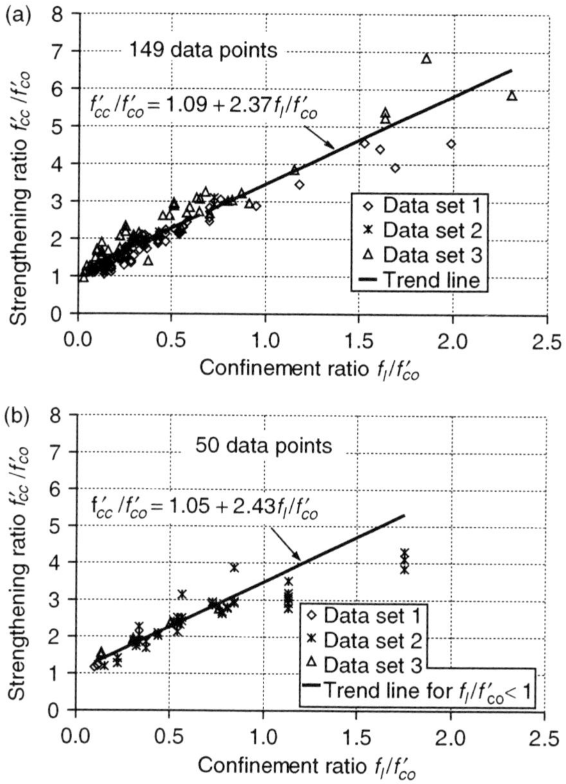

Figure 6.13 Comparison of compressive strength between adhesive-bonded and unbonded FRP-confined concrete specimens: (a) concrete confined by wrapped FRP sheets or wound strands; (b) concrete confined by FRP tubes

a comprehensive assessment of existing strength models using data set 1. This data set is also the largest of the three data sets and the least affected by the axial stiffness of the FRP as the vast majority of the results in data set 1 are for FRP-wrapped specimens with fibres only in the hoop direction. Comparisons of existing strength models with set 1 data are given in Figure 6.14. The statistical performance of each model is given in Table 6.4. In plotting the predictions of Mirmiran *et al.*'s (1998a) model, Samaan *et al.*'s (1998) model, which was proposed by the same group, was used to predict $f'_{cc2:1}$ in Equation 6.4.

Most of the test data lie in the bottom left corners of Figures 6.14(a) and 6.14(b) where the concrete axial compressive strength is less than tripled by FRP confinement. This is the important region for strengthening applications, as it is unlikely that the axial compressive strength of concrete needs to be more than tripled in a strengthening situation. Figure 6.14 shows that some of the models give close predictions of the test results but others do not.

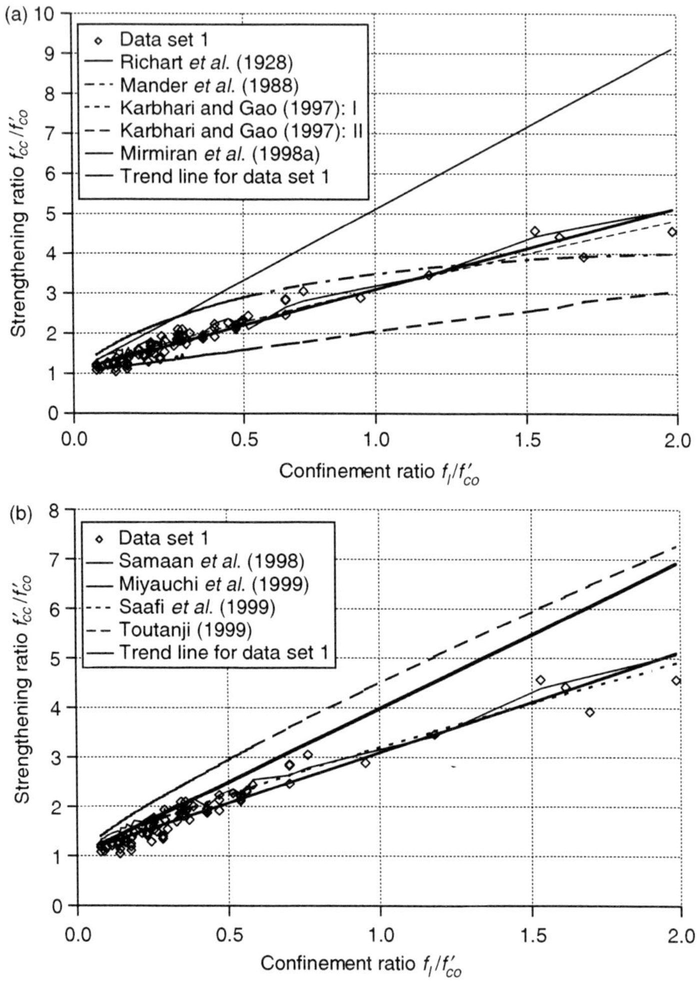

Figure 6.14 Comparison of strength models for FRP-confined concrete with test results: (a) models of Richart *et al.*, Mander *et al.*, Karbhari and Gao (I and II) and Mirmiran *et al.*; (b) models of Samaan *et al.*, Miyauchi *et al.*, Saafi *et al.* and Toutanji

Of the nine models, Karbhari and Gao's model I and Saafi *et al.*'s model give very similar predictions and perform the best (Table 6.4 and Figure 6.14); they are very close to the trend line of the set 1 data. Richart *et al.*'s and Mander *et al.*'s models, which are commonly used for steel-confined concrete, do not perform well: the former is significantly unconservative for all cases while the latter is significantly unconservative for the range of relatively low degree of confinement but becomes over-conservative for the range of high degree of confinement. The model of Toutanji also significantly overestimates the strength of FRP-confined

Table 6.4 Statistical performance of strength models for FRP-confined concrete as assessed by data set 1

Model	Test-to-predicted f'_{cc} ratio		
	Average	Standard deviation	Coefficient of variation (%)
Richart *et al.* (1928)	0.74	0.09	11.6
Mander *et al.* (1988)	0.77	0.11	14.3
Karbhari and Gao (1997): I	0.96	0.09	9.5
Karbhari and Gao (1997): II	1.27	0.20	15.3
Mirmiran *et al.* (1998a)	0.91	0.11	11.4
Samaan *et al.* (1998)	0.90	0.09	10.1
Miyauchi *et al.* (1999)	0.87	0.08	9.3
Saafi *et al.* (1999)	0.93	0.09	9.4
Toutanji (1999)	0.74	0.06	8.6

concrete. Miyauchi *et al.*'s model is the only model proposed for FRP-confined concrete which has a linear relationship between the strengthening ratio f'_{cc}/f'_{co} and the confinement ratio f_l/f'_{co}, but is unconservative. However, Karbhari and Gao's model II is over-conservative.

In addition to a comparison of available strength models with data set 1, Lam and Teng (2001a) also compared their predictions with data set 2. The concussions are similar to those given above.

6.4.6 Lam and Teng's Model

As the experimental data show that the strength of FRP-confined concrete f'_{cc} is linearly related to the lateral confining pressure f_l (Figures 6.12 and 6.13), the additional complexity in representing this relationship as done by most of the recent models (Karbhari and Gao 1997, Mirmiran *et al.* 1998a, Samaan *et al.* 1998, Saafi *et al.* 1999, Toutanji 1999) appears to be unnecessary (Lam and Teng 2001a). Lam and Teng (2001a) found that Equation 6.3 with k_1 equal to 2.15 gives the closest predictions overall for the set 1 data as shown in Table 6.5,

Table 6.5 Statistical performance of Lam and Teng's model as assessed by data set 1

Model	Test-to-predicted f'_{cc} ratio		
	Average	Standard deviation	Coefficient of variation (%)
Lam and Teng (2001a)	1.03	0.10	9.4
$\frac{f'_{cc}}{f'_{co}} = 1 + 2.15\frac{f_l}{f'_{co}}$	1.00	0.09	9.1

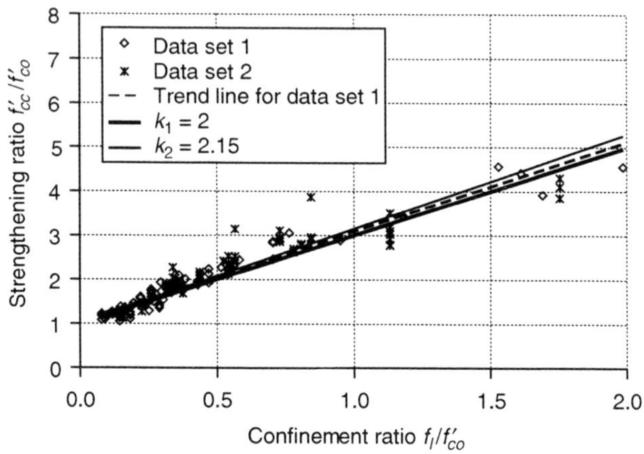

Figure 6.15 Comparison of Lam and Teng's strength model for FRP-confined concrete with test results

but the predictions are slightly on the unconservative side for the region of high degree of confinement (i.e. $f_l/f'_{co} > 1$) (Figure 6.15). The following equation for the axial compressive strength of FRP-confined concrete was therefore proposed by Lam and Teng (2001a) for design use:

$$\frac{f'_{cc}}{f'_{co}} = 1 + 2\frac{f_l}{f'_{co}} \tag{6.9}$$

This equation errs slightly on the conservative side and is simpler than all other existing models developed specifically for FRP-confined concrete. Since the equation was developed using set 1 data of their database, Lam and Teng (2001a) suggested that in design application, the tensile strength of the FRP in the hoop direction should be determined using flat-coupon tests according to ASTM D3039/D3039M-95a (1995) or a similar method. Nevertheless, Lam and Teng noted that ring-splitting tests (e.g. ASTM D2290-92 1992) have been quite commonly used by researchers for prefabricated FRP tubes and the use of Equation 6.9 with FRP tensile strengths from ring-splitting tests is also acceptable, because set 2 data with FRP tensile strengths from ring tests are conservatively predicted by Equation 6.9 if $f_l/f'_{co} < 1$ (Figure 6.15).

6.5 Stress–Strain Behaviour of FRP-confined Concrete

6.5.1 General

An understanding of the stress–strain behaviour of FRP-confined concrete is also necessary for the design of FRP strengthening measures for a column, particularly when the column is subject to combined bending and axial load and when its seismic ductility is of concern. Although many stress–strain models are available for steel-confined concrete (e.g. Ahmad and Shah 1982, Mander *et al.* 1988, Cusson and Paultre 1995) and one of them (Mander *et al.* 1988) has been used for FRP-confined concrete, these models are not strictly applicable to FRP-confined concrete as they do not reflect the bilinear behaviour observed in tests on FRP-confined concrete. Consequently, a number of stress–strain models have been proposed specifically for FRP-confined concrete (Lam and Teng 2001b). In this section, existing axial stress–strain models for FRP-confined concrete are reviewed and compared with experimental data based on the work of Lam and Teng (2001b). Their recently proposed model is also introduced and recommended for design use.

It should be mentioned that some of the existing models include predictions of axial stress–hoop strain relationships. This aspect is not addressed here, so only axial stress–strain relationships are discussed, and they are generally referred to as stress–strain relationships only for simplicity.

6.5.2 Review of Stress–Strain Models

Early investigations showed that for steel confined concrete, the axial compressive strain ε_{cc} at the peak axial stress can be related to the lateral confining pressure (Richart *et al.* 1929) by

$$\varepsilon_{cc} = \varepsilon_{co}\left(1 + k_2\frac{f_l}{f'_{co}}\right) \tag{6.10}$$

where ε_{co} is the axial strain of the unconfined concrete at its peak stress and k_2 is the strain enhancement coefficient. Richart *et al.* (1929) suggested $k_2 = 5k_1$ for steel-confined concrete, where k_1 is the confinement effectiveness coefficient for strength enhancement defined by Equation 6.3 and equals 4.1 for steel confinement.

The first stress–strain model for FRP-confined concrete appears to have been proposed by Fardis and Khalili (1982), who suggested that

$$\sigma_c = \frac{E_c\varepsilon_c}{1 + \varepsilon_c\left(E_c/f'_{cc} - 1/\varepsilon_{cc}\right)} \tag{6.11}$$

where E_c is the elastic modulus of the unconfined concrete, and σ_c and ε_c are the axial stress and axial strain of the confined concrete respectively. They also suggested that the axial strain of the confined concrete at peak stress can be predicted by

$$\varepsilon_{cc} = 0.002 + 0.0005\frac{2E_{frp}t_{frp}}{df'_{co}} \tag{6.12}$$

where E_{frp} is the elastic modulus in the hoop direction of the confining FRP.

A model originally proposed by Mander *et al.* (1988) for steel-confined concrete has also been used by various researchers (e.g. Saadatmanesh *et al.*

1994, Priestley *et al.* 1996) to predict the stress–strain behaviour of FRP-confined concrete. This model is based on an equation suggested by Popovics (1973). The axial stress of confined concrete is given by

$$\sigma_c = \frac{f'_{cc} x r}{r - 1 + x^r} \tag{6.13}$$

where f'_{cc} is given by Equation 6.6, $x = \varepsilon_c/\varepsilon_{cc}$, and $r = E_c/(E_c - E_{sec})$ with $E_{sec} = f'_{cc}/\varepsilon_{cc}$. The axial strain ε_{cc} at the peak stress f'_{cc} is found by rearranging Equation 6.10 and assuming that $k_2 = 5k_1$ as suggested by Richart *et al.* (1929). This is given by

$$\varepsilon_{cc} = \varepsilon_{co}\left[1 + 5\left(\frac{f'_{cc}}{f'_{co}} - 1\right)\right] \tag{6.14}$$

The stress–strain curve of Equation 6.13 terminates when the ultimate axial strain of the confined concrete, ε_{cu}, is reached. This ultimate strain is defined as the axial strain at which the first hoop fractures (Scott *et al.* 1982). Mander *et al.* (1988) suggested that this ultimate axial strain be found using the energy balance concept as expressed in the following equation:

$$U_{sh} = U_{cc} + U_{sc} - U_{co} \tag{6.15}$$

where U_{sh}, U_{cc}, U_{sc} and U_{co} are the ultimate strain energies per unit volume of the confining reinforcement, the confined concrete, the longitudinal steel reinforcement, and the unconfined concrete respectively. Based on this concept, a conservative estimate for the ultimate axial strain of steel-confined concrete (Priestley *et al.* 1996) can be found as

$$\varepsilon_{cu} = 0.004 + \frac{1.4\rho_{sh} f_{yh} \varepsilon_{su}}{f'_{cc}} \tag{6.16}$$

where ρ_{sh} is the volumetric ratio and f_{yh} the yield strength of transverse steel reinforcement, and ε_{su} is the strain of the steel at maximum tensile stress. Furthermore, Seible *et al.* (1997) suggested that for FRP-confined concrete,

$$\varepsilon_{cu} = 0.004 + \frac{2.8\rho_{frp} f_{frp} \varepsilon_{frp,rup}}{f'_{cc}} \tag{6.17}$$

where $\varepsilon_{frp,rup}$ is the ultimate tensile strain of FRP in the hoop direction. The stress–strain model of Mander *et al.* (1988) for steel-confined concrete is illustrated in Figure 6.16. It should be noted that for FRP-confined concrete, in most cases the ultimate axial strain and the strain at the ultimate axial compressive strength (i.e. peak axial stress) are the same. They are different only when the effect of FRP confinement is small.

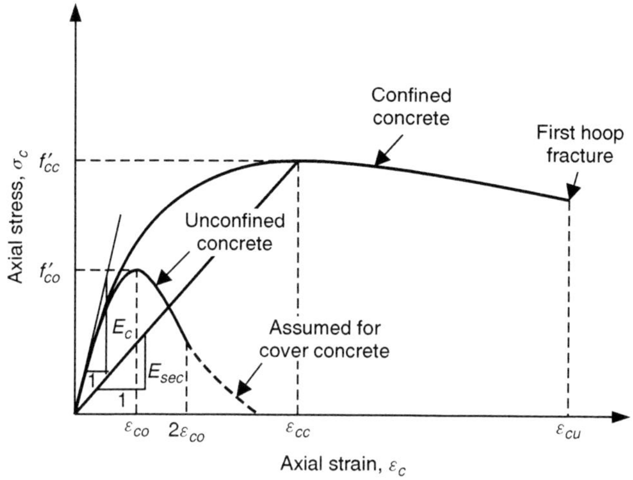

Figure 6.16 Mander *et al.*'s stress–strain model for steel-confined concrete

As extensive experimental results have shown that the stress–strain behaviour of FRP-confined concrete is significantly different from that of steel-confined concrete, a number of stress–strain models have been proposed for FRP-confined concrete in the past few years (Lam and Teng 2001b). While some of these models have been based on test data of concrete-filled FRP tubes (e.g. Samaan *et al.* 1998, Saafi *et al.* 1999), most of the models have been based mainly on tests on FRP-wrapped concrete specimens (e.g. Karbhari and Gao 1997, Miyauchi *et al.* 1999, Spoelstra and Monti 1999, Toutanji 1999, Xiao and Wu 2000). These models are summarised in Table 6.6. It should be noted that models with dimensional inconsistency are given here with stresses in MPa and lengths in mm. It should be further noted that model I of Karbhari and Gao (1997) is not a complete model as it only defines the ultimate strain point and that the model of Spoelstra and Monti (1999) requires an iterative procedure to generate the stress–strain curve.

It can be seen from Table 6.6 that the models of Saafi *et al.* and Toutanji, developed for concrete-filled FRP tubes and FRP-wrapped concrete respectively, are similar in form. Of the seven models given in Table 6.6, four (Karbhari and Gao 1997, Samaan *et al.* 1998, Saafi *et al.* 1999 and Toutanji 1999) feature monotonically increasing stress–strain curves only, with the axial strain at peak stress, ε_{cc}, coinciding with the ultimate axial strain, ε_{cu}, whilst the rest (Miyauchi *et al.* 1999, Spoelstra and Monti 1999 and Xiao and Wu 2000) include the possibility of a descending branch after the stress–strain curve peaks at f'_{cc}, with the ultimate axial strain ε_{cu} being greater than the axial strain at peak stress, ε_{cc}. In Miyauchi *et al.*'s model, the stress–strain curve can be of either the increasing type or the decreasing type. Spoelstra and Monti's model predicts a concrete stress below f'_{co} at ultimate axial strain for $f_l/f'_{co} < 0.07$ and for the limiting case of unconfined concrete, this concrete stress is only 20 % of f'_{co} occurring at an ultimate

Table 6.6 Stress–strain models for FRP-confined concrete

Reference		Stress–strain model	Parameters
Karbhari and Gao (1997)	Model I	$\varepsilon_{cc} = \varepsilon_{co} + 0.01\dfrac{2f_{frp}t_{frp}}{df'_{co}}$, f'_{cc} defined in Table 6.2	
	Model II (Figure 6.17)	At point A, $\sigma_c = f'_{co} + 4.1f'_{co}\nu_c\dfrac{2t_{frp}}{d}\dfrac{E_{frp}}{E_c}$, $\varepsilon_c = \dfrac{f_c}{E_{eff}}$	
		At point B, $\sigma_c = f'_{cc} = f'_{co} + 3.1f'_{co}\nu_c\dfrac{2t_{frp}}{d}\dfrac{E_{frp}}{E_c} + \dfrac{2f_{frp}t_{frp}}{d}$ (Equation 6.5)	$E_{eff} = E_c\dfrac{A_c}{A_c + A_{frp}} + E_{frp}\dfrac{A_{frp}}{A_c + A_{frp}}$
		$\varepsilon_c = \varepsilon_{cc} = 1 - \dfrac{1.004[1 - f'_{co}/E_{eff} - 4.1f'_{co}\nu_c(2t_{frp}/d)(E_{frp}/(E_cE_{eff}))]}{(1+\varepsilon_{frp,rup})^2}$	
Samaan *et al.* (1998) (Figure 6.18)	$\sigma_c = \dfrac{(E_c - E_2)\varepsilon_c}{\{1 + [(E_c - E_2)\varepsilon_c/f_o]^n\}^{1/n}} + E_2\varepsilon_c$		$E_c = 3950\sqrt{f'_c}$
	$\varepsilon_{cc} = \dfrac{f'_{cc} - f_o}{E_2}$		$E_2 = 245.61f_{co}^{0.2} + 1.3456\dfrac{E_{frp}t_{frp}}{d}$ $f_o = 0.872f'_{co} + 0.371f_l + 6.258$ f'_{cc} defined in Table 6.2 $n = 1.5$

Miyauchi *et al.* (1999)	Increasing type	$\sigma_c = f'_{co}[2(\varepsilon_c/\varepsilon_{co}) - (\varepsilon_c/\varepsilon_{co})^2] \quad (0 \leqslant \varepsilon_c \leqslant \varepsilon_t)$ $\sigma_c = f'_{cu} - \lambda(\varepsilon_{cu} - \varepsilon_c) \quad (\varepsilon_t \leqslant \varepsilon_c \leqslant \varepsilon_{cu})$	$\varepsilon_t = \varepsilon_{co} - \lambda\varepsilon_{co}^2/(2f'_{co})$ $\lambda = \{-2f'_{co}(\varepsilon_{cu} - \varepsilon_{co}) + [4f'_{co}(f'_{co}\varepsilon_{cu}^2 - 2f'_{co}\varepsilon_{co}\varepsilon_{cu} + f'_{cu}\varepsilon_{co}^2)]^{0.5}\}/\varepsilon_{co}^2$
	Decreasing type	$\sigma_c = f'_{co}[2(\varepsilon_c/\varepsilon_{co}) - (\varepsilon_c/\varepsilon_{co})^2] \quad (0 \leqslant \varepsilon_c \leqslant \varepsilon_{co})$ $\sigma_c = f'_{co} + (\varepsilon_c - \varepsilon_{co})(f'_{cu} - f'_{co})/(\varepsilon_{cu} - \varepsilon_{co})$ $(\varepsilon_{co} \leqslant \varepsilon_c \leqslant \varepsilon_{cu})$	f'_{cu} = axial stress of confined concrete at ultimate strain ε_{cu} For increasing type, $f'_{cu} = f'_{cc}$ (defined in Table 6.2) and $\varepsilon_{cu} = \varepsilon_{cc} = \varepsilon_{co}$ $\left[1.0 + (15.87 - 0.093f'_{co})(f_l/f'_{co})^{0.246+0.0064f'_{co}}\right]$
Saafi *et al.* (1999)		$\sigma_c = \dfrac{E_c\varepsilon_c}{1 + (E_c/f_{ua} - 2/\varepsilon_{ua} + E_{ua}E_c\varepsilon_{ua}/f_{ua}^2)\varepsilon_c + (1/\varepsilon_{ua}^2 - E_{ua}E_c/f_{ua}^2)\varepsilon_c^2} \quad (0 \leqslant \varepsilon_h \leqslant 0.002)$	$f_{ua} = f'_{co}\left[1 + 0.0213\left(\dfrac{E_{frp}t_{frp}}{df'_{co}}\right)^{0.84}\right]$
		$\sigma_c = f'_{co}\left[1 + 2.2\left(\dfrac{2E_{frp}t_{frp}\varepsilon_h}{df'_{co}}\right)^{0.84}\right]$ and	ε_h = hoop strain of confined concrete
			$\varepsilon_{ua} = \varepsilon_{co}\left[1 + 0.0783\left(\dfrac{E_{frp}t_{frp}}{df'_{co}}\right)^{0.84}\right]$
		$\varepsilon_c = \varepsilon_{co}\left[1 + (537\varepsilon_h + 2.6)\left(\dfrac{\sigma_c}{f'_{co}} - 1\right)\right] \quad (0.002 \leqslant \varepsilon_h \leqslant \varepsilon_{frp,rup})$	$E_{ua} = 0.272\dfrac{f'_{co}}{\varepsilon_{co}}$
Spoelstra and Monti (1999)		$f'_{cu} = f'_{co}\left(0.2 + 3\sqrt{\dfrac{f_l}{f'_{co}}}\right)$ and $\varepsilon_{cu} = \varepsilon_{co}\left(2 + 1.25\dfrac{E_c}{f'_{co}}\varepsilon_{frp,rup}\sqrt{\dfrac{f_l}{f'_{co}}}\right)$	

(continued overleaf)

Table 6.6 *(continued)*

Reference	Stress–strain model	Parameters
Toutanji (1999)	$\sigma_c = \dfrac{E_c \varepsilon_c}{1 + (E_c/f_{ua} - 2/\varepsilon_{ua} + E_{ua} E_c \varepsilon_{ua}/f_{ua}^2)\varepsilon_c + (1/\varepsilon_{ua}^2 - E_{ua} E_c/f_{ua}^2)\varepsilon_c^2}$ $(0 \leqslant \varepsilon_h \leqslant 0.002)$	$f_{ua} = f'_{co}\left[1 + 0.0178\left(\dfrac{2E_{frp}t_{frp}}{df'_{co}}\right)^{0.85}\right]$
	$\sigma_c = f'_{co}\left[1 - 3.5\left(\dfrac{2E_{frp}t_{frp}\varepsilon_h}{df'_{co}}\right)^{0.85}\right]$ and	$\varepsilon_{ua} = \varepsilon_{co}\left[1 + 0.0448\left(\dfrac{2E_{frp}t_{frp}}{df'_{co}}\right)^{0.85}\right]$
	$\varepsilon_c = \varepsilon_{co}\left[1 + (310.57\varepsilon_h + 1.9)\left(\dfrac{\sigma_c}{f'_{co}} - 1\right)\right]$ $(0.002 \leqslant \varepsilon_h \leqslant \varepsilon_{frp,rup})$	$E_{ua} = 0.3075\dfrac{f'_{co}}{\varepsilon_{co}}$
Xiao and Wu (2000)	$\sigma_c = E_c \varepsilon_c + 2\nu_c f_r$ and	
	$\varepsilon_c = -\dfrac{1 + (C_j/E_c)(1 - \nu_c - 2\nu_c^2)}{\nu_c}\varepsilon_h$ (before reaching f'_{co})	$C_j = \dfrac{2t_{frp}}{d}E_{frp}$
	$\sigma_c = \alpha f'_{co} + k f_r$ and	$f_r = -C_j \varepsilon_h$ (ε_h defined as negative)
	$\varepsilon_c = \dfrac{\varepsilon'_{ro} - \varepsilon_h}{\nu'_c}$ (after reaching f'_{co})	$\alpha = 1.1, \quad k = 4.1 - 0.75\dfrac{f'^2_{co}}{C_j},$
		$\varepsilon'_{ro} = -0.0005, \quad \nu'_c = 7\left(\dfrac{f'_{co}}{C_j}\right)^{0.8}$

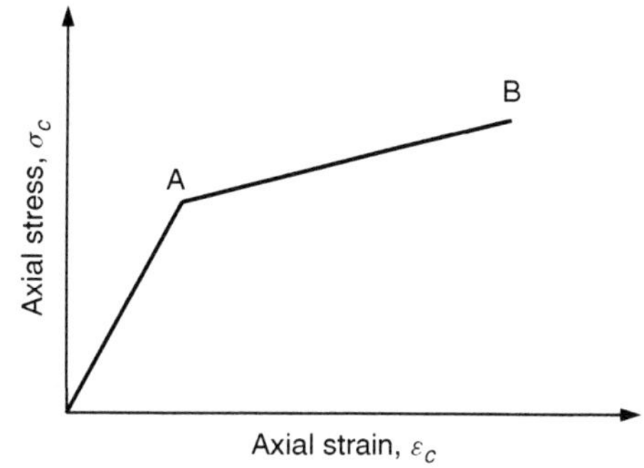

Figure 6.17 Karbhari and Gao's stress–strain model for FRP-confined concrete

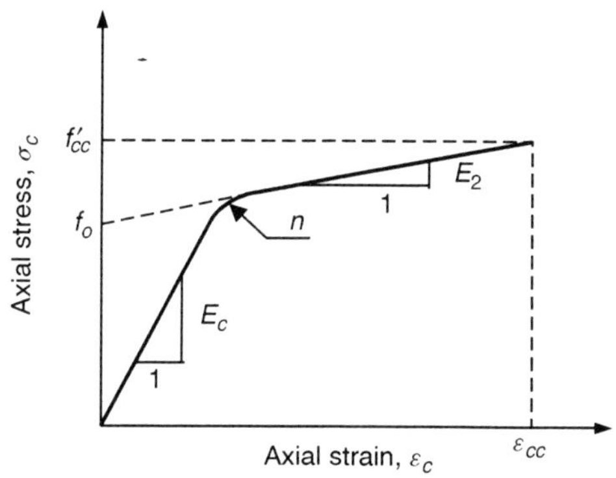

Figure 6.18 Samaan *et al.*'s stress–strain model for FRP-confined concrete. Reproduced from Samaan *et al.* (1998) by permission of American Society of Civil Engineers

strain $\varepsilon_{cu} = 0.004$. Xaio and Wu's (2000) model predicts a descending branch after reaching f'_{co} for $C_j/f'^2_{co} < 0.183$.

6.5.3 Experimental Data

Many tests have been conducted on the stress–strain behaviour of FRP-confined concrete as reviewed by Lam and Teng (2001b). Test curves are generally presented in graphs, so the development of a database containing all details of test stress–strain curves is extremely difficult if not completely impossible. Fortunately, owing to the bilinear nature of stress–strain curves of sufficiently confined concrete, each curve can be closely defined by a few characteristic points. Lam and Teng (2001b) developed a large test database which contains the axial strain of confined concrete, ε_{cc}, at peak stress for 171 cases (each case may represent the average of up to five nominally identical specimens), and for a smaller number of cases, the intercept of the stress axis by the second linear portion of the stress–strain curve and hoop strain in the FRP at rupture. For all these stress–strain curves, the strain at peak stress coincides with the ultimate strain and is referred to as the ultimate strain in the following discussion, as the few curves with a descending branch (Demers and Neale 1994, Mastrapa 1997, Xiao and Wu 2000) were excluded by Lam and Teng from the database. When necessary, this database is referred to as the strain database to differentiate it from the strength database (Lam and Teng 2001a) discussed in Section 6.4. All tests covered by the strain database are also covered by the strength database.

The first two parameters in the strain database together with the compressive strength of confined concrete define the second linear portion of the stress–strain curve. An analysis of experimental compressive strengths of FRP-confined concrete is presented in Section 6.4.4. As for that analysis, the test results in the strain database for the analysis of stress–strain curves were also divided into three data sets (Lam and Teng 2001b): sets 1 and 2 are for those with FRP material properties from flat-coupon tests and ring-splitting tests respectively by researchers, while set 3 is for the rest.

6.5.4 Trends of Experimental Data

6.5.4.1 Ultimate Axial Strain

The ultimate axial strain of FRP-confined concrete is one of the key parameters for its stress–strain curve and can be related to the lateral confining pressure f_l (Karbhari and Gao 1997, Lam and Teng 2001b). Lam and Teng (2001b) observed a roughly linear relationship between the ultimate axial strain and the confinement ratio for both data sets 1 and 2, with the exception of four data points in data set 1 reported by Karbhari and Gao (1997), although this is accompanied by significant scatters (Figure 6.19). In Figure 6.19, the ultimate axial strain is normalised by the strain of unconfined concrete at peak stress, ε_{co}, which is assumed to have the commonly accepted value of 0.002 throughout the data interpretation in this section. Data set 3 shows a poorer correlation with a linear trend, which is attributable to uncertainty in the material properties. These plots show that the normalised ultimate strain of confined concrete also depends significantly on other parameters. This is in contrast with the situation for the strength of FRP-confined concrete, where the strengthening ratio is predominantly dependent on the confinement ratio (Figure 6.12).

Lam and Teng (2001b) also examined the strain data trends separately for FRP-wrapped concrete and concrete-filled FRP tubes using those data with reliable material properties (i.e. data sets 1 and 2). The trends were found to be similar to those seen in Figure 6.19, as all the set 1 data are for FRP-wrapped specimens, although in two of the cases, the FRP and the concrete were separated by foil, whilst the set 2 data are almost all for concrete-filled tubes (four test results for FRP-wrapped specimens).

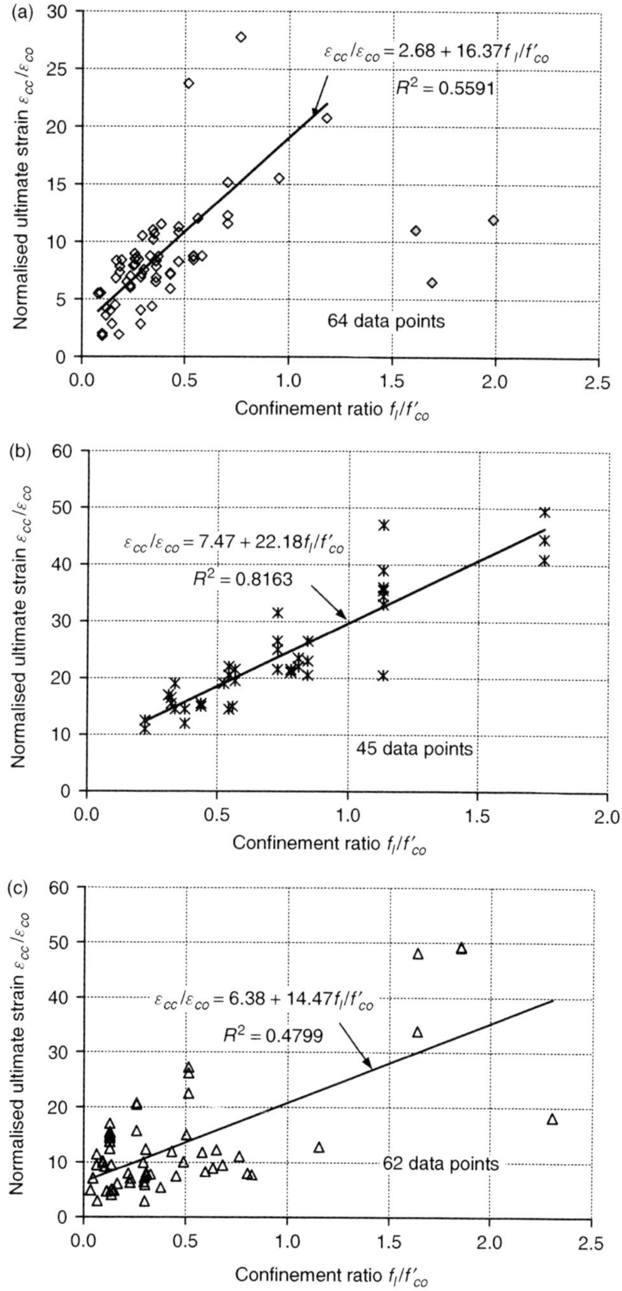

Figure 6.19 Effect of source of FRP properties on trends of strain data: (a) data set 1, FRP properties from flat-coupon tensile tests; (b) data set 2, FRP properties from ring-splitting tests; (c) data set 3, FRP properties from manufacturers or unspecified sources

Spoelstra and Monti (1999) noted that the deformability of GFRP-confined concrete is almost twice that of CFRP-confined concrete. Lam and Teng (2001b) studied in detail the effect of FRP material type on the ultimate axial strain of FRP-confined concrete using data sets 1 and 2. They related the ultimate axial strain of confined concrete to the confining pressure as in many existing models and defined the strain enhancement coefficient k_2 by the following equation:

$$\frac{\varepsilon_{cc}}{\varepsilon_{co}} = 2 + k_2 \frac{f_l}{f'_{co}} \tag{6.18}$$

Given that ε_{cc} for concrete sufficiently confined by FRP is the ultimate strain, Equation 6.18 predicts an ultimate strain of 0.004 for unconfined concrete.

Table 6.7 shows the performance of specimens with different FRP materials in terms of the strain enhancement coefficient k_2 defined by Equation 6.18. It can be seen that the strain enhancement coefficient depends strongly on the type of FRP. The strain enhancement coefficients for E-glass, S-glass and AFRP-confined concrete are similar and are much larger than those for high-modulus CFRP-confined concrete, with those for normal CFRP-confined concrete in between.

Lam and Teng (2001b) also plotted the normalised ultimate axial strain against the confinement ratio for CFRP-confined concrete and GFRP-confined concrete separately (Figure 6.20). All the data shown in Figure 6.20(a) are for concrete

Table 6.7 Effect of type of FRP on strain enhancement coefficient

Type of FRP	Type of specimen	Tensile test method	No. of data points	Ultimate tensile strain of FRP, $\varepsilon_{frp,rup}$		$k_2 = \frac{\varepsilon_{cc}/\varepsilon_{co} - 2}{f_l/f'_{co}}$	
				Mean	Standard deviation	Mean	Standard deviation
HM carbon	Wrapping	Coupons	4	0.003	0	6.4	5.1
Carbon	Wrapping	Coupons	37	0.014	0.002	16.3	7.1
HM carbon	Wrapping[a]	Coupons	1	0.003	–	-2.0	–
Carbon	Wrapping[a]	Coupons	1	0.013	–	13.8	–
Carbon and glass	Wrapping	Coupons	1	0.013	–	31.7	–
E-glass	Wrapping	Coupons	9	0.020	0.002	26.7	5.6
S-glass	Wrapping	Rings	4	0.029	0	33.7	4.8
E-glass	Tube	Rings	37	0.018	0.005	32.6	9.0
S-glass	Tube	Rings	4	0.029	0	30.0	2.2
Aramid	Wrapping	Coupons	7	0.022	0.007	29.9	8.9
	Total		105	0.017	0.006	24.7	11.4

Notes: HM = high modulus; four data points from Karbhari and Gao (1997) were excluded.
[a]With the FRP and the concrete separated by foil.

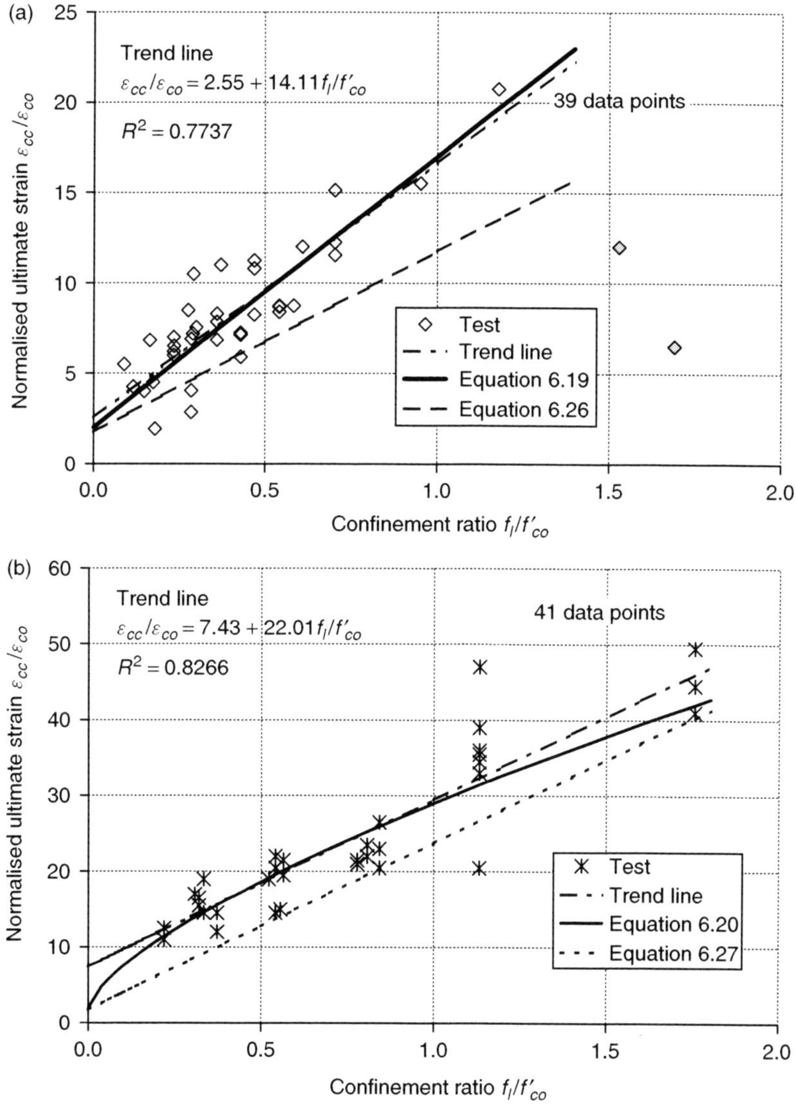

Figure 6.20 Effect of type of FRP on the ultimate strain of FRP-confined concrete: (a) CFRP-wrapped concrete specimens; (b) concrete-filled GFRP tubes

cylinders wrapped with normal CFRP, with the FRP material properties from flat-coupon tests by researchers. On the other hand, the data shown in Figure 6.20(b) are all for concrete-filled E-glass or S-glass FRP tubes, with the FRP material properties from ring tests by researchers. Both Figures 6.20(a) and 6.20(b) show a better correlation between $\varepsilon_{cc}/\varepsilon_{co}$ and f_l/f'_{co} than that observed in Figure 6.19.

This indicates that the ultimate axial strain of FRP-confined concrete depends on not only the confining pressure, but also the type of FRP. This is in contrast to the case of compressive strength of FRP-confined concrete, which has been shown to depend little on the type of FRP (Lam and Teng 2001a). The ultimate axial strain of normal CFRP-wrapped concrete can be approximated by (Figure 6.20a)

$$\frac{\varepsilon_{cc}}{\varepsilon_{co}} = 2 + 15\frac{f_l}{f'_{co}} \tag{6.19}$$

whilst that of concrete-filled GFRP tubes can be approximated by (Figure 6.20b)

$$\frac{\varepsilon_{cc}}{\varepsilon_{co}} = 2 + 27\left(\frac{f_l}{f'_{co}}\right)^{0.7} \tag{6.20}$$

It should be mentioned that the two data points (shaded) in Figure 6.20(a) reported by Karbhari and Gao (1997) were excluded in producing the trend line.

6.5.4.2 Intercept of Stress Axis by Second Linear Portion

The intercept of the stress axis by the second linear portion f_o (Figure 6.18) can be used as a parameter to characterise the bilinear stress–strain curve of FRP-confined concrete as has been done by Samaan *et al.* (1998) (Figure 6.18 and Table 6.6) who proposed that

$$f_o = 0.872 f'_{co} + 0.371 f_l + 6.258 \text{ (MPa)} \tag{6.21}$$

based on their test results. Lam and Teng's (2001b) measured values of f_o from 60 stress–strain curves reported with the test results in data sets 1 and 2 of their strain database and found that for most of these curves, the ratio of f_o to f'_{co} falls between 0.9 and 1.2 (Figure 6.21), with an average of 1.08 and a standard deviation of 0.11. A small dependency of f_o/f'_{co} on f_l/f'_{co} may be seen in Figure 6.21, but the data are so scattered that this small dependency is not worthy of much attention.

6.5.4.3 FRP Hoop Rupture Strain

Xiao and Wu (2000) and Shahawy *et al.* (2000) reported that for FRP-wrapped circular concrete specimens, the hoop strain of the FRP at rupture did not reach the ultimate tensile strain of FRP obtained from flat-coupon tensile tests. Shahawy *et al.* (2000) suggested that ring-splitting tests provide a closer estimate of the hoop rupture strain of FRP. Lam and Teng (2001a) found from their strength database for FRP-confined concrete that the observed confinement effectiveness coefficient k_1 is smaller for data set 1 (all for FRP-wrapped concrete) than for data set 2 (almost all for concrete-filled FRP tubes). They attributed this

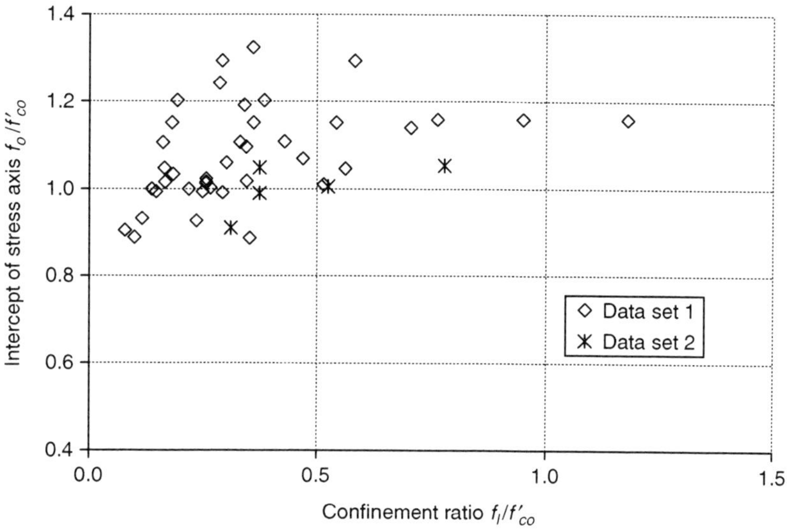

Figure 6.21 Intercept of the stress axis by the linear second portion versus confinement ratio

to the lower FRP rupture strain of an FRP wrap than that from flat-coupon tests.

FRP strains at hoop rupture were reported by researchers for 79 cases of the sets 1 and 2 data in Lam and Teng's (2001b) strain database. They found from these data that the ratio of FRP hoop rupture strain $\varepsilon_{h,rup}$ to FRP rupture strain $\varepsilon_{frp,rup}$ from material tests ranges from 0.24 to 1.35, with an average of 0.79 and a standard deviation of 0.28. For the 44 cases with FRP properties from flat-coupon tests, this ratio is between 0.24 and 0.90, with an average of 0.64 and a standard deviation of 0.20. For the 35 results with FRP properties from ring-splitting tests, the ratio is between 0.61 and 1.35, with an average of 0.99 and a standard deviation of 0.23. The confinement ratio does not affect this ratio significantly (Figure 6.22). These statistics lend support to the suggestion (Shahawy *et al.* 2000) that ring-splitting tests provide a closer estimate of the rupture strength in the hoop direction of an FRP tube.

Whilst it may be possible to develop improved confinement models for FRP-confined concrete by defining the FRP tensile rupture strain/strength in the hoop direction using actual strains measured on circular concrete specimens, such an approach can only be useful if the relationship between flat-coupon tests, ring tests and tensile strengths of FRP in circular concrete specimens is well established for both FRP-wrapped specimens and concrete-filled FRP tubes. Such an attempt may not succeed as measurements of FRP hoop rupture strains tend to show such large scatters as to obscure any relationships between the sought parameters, as already noted above for the existing data.

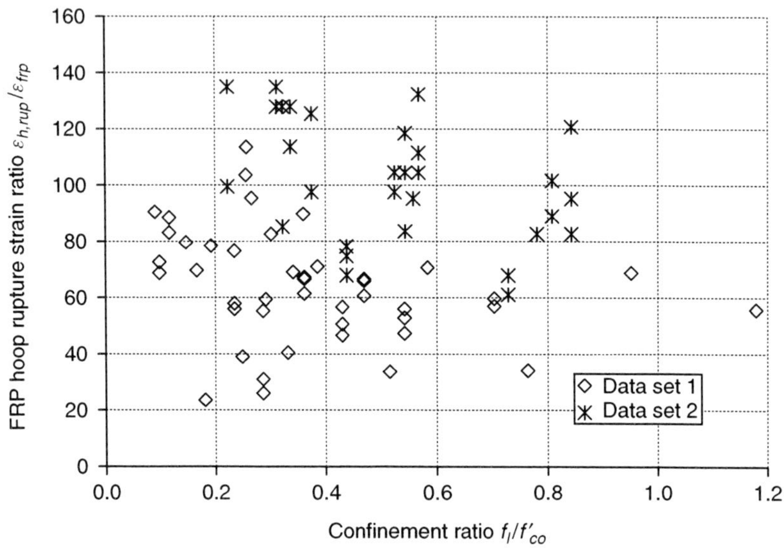

Figure 6.22 FRP hoop rupture strain ratio versus confinement ratio

6.5.5 Performance of Existing Stress–Strain Models

6.5.5.1 Ultimate Axial Strain

All stress–strain models for FRP-confined concrete provide predictions of the ultimate axial strain. Table 6.8 shows the findings from an assessment of the models reviewed in Section 6.5.2 by Lam and Teng (2001b) using their strain database. In their assessment, it was assumed that $E_c = 4730\sqrt{f'_{co}}$ (MPa) and $\nu_c = 0.18$ when these values were needed. For Xiao and Wu's model, the FRP strain at hoop rupture was taken to be half the FRP rupture strain from flat-coupon tests following their suggestion. Table 6.8(a) is for CFRP-wrapped concrete cylinders (high-modulus CFRP excluded), whilst Table 6.8(b) is for concrete-filled E-glass and S-glass FRP tubes.

As can be seen from Table 6.8, most of the models overestimate the ultimate axial strain of CFRP-confined concrete, but all the models underestimate that of GFRP-confined concrete. For CFRP-confined concrete, the model of Fardis and Khalili performs the best. For GFRP-confined concrete, the models of Saafi *et al.* and Spoelstra and Monti perform the best. A common flaw in all these models is that the important effect of FRP type is not taken into account.

6.5.5.2 Stress–Strain Curves

Lam and Teng (2001b) compared existing stress–strain models with test curves supplied by Xiao (2001) for the tests described in Xiao and Wu (2000). Their tests were conducted on concrete cylinders 152 mm in diameter and 305 mm

Table 6.8 Ultimate axial strain of FRP-confined concrete: accuracy of existing models

(a) CFRP-wrapped concrete cylinders (39 data points)

Model	Test-to-predicted ε_{cc} ratio		
	Average	Standard deviation	Coefficient of variation (%)
Fardis and Khalili (1982)	1.03	0.33	32
Seible *et al.* (1997)	0.82	0.31	38
Karbhari and Gao (1997): I	2.65	0.77	29
Karbhari and Gao (1997): II	0.70	0.37	53
Samaan *et al.* (1998)	0.66	0.30	46
Miyauchi *et al.* (1999)	0.92	0.24	26
Saafi *et al.* (1999)	0.89	0.40	45
Spoelstra and Monti (1999)	0.77	0.26	33
Toutanji (1999)	1.32	0.51	39
Xiao and Wu (2000)	1.26	0.48	38

(b) Concrete-filled GFRP tubes (41 data points)

Model	Test-to-predicted ε_{cc} ratio		
	Average	Standard deviation	Coefficient of variation (%)
Fardis and Khalili (1982)	2.31	0.64	28
Seible *et al.* (1997)	1.52	0.42	28
Karbhari and Gao (1997): I	5.20	0.92	18
Karbhari and Gao (1997): II	1.40	0.36	26
Samaan *et al.* (1998)	1.44	0.39	27
Miyauchi *et al.* (1999)	1.93	0.37	19
Saafi *et al.* (1999)	1.23	0.47	33
Spoelstra and Monti (1999)	1.28	0.30	24
Toutanji (1999)	1.90	0.65	34
Xiao and Wu (2000)	2.15	0.50	23

in height wrapped with one to three layers of CFRP. The tensile strength and elastic modulus of the CFRP used in these tests were 1577 MPa and 105 000 MPa respectively, which were determined from flat-coupon tensile tests and calculated on the basis of a nominal thickness of 0.381 mm per layer. The concrete cylinder compressive strength ranged from 33.7 to 55.2 MPa in these tests.

Four stress–strain curves were selected for comparison in Lam and Teng (2001b), based on the single criterion that their experimental confinement effectiveness coefficients k_1 are around two, and thus similar to the average value of k_1 for data set 1 of Lam and Teng's (2001a) strength database. Figure 6.23 shows comparisons between various stress–strain models and test results. In plotting the predicted stress–strain curves, the same assumptions as used for ultimate strain comparisons above were adopted. For Xiao and Wu's model, the FRP strain at hoop rupture was again taken to be 50 % of the FRP rupture strain from flat-coupon tests.

From Figure 6.23, it can be seen that large differences exist between different models and most models are unsatisfactory. Overall, the models of Xaio and Wu, Saafi *et al.* and Miyauchi *et al.* provide closer predictions than the others.

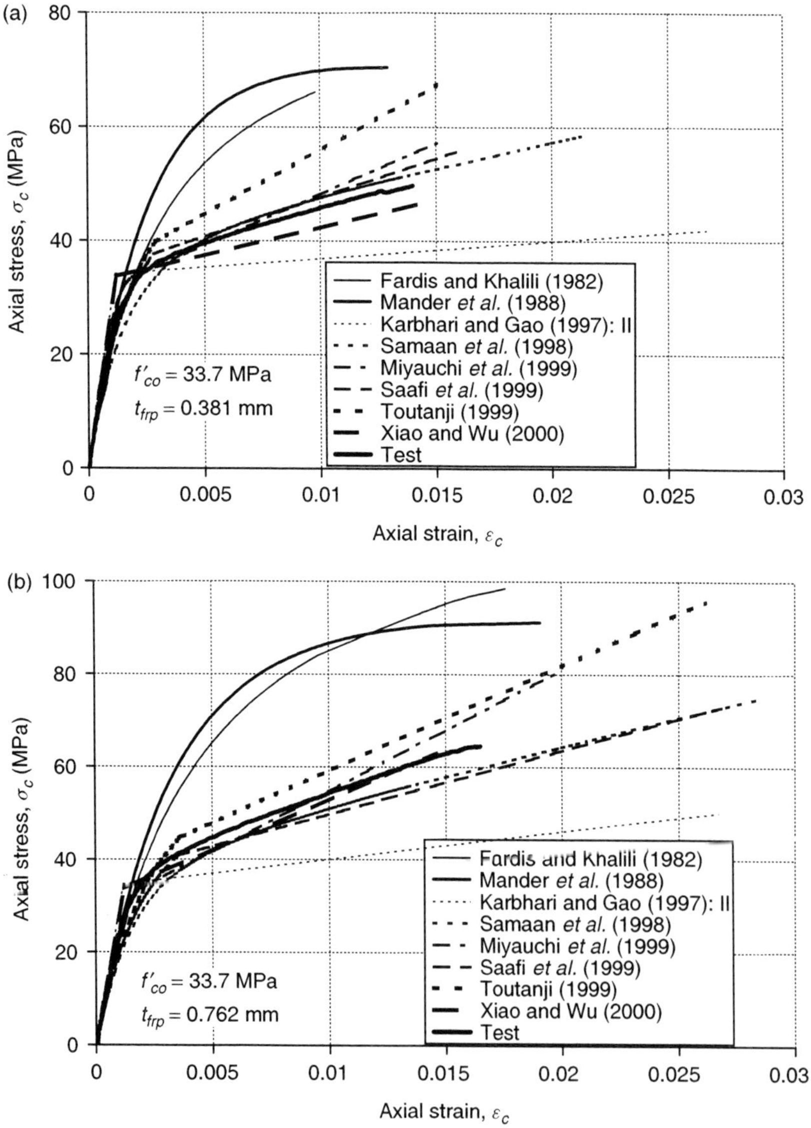

Figure 6.23 Theoretical stress–strain models versus test results

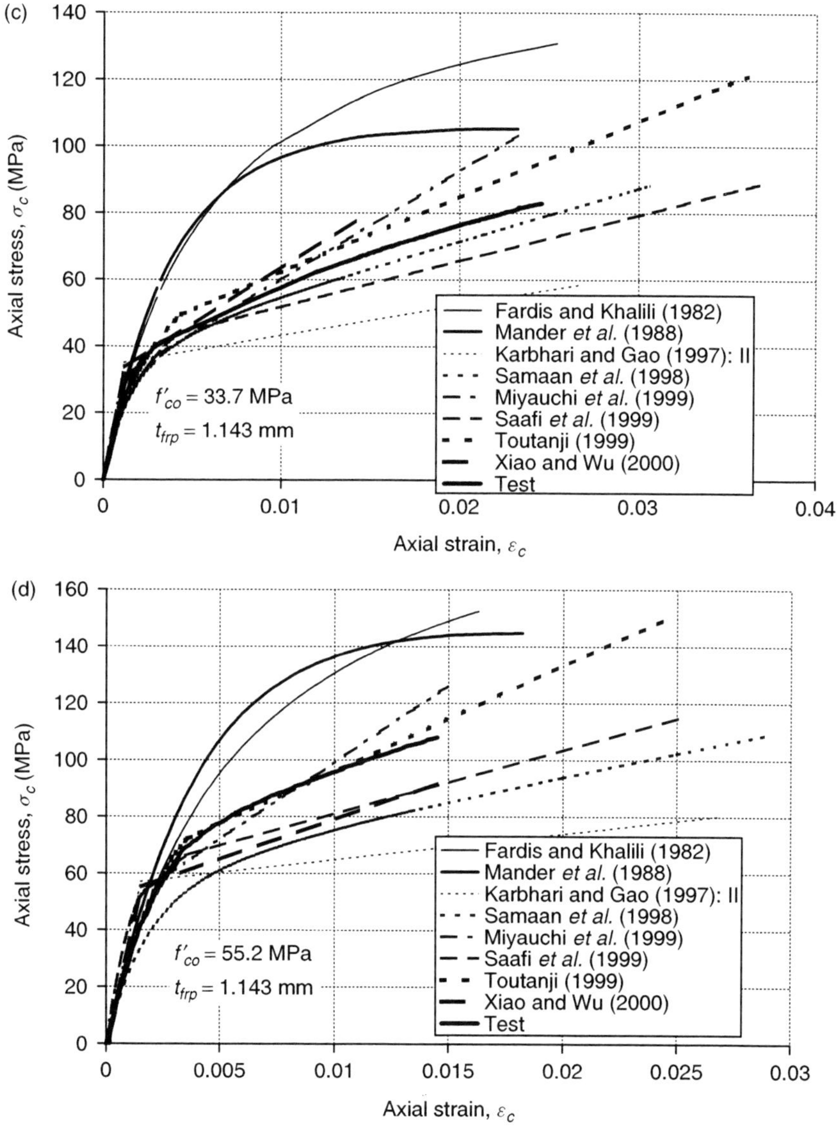

Figure 6.23 *(continued)*

6.5.6 Lam and Teng's Stress–Strain Model

6.5.6.1 Best-fit Stress–Strain Model

Based on a review of previous models and their large strain database, Lam and Teng (2001b) recently proposed a simple but accurate stress–strain model.

This model is based on the following assumptions: (a) the first portion of the stress–strain curve is parabolic (but its shape is clear to being linear); (b) the second portion of the stress–strain curve is linear; (c) the first portion meets the second portion smoothly (i.e. there is no change in slope between the two portions when they meet); (d) the stress axis is intercepted by the linear second portion at the compressive strength of unconfined concrete; and (e) the linear second portion ends at a point where both the compressive strength and the ultimate axial strain are reached. These assumptions were derived from careful observations of test results and can be easily justified by examining the discussions presented earlier in this section.

Based on these assumptions, Lam and Teng (2001b) established the following stress–strain model for FRP-confined concrete:

$$\sigma_c = E_c\varepsilon_c - \frac{(E_c - E_2)^2}{4f'_{co}}\varepsilon_c^2 \quad \text{for } 0 \leqslant \varepsilon_c \leqslant \varepsilon_t \tag{6.22a}$$

and

$$\sigma_c = f'_{co} + E_2\varepsilon_c \quad \text{for } \varepsilon_t \leqslant \varepsilon_c \leqslant \varepsilon_{cc} \tag{6.22b}$$

Equation 6.22 defines a stress–strain curve (Figure 6.24) consisting of a parabolic first portion and a linear second portion with a smooth transition at ε_t which is given by

$$\varepsilon_t = \frac{2f'_{co}}{(E_c - E_2)} \tag{6.23}$$

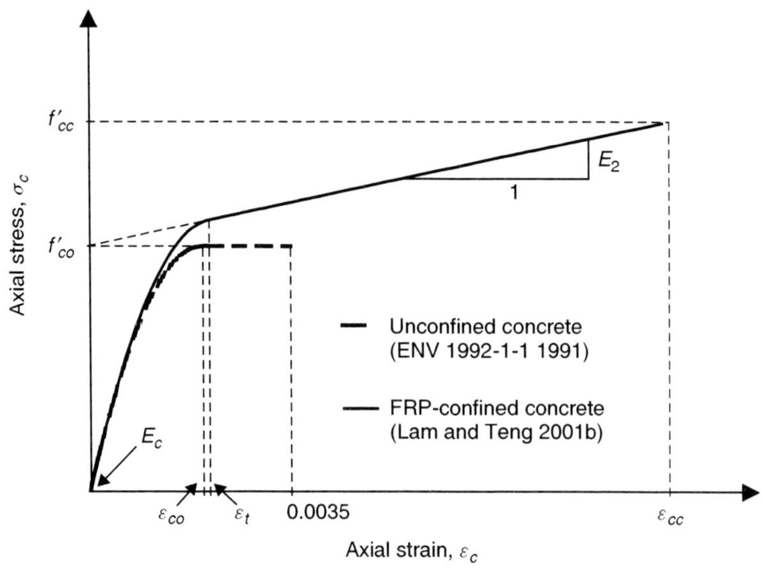

Figure 6.24 Lam and Teng's stress–strain model for FRP-confined concrete

where E_2 is the slope of the linear second portion, given by

$$E_2 = \frac{f'_{cc} - f'_{co}}{\varepsilon_{cc}} \tag{6.24}$$

To define E_2, the axial compressive strength and the ultimate axial strain of the confined concrete need to be known. Lam and Teng (2001b) proposed Equation 6.9 for the former, and as a result, the slope of the linear second portion of the stress–strain curve E_2 is simplified to

$$E_2 = \frac{2f_l}{\varepsilon_{cc}} \tag{6.25}$$

For the ultimate axial strain, Lam and Teng (2001b) proposed Equation 6.19 for normal CFRP-wrapped concrete and Equation 6.20 for GFRP tube-confined concrete respectively, based on the observed dependency of ultimate axial strain on type of FRP as discussed earlier. The two ultimate strain equations also make use of FRP tensile strengths from different tests: flat-coupon tests for Equation 6.19 and ring-splitting tests for Equation 6.20 respectively. These requirements are a direct consequence of Lam and Teng's (2001b) strain database (Table 6.7). Table 6.9 shows that Equations 6.19 and 6.20 provide close predictions of test results.

Lam and Teng's (2001b) model is the first stress–strain model for FRP-confined concrete which takes into account the effect of FRP type. Although Lam and Teng (2001b) were able to propose expressions for ultimate axial strains only for two types of FRP-confined concrete, owing to the limited test results available, the stress–strain model is generally applicable to other types of FRP once test data allow the establishment of a suitable ultimate axial strain expression. Future fine-tuning of expressions for the compressive strength and the ultimate axial strain can be easily accommodated by the model. The only limitation to the model is that the stress–strain curve has an ascending second portion as may be ensured by a sufficient amount of FRP. Figure 6.25 shows comparisons between Lam and Teng's (2001b) stress–strain model and the test curves from Xiao (2001). The close agreement between the model and the test results is evident.

Table 6.9 Comparison of Equations 6.19 and 6.20 with test results

Equation	Experimental data			Test-to-predicted ε_{cc} ratio		
	Data set	Type of specimen	No. of data points	Average	Standard deviation	Coefficient of variation (%)
6.19	1	CFRP wrapped	39	1.01	0.31	31
6.20	2	GFRP tubes	41	1.03	0.18	17

6.5.6.2 Stress–Strain Model for Design Use

Lam and Teng (2001b) appear to be the first to have discussed the need for a stress–strain model for design use which provides a conservative prediction of the ultimate strain. On the other hand, they suggested that their strength model (Equation 6.9) can be directly used in design as it merely relates the confined strength to the unconfined strength of concrete, and as such does not

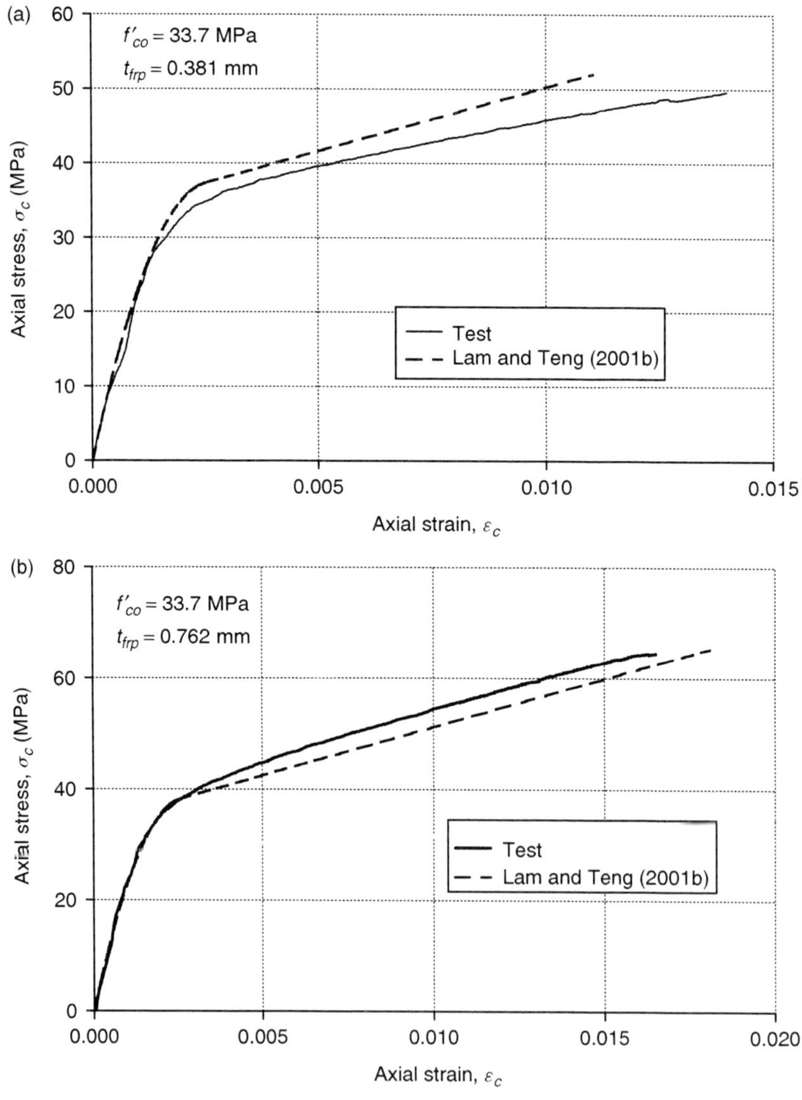

Figure 6.25 Lam and Teng's stress–strain model versus test results

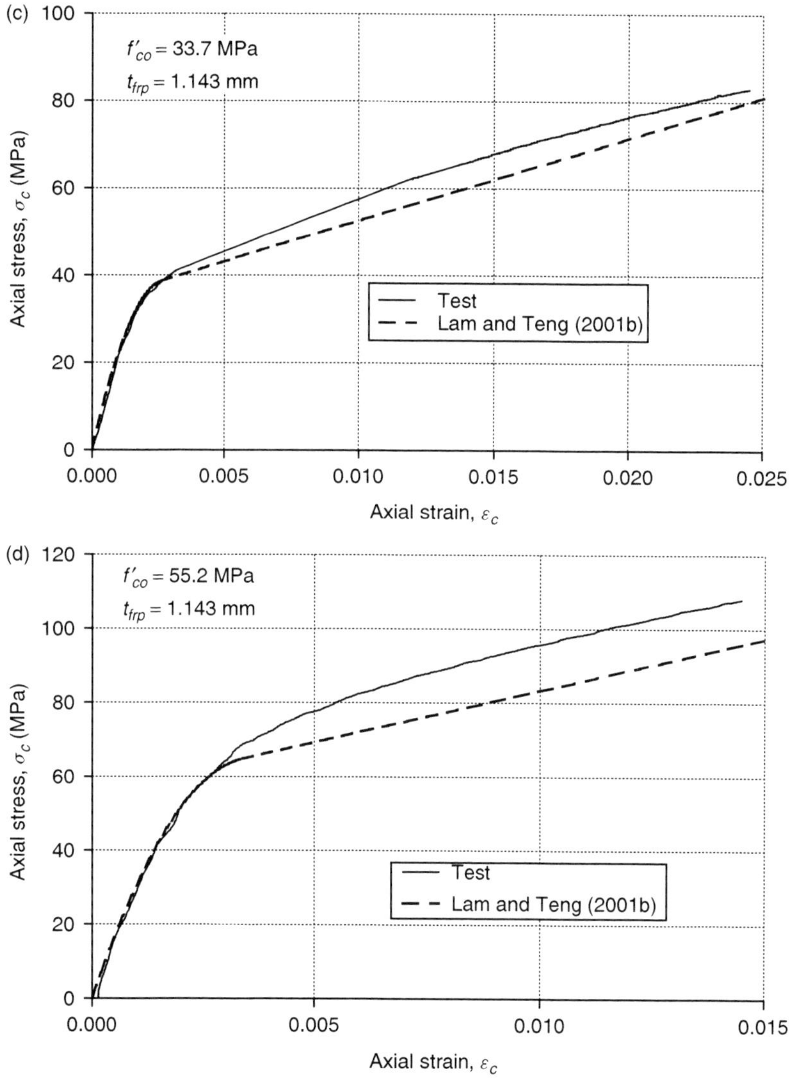

Figure 6.25 (*continued*)

erode the safety margin embedded in unconfined concrete strength in established design approaches. As shown in Section 6.9, the strength predictions of columns subject to either axial or eccentric loads are in fact not sensitive to ultimate strain values, but the ductility level of a column is likely to be over-predicted using a best-fit stress–strain model. As ductility demands are important for moment redistribution in indeterminate structures and in seismic design, such over-predictions must be prevented by a lower bound value for the ultimate strain. Therefore, a lower bound stress–strain model should be adopted in design against both static and seismic loads.

Lam and Teng (2001b) recommended that for design use, the ultimate axial strain of confined concrete in their stress–strain model be given by the following lower bound formulae:

$$\frac{\varepsilon_{cc}}{\varepsilon_{co}} = 1.75 + 10\left(\frac{f_l}{f'_{co}}\right) \quad \text{for CFRP-wrapped concrete} \tag{6.26}$$

$$\frac{\varepsilon_{cc}}{\varepsilon_{co}} = 1.75 + 22\left(\frac{f_l}{f'_{co}}\right) \quad \text{for concrete-filled GFRP tubes} \tag{6.27}$$

These two lower bound equations are seen to provide conservative predictions in most cases (Figure 6.20). An important advantage of this lower bound stress–strain model is that it reduces to the idealised stress–strain curve for unconfined concrete adopted by Eurocode 2 (ENV 1992-1-1 1991) for design use (Figure 6.24), making it particularly attractive for application in design.

6.6 FRP-confined Rectangular Concrete Columns

6.6.1 General

In Sections 6.4 and 6.5, the compressive strength and stress–strain models for FRP-confined concrete were thoroughly discussed. These models are based on test results on FRP-confined circular specimens and are thus directly applicable in predicting the strength gains of concrete in circular RC columns. They are, however, not directly applicable to rectangular RC columns in which the concrete is not uniformly confined by an FRP jacket. As a result, the effectiveness of confinement is much reduced (Mirmiran *et al.* 1998a). Owing to the nonuniformity of confinement, the compressive strength of confined concrete varies over the section. The commonly accepted approach to deal with confined concrete in noncircular sections is to find an equivalent model in terms of the average axial stress (equals load divided by cross-sectional area). As this is the common approach, the word 'average' is generally omitted though implicitly implied. In this section, such models for the axial compressive strength of confined concrete in rectangular columns based on tests on square and rectangular specimens are discussed. Two earlier models are discussed first, followed by a detailed discussion of the recent model of Lam and Teng (2001c) which is recommended for design use in this chapter.

6.6.2 Existing Models

6.6.2.1 General Equations

A rectangular column with rounded corners is shown in Figure 6.26, where the width is denoted by b and is assumed to be smaller than the depth h in this

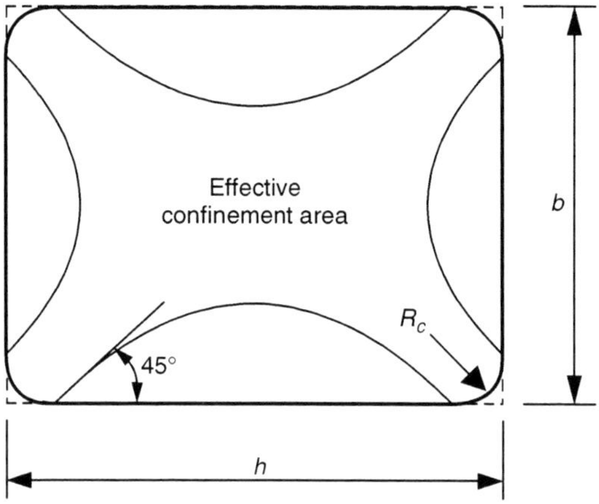

Figure 6.26 Effectively confined concrete in a rectangular column

chapter for convenience. The aspect ratio h/b defines the shape of a rectangular section. The corner radius is denoted by R_c. As corner rounding is limited by the presence of internal steel reinforcement, the value of R_c is generally small.

Studies on steel-confined concrete (Mander *et al.* 1988, Cusson and Paultre 1995) have led to the simple proposition that for rectangular columns, the concrete contained by four second-degree parabolas as shown in Figure 6.26 is fully confined whilst the confinement to the rest is negligible. These parabolas intersect the edges at 45°. This proposition has been applied to FRP-confined rectangular concrete columns (e.g. Restrepol and De Vino 1996, Harries *et al.* 1998, Chaallal and Shahawy 2000). This proposition reflects the fact that FRP confinement is less effective for a rectangular section than for a circular section in which all concrete is confined to the same degree.

In terms of the entire section, an effective confining pressure f_l' can be defined as a function of the shape through the use of a shape factor k_s as

$$f_l' = k_s f_l \tag{6.28}$$

where f_l is the confining pressure provided by an FRP jacket of the same thickness to an equivalent circular column of diameter D and can thus be evaluated by

$$f_l = \frac{2 f_{frp} t_{frp}}{D} \tag{6.29}$$

In order to determine the effective confining pressure using Equations 6.28 and 6.29, both the shape factor and the diameter of the equivalent circular column need to be defined. Once these two parameters are defined, the strength of FRP-confined concrete in rectangular columns can be predicted using the compressive strength models for FRP-confined concrete developed on the basis of circular specimens (e.g. Lam and Teng 2001a) by replacing the confining pressure with the effective confining pressure. That is,

$$\frac{f'_{cc}}{f'_{co}} = 1 + k_1 k_s \frac{f_l}{f'_{co}} \tag{6.30}$$

where k_1 is the confinement effectiveness coefficient obtained from tests on FRP-confined circular specimens and is equal to two according to Lam and Teng (2001a).

6.6.2.2 Shape Factor

The proposition represented in Figure 6.26, which was originally established for steel-confined columns, has also been used by Restrepol and De Vino (1996), Harries *et al.* (1998) and Chaallal and Shahawy (2000) to derive the shape factor for FRP-confined rectangular columns with rounded corners. The shape factor is given by the ratio of the effective confinement area A_e to the total area of concrete enclosed by the FRP jacket, A_c, as follows:

$$k_s = \frac{A_e}{A_c} = \frac{1 - [(b - 2R_c)^2 + (h - 2R_c)^2]/(3A_g) - \rho_{sc}}{1 - \rho_{sc}} \tag{6.31}$$

where A_g is the gross area of column section with rounded corners, and ρ_{sc} is the cross-sectional area ratio of longitudinal steel. Equation 6.31 is referred to herein as the effective confinement area model for the shape factor. In Equation 6.31, the cross-sectional area of longitudinal steel outside the parabolas is deducted from the gross area twice, but the effect of this simplification is small and conservative. The gross area of the column section can be evaluated by

$$A_g = bh - (4 - \pi)R_c^2 \tag{6.32}$$

On the other hand, Mirmiran *et al.* (1998a) proposed a modified confinement ratio m for rectangular columns to replace the confinement ratio f_l/f'_{co} for circular columns to account for the shape effect. The modified confinement ratio can be written as

$$m = \left(\frac{2R_c}{h}\right)\frac{f_l}{f'_{co}} \tag{6.33}$$

where f_l is the confining pressure to an equivalent circular column. Using test data available to them, Mirmiran *et al.* (1998a) showed that for values of $m <$ 15 %, FRP confinement was ineffective in strengthening the concrete column, and no strength enhancement could be expected. They suggested that for values of $m >$ 15 %, the confinement model of Samaan *et al.* (1998) could be used. Their proposal is equivalent to defining the shape factor as follows:

$$k_s = \frac{2R_c}{h} \tag{6.34}$$

Equation 6.34 is herein referred to as the corner radius model for the shape factor.

6.6.2.3 Diameter of Equivalent Circular Column

In the studies of Restrepol and De Vino (1996), Harries *et al.* (1998) and Chaallal and Shahawy (2000), the diameter of the equivalent circular column for finding the nominal confining pressure f_l of FRP-confined rectangular columns has been left undefined, although according to the study of Cusson and Paultre (1995) for concrete columns confined by transverse steel reinforcement, D should be the average side length $(b+h)/2$. This definition was thus used by Lam and Teng (2001c) with the effective confinement area model.

Mirmiran *et al.* (1998a) were concerned with only square columns and defined D as the section side length, but it has been confirmed through private communication that D should be the longer side length in the case of a rectangular section (Mirmiran 2000). The use of the section depth h in Equation 6.34 is based on the same information.

6.6.3 Experimental Data

Only a limited amount of test data exists on FRP-confined square or rectangular plain concrete specimens subject to compression (Demers and Neale 1994, Picher *et al.* 1996, Demers *et al.* 1996, Harries *et al.* 1998, Mirmiran *et al.* 1998a, Rochette and Labossiere 2000). Lam and Teng (2001c) noted that tests available before their study covered square and rectangular concrete specimens with the section depth ranging from 152 mm to 203 mm, and the corner radius from 5 mm to 38 mm. The unconfined concrete strength of these specimens ranged from 32.2 MPa to 43.7 MPa. FRP confinement was provided by either wrapped FRP sheets or FRP tubes.

Lam and Teng (2001c) used these data to assess the validity of both the effective confinement area and the corner radius models. The data presented in the papers of Picher *et al.* (1996) and Demers *et al.* (1996) were not used because they were suspected to have been included in the database presented in a subsequent paper from the same group (Rochette and Labossiere 2000), although Lam and Teng (2001c) were not fully confident about this point owing to some small differences in the material properties given in these papers for apparently the same test specimens. In addition, the four test results presented by Harries *et al.* (1998) were also left out because the corner radii of the specimens were not reported. Thus, a total of 39 usable test results from the published literature were available to Lam and Teng.

These 39 test results were plotted by Lam and Teng (2001c) against the modified confinement ratio (= shape factor × confinement ratio of equivalent circular column) for the two different definitions of shape factor (Figure 6.27). In both cases, a roughly linear relationship can be observed, but the correlation is poor. Nevertheless, the effective confinement area model results in much less scatter than the corner radius model in the confinement effectiveness coefficient k_1 deduced using Equation 6.30 for small values of f_l' (Figure 6.28). The k_1 range

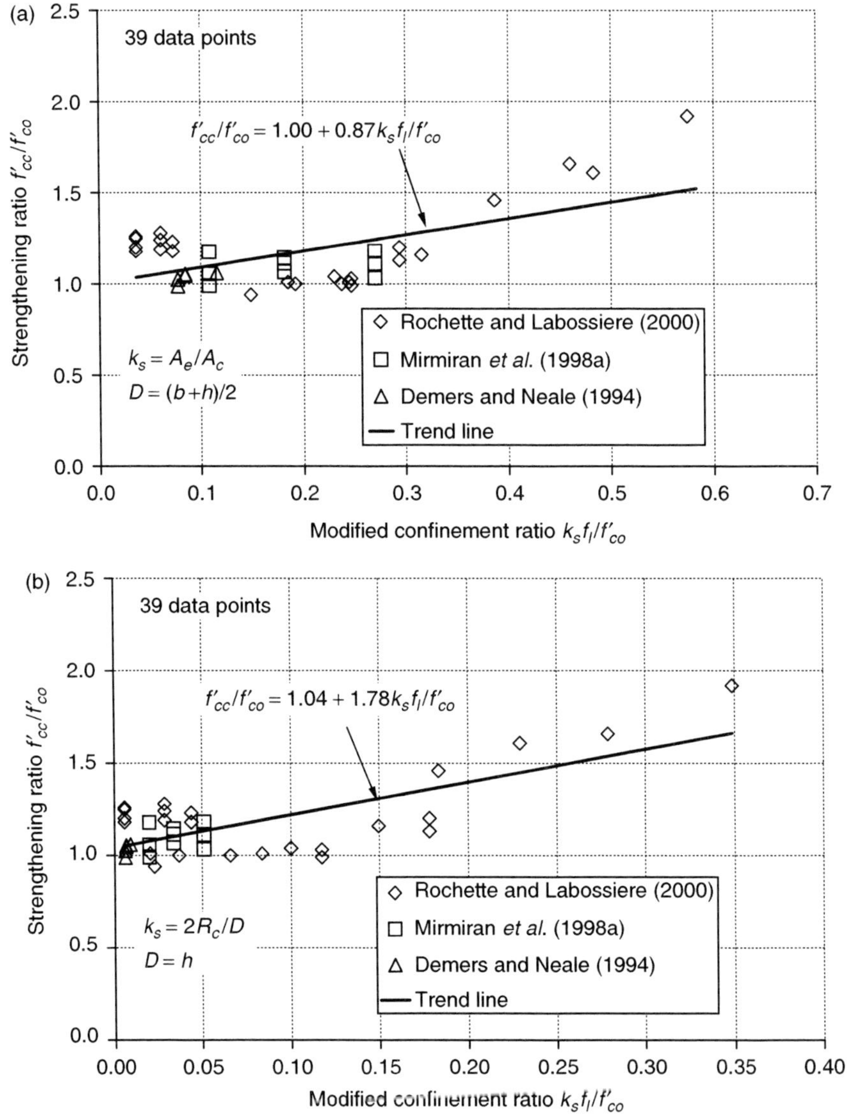

Figure 6.27 Strengthening ratio versus two different modified confinement ratios for FRP-confined rectangular concrete columns: (a) effectively confined area model; (b) corner radius model

is from −0.4 to 7.0 for the former (Figure 6.28a) but from −2.6 to 45 for the latter (Figure 6.28b). As most of these tests were conducted on specimens with a small amount of FRP and insignificant strength gains, these comparisons are inconclusive.

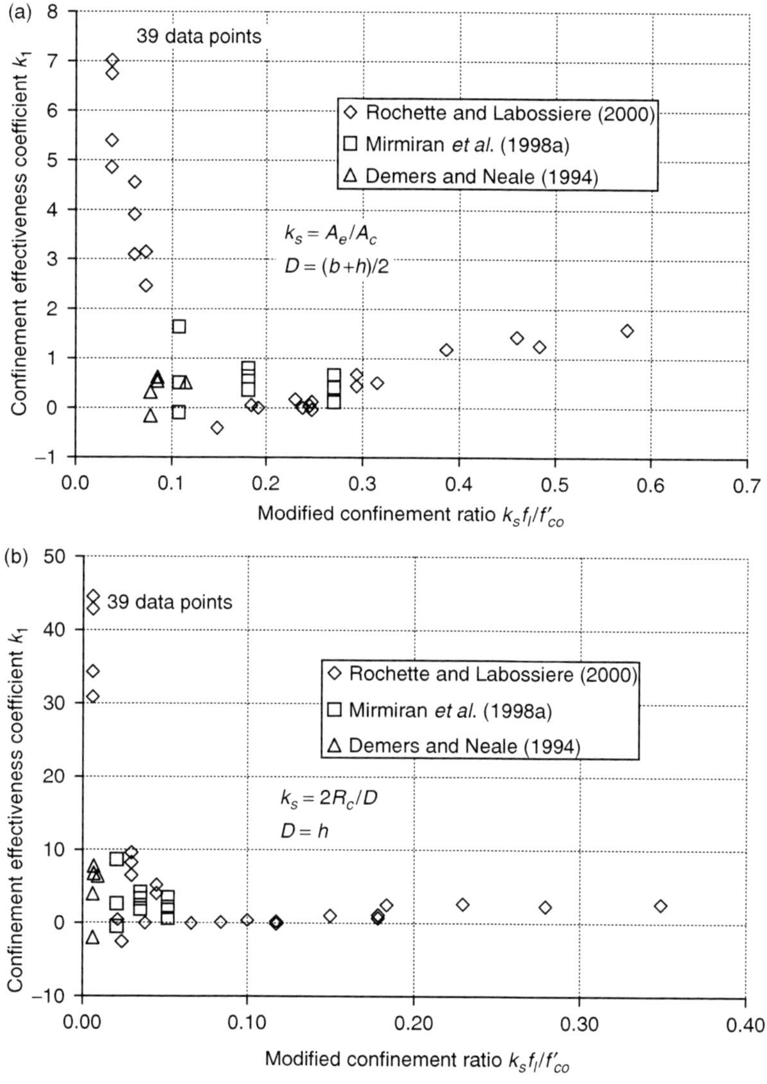

Figure 6.28 Confinement effectiveness coefficient versus two different modified confinement ratios for FRP-confined rectangular concrete columns: (a) effectively confined area model; (b) corner radius model

6.6.4 Lam and Teng's Model

Lam and Teng (2001c) carried out further tests on FRP-confined square and rectangular concrete specimens. The test programme included 10 square specimens and 2 rectangular specimens. The amount of FRP and corner radius were varied to cover various degrees of confinement. Two circular specimens were also tested

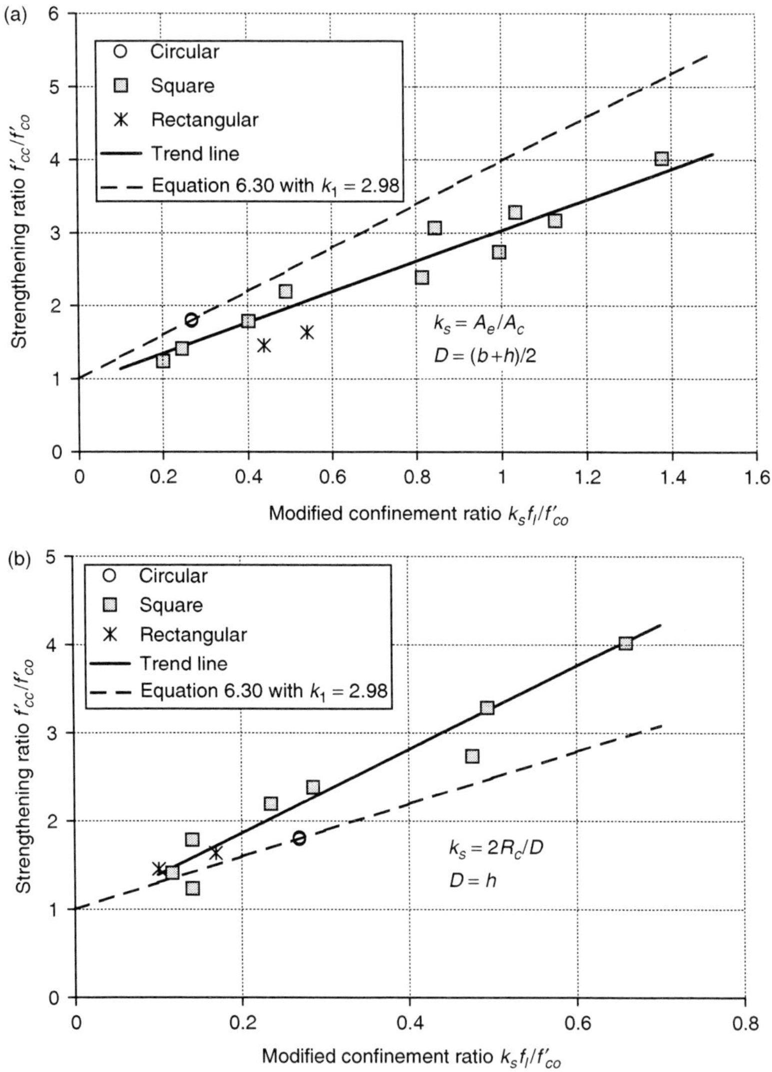

Figure 6.29 Performance of shape factor models for FRP-confined rectangular concrete columns as assessed by Lam and Teng's test data: (a) effectively confined area model; (b) corner radius model

which led to a value of 2.98 for the FRP confinement effectiveness coefficient k_1 for use with Equation 6.30 in predicting the test results of square and rectangular specimens. Figure 6.29 shows the test data against the predictions of the two shape factor models. The effective confinement area model is seen to be unconservative while the corner radius model is conservative and shows a larger scatter.

To achieve a better prediction of the compressive strength of FRP-confined concrete, Lam and Teng (2001c) proposed an improved effective confinement area model for the shape factor. In this model, the effective confinement area is contained by four parabolas as illustrated by Figure 6.30, with the initial slopes of the parabolas being the same of the adjacent diagonal lines. The initial slopes become one only for square sections. The effective confinement area ratio A_e/A_c is therefore given by

$$\frac{A_e}{A_c} = \frac{1 - [(b/h)(h - 2R_c)^2 + (h/b)(b - 2R_c)^2]/(3A_g) - \rho_{sc}}{1 - \rho_{sc}} \tag{6.35}$$

The shape factor k_s is given by the effective confinement area ratio divided by the aspect ratio, that is

$$k_s = \frac{b}{h}\frac{A_e}{A_c} \tag{6.36}$$

The diameter of the equivalent circular column D in Equation 6.29 is defined by Lam and Teng (2001c) as the diagonal distance of the section. This means that the equivalent circular section circumscribes the rectangular section. The equivalent confining pressure f_l is given by

$$f_l = \frac{2 f_{frp} t_{frp}}{\sqrt{h^2 + b^2}} \tag{6.37}$$

where $\sqrt{h^2 + b^2}$ is the diagonal length of the rectangular section.

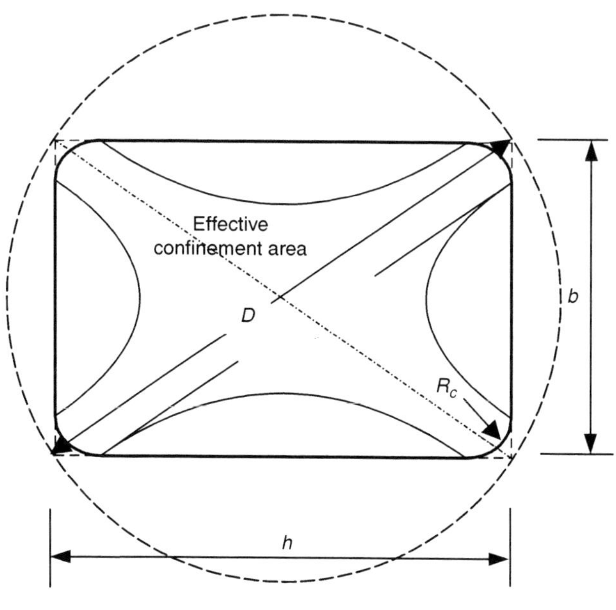

Figure 6.30 Lam and Teng's shape factor model for rectangular concrete columns

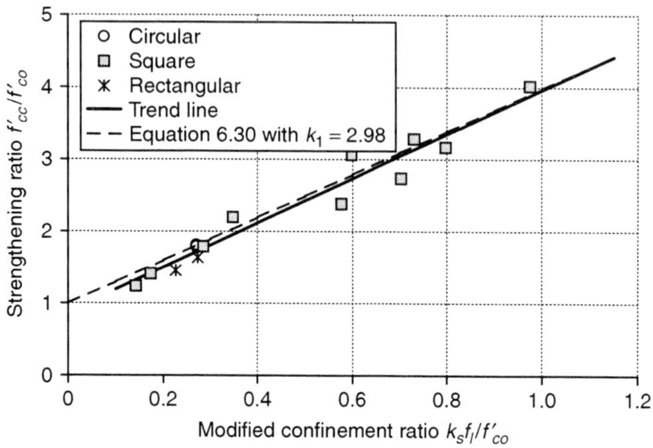

Figure 6.31 Performance of Lam and Teng's shape factor model for rectangular concrete columns as assessed by their own data

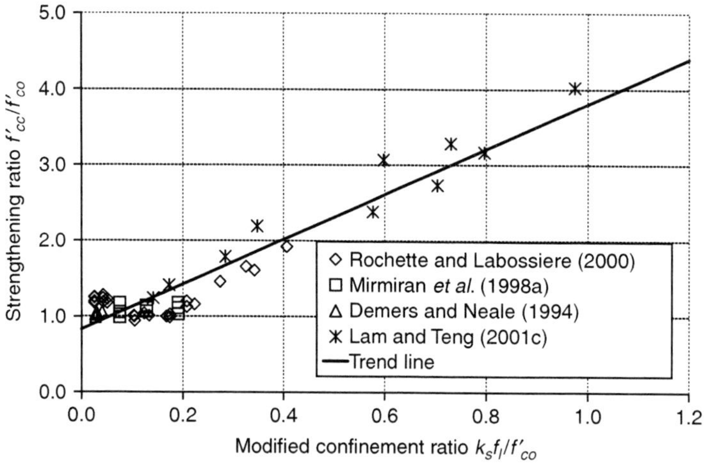

Figure 6.32 Performance of Lam and Teng's shape factor model for FRP-confined rectangular concrete columns as assessed by all test data

Figure 6.31 shows that predictions based on Lam and Teng's (2001c) model almost coincide with the best-fit line of the test data, indicating that this model is an accurate representation of the shape factor. Figure 6.32 shows that when all the available data are plotted using Lam and Teng's (2001c) model, a linear trend can be observed between the strengthening ratio f'_{cc}/f'_{co} and the modified confinement ratio $k_s f_l/f'_{co}$, indicating that the shape effect is appropriately represented, although there is some uncertainty at low degrees of confinement.

6.7 FRP-confined Elliptical Columns

As discussed in Section 6.2, rectangular columns can be transformed into elliptical columns for more effective FRP confinement. For this method to be widely used, a reliable strength model for FRP-confined elliptical columns is required. Teng and Lam (2001) presented the first study on FRP-confined elliptical concrete columns. Tests were carried out that covered both circular and elliptical specimens, with the aspect ratio (the ratio of the major to the minor axis length) a/b (Figure 6.9) ranging from 1 to 2.5, with or without CFRP wrapping. The test results showed that the confinement effectiveness decreases as the aspect ratio increases. Based on these test results, Teng and Lam (2001) proposed that the compressive strength of FRP-confined concrete in an elliptical column be predicted using Equation 6.30. In their model, the circular column for finding the equivalent confining pressure f_l is one with the same FRP volumetric ratio. The area of an ellipse is πab, and its circumference is closely approximated by $\pi[1.5(a+b)-\sqrt{ab}]$. Therefore, the FRP volumetric ratio for an elliptical column confined by an FRP jacket with a thickness of t_{frp} is given by

$$\rho_{frp} = \frac{t_{frp}[1.5(a+b)-\sqrt{ab}]}{ab} \tag{6.38}$$

The equivalent confining pressure is thus given by

$$f_l = \frac{\rho_{frp} f_{frp}}{2} = \frac{t_{frp}[1.5(a+b)-\sqrt{ab}]f_{frp}}{2ab} \tag{6.39}$$

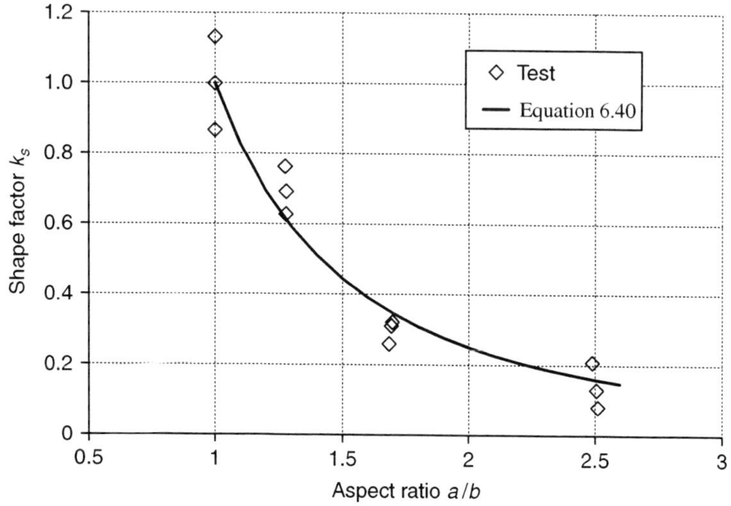

Figure 6.33 Comparison of Teng and Lam's shape factor model for FRP-confined elliptical columns with test data

The shape factor k_s is given by the following simple equation:

$$k_s = \left(\frac{b}{a}\right)^2 \tag{6.40}$$

which provides a close approximation of the test data (Figure 6.33).

6.8 Design Recommendation for Axially Loaded Columns

6.8.1 General Expression for the Axial Strength of FRP-confined Concrete

Based on the discussions in the preceding sections, the axial compressive strength of FRP-confined concrete can be related to that of unconfined concrete through

$$f'_{cc} = f'_{co} + k_1 k_g k_s f_l \tag{6.41}$$

where the coefficients k_1, k_g and k_s recommended for design use are detailed in Table 6.10. The gap factor k_g introduced to account for FRP confinement with significant gaps is only available for circular columns (Equation 6.8) but not for rectangular or elliptical columns, although such expressions are available for steel-confined rectangular columns (Mander *et al.* 1988, Cusson and Paultre 1995). For circular columns, Equation 6.8 shows that for a gap distance smaller than 5 % of the column diameter, the effect of gap on the lateral confining pressure is less than 5 %. In such cases, the gap effect is insignificant and can be ignored.

6.8.2 Ultimate Strength of FRP-confined RC Columns under Axial Load

According to BS 8110 (1997), the ultimate axial strength of short RC columns for design use is evaluated using

Table 6.10 k_1, k_g, k_s and f_l for circular and elliptical columns for design use

Column shape	Confinement effectiveness coefficient k_1	Gap factor k_g	Shape factor k_s	Lateral confining pressure f_l
Circular	2	$\frac{(1 - s_{frp}/(2d))^2}{1 - \rho_{sc}}$	1	$\frac{2 f_{frp} t_{frp}}{d}$
Rectangular	2	–	$\frac{b}{h}\frac{A_e}{A_c}$	$\frac{2 f_{frp} t_{frp}}{\sqrt{h^2 + b^2}}$
Elliptical	2	–	$\left(\frac{b}{a}\right)^2$	$f_l = \frac{t_{frp}\lfloor 1.5(a+b) - \sqrt{ab}\rfloor f_{frp}}{2ab}$

$$N_u = \frac{0.67}{\gamma_c} f_{cu} A_c + \frac{f_y}{\gamma_s} A_{sc} \tag{6.42}$$

where γ_c and γ_s are partial safety factors for concrete and steel respectively; A_c and A_{sc} are cross-sectional areas of concrete and longitudinal steel respectively; f_{cu} is the characteristic cube compressive strength of concrete determined using 150 mm cubes; and f_y is the characteristic yield strength of longitudinal steel. The term $0.67 f_{cu}$ represents the axial compressive strength of the unconfined concrete in a real RC column in terms of the cube strength f_{cu}. Therefore, based on Equation 6.41, if the column is confined using FRP composites, the axial compressive strength of the concrete in a column becomes

$$f'_{cc} = 0.67 f_{cu} + k_1 k_g k_s f_l \tag{6.43}$$

The design strength of an RC column according to BS 8110 (1997) thus becomes

$$N_u = \left(\frac{0.67}{\gamma_c} f_{cu} + k_1 k_g k_s \frac{f_l}{\gamma_{frp}} \right) A_c + \frac{f_y}{\gamma_s} A_{sc} \tag{6.44}$$

where γ_{frp} is the partial safety factor for the FRP, as specified in Section 1.6.

6.9 Design Recommendation for Eccentrically Loaded Columns

6.9.1 General

In practice, most columns are eccentrically loaded with their sections subjected to combined bending and axial compression. FRP confinement is effective not only for axially loaded columns but also for eccentrically loaded columns in enhancing their load-carrying capacity. As with axially loaded columns, only eccentrically loaded RC columns confined by an FRP jacket with the fibres or the main fibres in the hoop direction are dealt with here. The longitudinal strength of these FRP jackets is small and can be ignored. For such FRP-confined columns under eccentric loads, the analysis procedure is similar to that for conventional columns as may be found in numerous RC textbooks (e.g. Park and Paulay 1975, Kong and Evans 1987). The only difference from the analysis procedure FRP confinement introduces, over conventional columns, is the use of a suitable concrete stress–strain relationship that considers the confinement effect of the FRP.

6.9.2 Existing Studies on FRP-confined RC Columns under Eccentric Loads

Several studies have considered the theoretical behaviour of columns subjected to combined bending and axial compression and have produced theoretical bending

Table 6.11 Existing theoretical studies on FRP-confined RC columns under eccentric loads

Reference	Column section	Stress–strain model	Longitudinal FRP strength
Fardis and Khalili (1982)	Circular	Fardis and Khalili (1982)	Yes
Saadatmanesh *et al.* (1994)	Circular	Mander *et al.* (1988)	No
Nanni and Norris (1995)	Circular, rectangular	Modified form of Fardis and Khalili (1981)	No*
Mirmiran and Shahawy (1996)	Circular	Modified form of Mander *et al.* (1988)	Yes
Mirmiran *et al.* (1999)	Square	Samaan *et al.* (1998)	Yes
Chaallal and Shahawy (2000)	Rectangular	Samaan *et al.* (1998)	Yes

*Some of the specimens had FRP with a small longitudinal strength

moment–axial load interaction diagrams ($M-N$ diagrams). All studies employed stress–strain curves for confined concrete based on tests of axially loaded FRP or steel-confined circular concrete specimens. This direct use of stress–strain curves for uniformly confined concrete to eccentrically loaded columns implies the assumption of a nonuniform confining pressure in an eccentrically loaded column which varies from zero at the neutral axis position to a maximum value at the extreme compression fibre of concrete. The existing studies are summarised in Table 6.11. It may be noted that some of these studies (Mirmiran and Shahawy, 1996; Mirmiran *et al.*, 1999) were carried out in the context of concrete-filled FRP tubes as new columns.

A small number of experimental studies also exist, but in all of them the FRP had longitudinal strength. These studies include Mirmiran *et al.* (1998b) and Mirmiran *et al.* (1999) on tests of square FRP tubes filled with concrete, Chaallal and Shahawy (2000) on tests of rectangular RC columns wrapped with FRP and Mirmiran *et al.* (2000) on tests of circular FRP tubes filled with concrete. As may be noted, all these studies were carried out by the same research group. It appears that no experimental study has investigated the monotonic behaviour of eccentrically loaded columns wrapped with FRP possessing strength in only the transverse direction.

6.9.3 Ultimate Strength of FRP-confined RC Columns under Eccentric Loads

Only circular sections are considered here, but the general procedure is applicable to other columns. The restriction to circular columns is due mainly to the fact that the stress–strain curve of FRP-confined concrete for noncircular sections has not been well defined.

For the analysis of a circular section subject to combined bending and axial compression as shown in Figure 6.34, complete composite action between concrete and FRP is assumed and the tensile strength of concrete is ignored. Plane

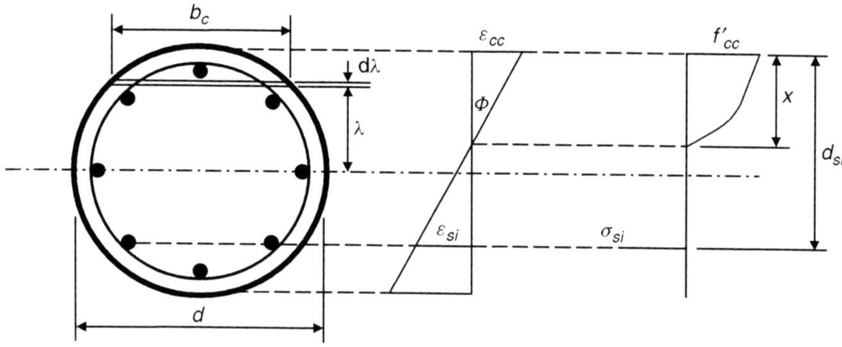

Figure 6.34 Strains and stresses over column depth at ultimate limit state

sections are assumed to remain plane. These are the same assumptions as adopted for beams in Section 3.4. The axial load N and the bending moment M at any stage of loading carried by the section with the reference axis going through the centre of the section are found by integrating the stresses over the section:

$$N = \int_{\lambda=d/2-x}^{d/2} \sigma_c b_c \, d\lambda + \sum_{i=1}^{n} (\sigma_{si} - \sigma_c) A_{si} \tag{6.45}$$

$$M = \int_{\lambda=d/2-x}^{d/2} \sigma_c b_c \lambda \, d\lambda + \sum_{i=1}^{n} (\sigma_{si} - \sigma_c) A_{si} \left(\frac{d}{2} - d_{si} \right) \tag{6.46}$$

where b_c is the width of the section at a distance λ from the reference axis, σ_{si} is the stress in the ith layer of longitudinal steel reinforcement that can be calculated from Equations 3.10 or 3.11 of Section 3.4, and A_{si} is the corresponding cross-sectional area of the longitudinal steel reinforcement. Compressive stresses are taken to be positive as in Section 3.4. The stress of concrete, σ_c, in the compression zone is recommended to be given by Lam and Teng's (2001b) stress–strain model for confined concrete (Equation 6.22) with an appropriate partial safety factor and is equal to zero in the tension zone.

Lam and Teng's (2001b) stress–strain model for design use, in terms of the concrete cube strength, is as follows:

$$\sigma_c = E_c \varepsilon_c - \frac{(E_c - E_2)^2}{4(0.67 f_{cu}/\gamma_c)} \varepsilon_c^2 \quad \text{for } 0 \leqslant \varepsilon_c \leqslant \varepsilon_t \tag{6.47a}$$

$$\sigma_c = 0.67 f_{cu}/\gamma_c + E_2 \varepsilon_c \quad \text{for } \varepsilon_t \leqslant \varepsilon_c \leqslant \varepsilon_{cc} \tag{6.47b}$$

where

$$\varepsilon_{cc} = 1.75 \varepsilon_{co} + 10 \frac{f_l}{0.67 f_{cu}} \varepsilon_{co} \tag{6.48}$$

$$E_c = 1000\frac{0.67f_{cu}}{\gamma_c}, \quad E_2 = \frac{2f_l}{\gamma_{frp}\varepsilon_{cc}} \tag{6.49}$$

$$\varepsilon_t = \frac{2(0.67f_{cu})}{\gamma_c(E_c - E_2)}, \quad f_l = \frac{2f_{frp}t_{frp}}{d}, \quad \varepsilon_{co} = 0.002 \tag{6.50}$$

In the above, the term $0.67f_{cu}$, representing the axial compressive strength of the unconfined concrete in a real RC column, should be replaced by $0.85f_c'$ if the concrete strength is defined by its cylinder compressive strength. The partial safety factors are as specified in Section 1.6. When no FRP confinement exists, this stress–strain curve reduces to the idealised stress–strain curve of unconfined concrete for design use recommended by Eurocode 2 (ENV 1992-1-1 1991) which is as follows in terms of the cube compressive strength:

$$\sigma_c = 1000\varepsilon_c(1 - 250\varepsilon_c)0.67f_{cu}/\gamma_c \quad \text{for } 0 \leqslant \varepsilon_c \leqslant 0.002 \tag{6.51a}$$

$$\sigma_c = 0.67f_{cu}/\gamma_c \quad \text{for } 0.002 \leqslant \varepsilon_c \leqslant 0.0035 \tag{6.51b}$$

Equations 6.45 and 6.46 are applicable at any stage of loading. The ultimate limit state of the column is reached when the strain at the extreme concrete compression fibre reaches the ultimate strain of FRP-confined concrete, signifying crushing of concrete due to FRP rupture (Figure 6.34). This ultimate strain is defined by Equations 6.26 and 6.27. At the ultimate limit state, the axial load, the moment and the neutral axis depth in Equations 6.45 and 6.46 may be taken as the three basic parameters, and once one of them is known or M_u and N_u are related by a specified load eccentricity, the values of the two remaining parameters can be easily determined. Numerical integration over the section can be carried out by the layer method in which the column section is divided into many small horizontal layers. For example, the axial load–bending moment interaction curves can be easily generated by finding the values of M_u and N_u corresponding to different neutral axis depths.

6.9.4 Moment–Curvature Analysis of FRP-confined RC Columns

A moment–curvature analysis can also be carried out easily. For a given axial load, moment–curvature plots can be generated by specifying a sequence of suitable strain values for the extreme compression fibre of concrete, $\varepsilon_{c,max}$, up to its ultimate value, ε_{cc}. For each strain value, the neutral axis depth is varied until the resultant axial force acting on the section, calculated from Equation 6.45, equals the applied axial load. Once the neutral axis position has been determined, the moment can be evaluated using Equation 6.46 and the curvature can be calculated by the following expression:

$$\Phi = \frac{\varepsilon_{c,max}}{x} \tag{6.52}$$

6.9.5 Typical Interaction Curves

Figure 6.35 shows typical interaction curves for a circular section with the following geometric and material properties: diameter $d = 1500$ mm, concrete cover to the longitudinal bars = 50 mm, concrete cube strength $f_{cu} = 40$ MPa, bar diameter = 40 mm, steel yield stress = 460 MPa. Altogether 32 bars are distributed evenly around the circumference. The column is wrapped by either 5 or 10 layers of CFRP formed from Tonen Forca Towsheets which are carbon fibre sheets produced in Japan. The ultimate tensile strength specified by the manufacturer (35 500 kg/cm^2 = 3483 MPa) based on a sheet thickness of 0.165 mm was adopted in the calculations. All partial safety factors were set to unity and the confinement effect of any transverse steel reinforcement was ignored.

Strengthening with 5 and 10 layers of CFRP provides a confinement ratio (f_l/f'_{co}) of 0.14 and 0.29 respectively, whilst the corresponding strengthening ratios (f'_{cc}/f'_{co}) are 1.29 and 1.57. These interaction curves shown in Figure 6.35 demonstrate that the maximum benefit of FRP confinement occurs when the section fails in pure compression for which the axial load capacity of the unstrengthened column $N_{uo} = 64\,780$ kN, but confinement is much less beneficial when the column fails in pure bending for which the moment capacity of the unstrengthened column $M_{uo} = 10\,277$ kN m. The effect of confinement increases with the level of axial load and the amount of FRP. Wrapping with 5 layers of CFRP leads to a 20 % increase in the axial load capacity but only 3.8 % increase in the bending moment capacity. The corresponding increases for 10 layers of CFRP are 41 % and 5.5 % respectively. This lack of increase in the

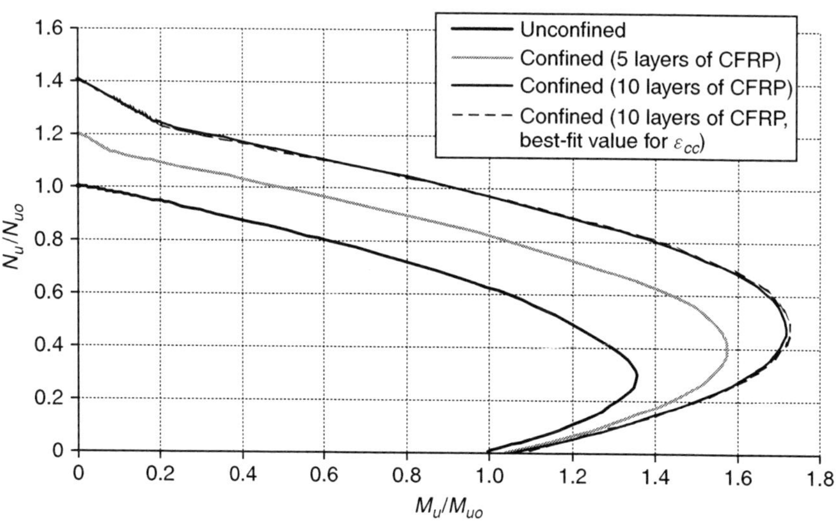

Figure 6.35 Bending moment–axial load interaction curves for a circular RC column

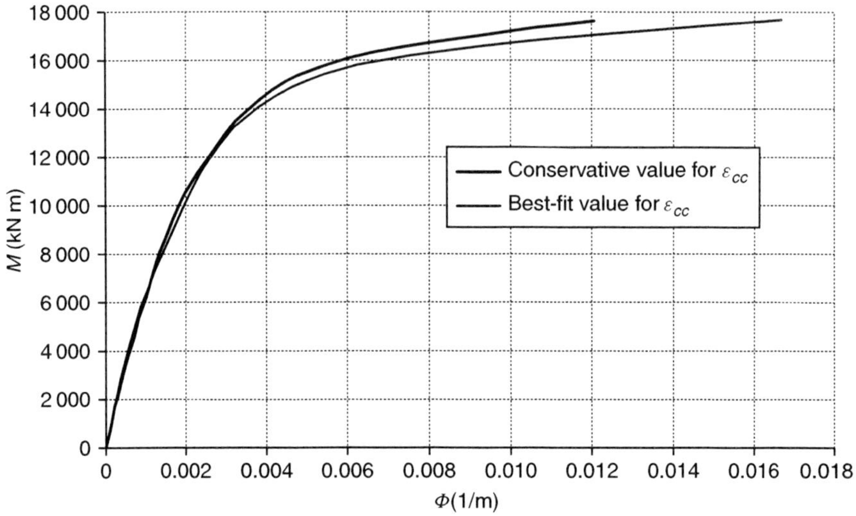

Figure 6.36 Moment–curvature curves for an FRP-confined circular RC column

moment capacity is an advantage in seismic retrofit as discussed in Chapter 7, as unwanted increase in flexural capacity will endanger adjacent structural elements and make brittle failure modes such as shear failure more likely.

The sharp slope change in the interaction curves at high axial loads for the wrapped columns (Figure 6.35) starts when the neutral axis begins to fall outside the column section. As the neutral axis depth is further increased, the parabolic portion of the confined concrete stress–strain curve gradually moves outside the column, with stresses over the section eventually governed entirely by the linear second portion of the stress–strain curve, at the beginning of the linear portion of the interaction curve at high axial loads.

For the case of wrapping with 10 layers of CFRP, another interaction curve was generated with the ultimate strain of concrete predicted by Lam and Teng's (2001b) best-fit model for CFRP-wrapped concrete. It can be seen from Figure 6.35 that this increase in the ultimate concrete strain has little effect on the load-carrying capacity, although it significantly affects the ultimate curvature value (Figure 6.36) which is important in seismic design. For seismic design, only the conservative model is appropriate.

6.10 Conclusions

This chapter has been concerned with the strengthening of RC columns under either axial or eccentric loads using FRP composites. Three methods for column strengthening have been identified: (a) wrapping; (b) filament winding; and

(c) prefabricated shell jacketing. All three methods provide external confinement to the concrete, thereby enhancing the axial compressive strength of the concrete. Failure of FRP-confined concrete is by tensile rupture of the FRP in the hoop direction. The stress–strain curve of FRP-confined concrete exhibits a bilinear shape.

The strength of FRP-confined concrete has been found to be lower than its steel-confined counterpart for the same amount of hoop confinement in terms of the supplied hoop force. As a result, a number of strength models have been developed specifically for FRP-confined concrete based on tests on circular specimens. These models have been compared with a test database compiled by Lam and Teng (2001a) to assess their performance. The recently proposed model of Lam and Teng (2001a) has been recommended for design use due to its simplicity and accuracy. In Lam and Teng's model, the strength of FRP-confined concrete is linearly related to the lateral confining pressure provided by the FRP.

Stress–strain curves of FRP-confined concrete exhibit a bilinear shape with an ascending second portion, provided the concrete is sufficiently confined. If, however, the effect of confinement is very small owing to a small amount of FRP and/or unfavourable sectional shapes, the stress–strain curve may feature a descending branch. Since in the latter case, only insignificant strength enhancement can be expected, only the former case has been examined in detail in this chapter. Available stress–strain models for FRP-confined concrete have been reviewed and assessed using a test database built by Lam and Teng (2001b). Most of these models give inaccurate predictions of the ultimate concrete strain and/or the shape of the stress–strain curve, and all of them do not account for the dependency of ultimate concrete strain on the type of FRP as shown by test data. The simple and accurate model recently developed by Lam and Teng (2001b), consisting of a parabolic first portion which meets smoothly with a linear second portion and differentiating between different types of FRP, has been recommended for design use, in which lower bound values of the ultimate concrete strain adopted are used for conservative predictions. This lower bound stress–strain curve reduces to the idealised stress–strain curve for unconfined concrete recommended for design use by Eurocode 2 (ENV 1992-1-1 1991), making it particularly attractive for practical application.

Strength and stress–strain models based on tests of circular specimens cannot be applied directly to columns of noncircular sections, as in the latter the confining pressure provided by the FRP is nonuniform. A number of studies have been carried out on FRP-confined rectangular columns, leading to several approaches for predicting the axial compressive strength. Lam and Teng's (2001c) model is the only one which provides accurate predictions of test results and has been recommended for design use. There has been only one study on the axial compressive strength of FRP-confined elliptical concrete columns (Teng and Lam 2001), which nevertheless led to an accurate strength model suitable for design use. The need to consider elliptical columns exists because rectangular columns can be transformed into elliptical ones for more effective FRP confinement.

Finally, design recommendations in the BS 8110 format have been advanced for predicting the ultimate strengths of FRP-confined RC columns subject to either axial or eccentric loads. In the design recommendation for axially loaded columns, a general expression for the axial compressive strength of FRP-confined concrete is presented, with the effects of column shapes and strap spacings in the case of partially confined concrete accounted for through the introduction of modification factors. The design recommendation for eccentrically loaded columns considers only circular sections as the stress–strain models for noncircular sections are not yet well defined, although the general procedure is applicable to noncircular sections. Based on this design recommendation, moment–axial load interaction curves and moment–curvature curves can be easily generated.

References

ACI 440R-96 (1996) *State-of-the-Art Report on Fiber Reinforced Plastic (FRP) Reinforcement for Concrete Structures*, American Concrete Institute (ACI) Committee 440, Farmington Hills, Michigan, USA.

Ahmad, S. H., Khaloo, A. R. and Irshaid, A. (1991) 'Behavior of concrete spirally confined by fibreglass filaments', *Magazine of Concrete Research*, Vol. 43, No. 156, pp. 143–148.

Ahmad, S. H. and Shah, S. P. (1982) 'Stress–strain curves of concrete confined by spiral reinforcement', *ACI Journal*, Vol. 79, No. 6, pp. 484–490.

ASTM D2290-92 (1992) *Standard Test Method for Apparent Tensile Strength of Ring or Tubular Plastics and Reinforced Plastics by Split Disk Method*, American Society for Testing and Materials (ASTM), Philadelphia, USA.

ASTM D3039/D3039M-95a (1995) *Standard Test Method for Tensile Properties of Polymer Matrix Composite Materials*, American Society for Testing and Materials (ASTM), Philadelphia, USA.

Ballinger, C., Maeda, T. and Hoshijima, T. (1993) 'Strengthening of reinforced concrete chimneys, columns and beams with carbon fiber reinforced plastics', *Proceedings of the International Symposium on Fiber-Reinforced-Plastic Reinforcements for Concrete Structures, ACI SP-138*, pp. 243–248.

BS 8110 (1997) *Structural Use of Concrete, Part 1. Code of Practice for Design and Construction*, British Standards Institution, London, UK.

Chaallal, O. and Shahawy, M. (2000) 'Performance of fibre-reinforced polymer-wrapped reinforced concrete column under combined axial-flexural loading', *ACI Structural Journal*, Vol. 97, No. 4, pp. 659–668.

Cusson, D. and Paultre, P. (1995) 'Stress–strain model for confined high-strength concrete', *Journal of Structural Engineering, ASCE*, Vol. 121, No. 3, pp. 468–477.

Demers, M. and Neale, K. W. (1994) 'Strengthening of concrete columns with unidirectional composite sheets', *Development in Short and Medium Span Bridge Engineering' 94*, Canadian Society for Civil Engineering, Montreal, Canada, pp. 895–905.

Demers, M. and Neale, K. W. (1999) 'Confinement of reinforced concrete columns with fibre-reinforced composite sheets–an experimental study', *Canadian Journal of Civil Engineering*, Vol. 26, pp. 226–241.

Demers, M., Hebert, D., Gauthier, M., Labossiere, P. and Neale, K. W. (1996) 'The strengthening of structural concrete with an aramid woven fibre/epoxy resin composite', *Proceedings of the Second International Conference on Advanced Composite Materials in Bridges and Structures*, Montreal, Canada, pp. 435–442.

ENV 1992-1-1 (1991) *Eurocode 2: Design of Concrete Structures–Part 1: General Rules and Rules for Buildings*, CEN, Brussels, Belgium.

Fardis, M. N. and Khalili, H. (1981) 'Concrete encased in fiberglass-reinforced plastic', *ACI Journal*, Vol. 78, pp. 440–445.

Fardis, M. N. and Khalili, H. (1982) 'FRP-encased concrete as a structural material', *Magazine of Concrete Research*, Vol. 34, No. 122, pp. 191–202.

Harmon, T. G. and Slattery, K. T. (1992) 'Advanced composite confinement of concrete', *Advanced Composite Materials in Bridges and Structures*, Canadian Society for Civil Engineering, Montreal, Canada, pp. 299–306.

Harries, K. A., Kestner, J., Pessiki, S., Sause, R. and Ricles, J. (1998) 'Axial behavior of reinforced concrete columns retrofit with FRPC jackets', *Proceedings of the Second International Conference on Composites in Infrastructures (ICCI)*, Tucson, Arizona, USA, edited by H. Saadatmanesh and M. R. Ehsani, pp. 411–425.

Howie, I. and Karbhari, V. M. (1994) 'Effect of materials architecture on strengthening efficiency of composite wraps for deteriorating columns in the North-East', *Infrastructure: New Materials and Methods of Repair, Proceedings of the Third Materials Engineering Conference*, Material Engineering Division, ASCE, New York, USA, pp. 199–206.

Karbhari, V. M. and Gao, Y. (1997) 'Composite jacketed concrete under uniaxial compression–verification of simple design equations', *Journal of Materials in Civil Engineering, ASCE*, Vol. 9, No. 4, pp. 185–193.

Kong, F. K. and Evans, R. H. (1987) *Reinforced and Prestressed Concrete*, Third Edition, Chapman & Hall, London, UK.

Lam, L. and Teng, J. G. (2001a) 'Strength models for FRP-confined concrete', *Journal of Structural Engineering, ASCE*, submitted.

Lam, L. and Teng, J. G. (2001b) 'Stress–strain models for FRP-confined concrete', *to be published.*

Lam, L. and Teng, J. G. (2001c) 'FRP-confined rectangular concrete columns', *to be published.*

Mander, J. B., Priestley, M. J. N. and Park, R. (1988) 'Theoretical stress–strain model for confined concrete', *Journal of Structural Engineering, ASCE*, Vol. 114, No. 8, pp. 1804–1826.

Mastrapa, J. C. (1997) *The Effect of Construction Bond on Confinement with FRP Composites*, MSc thesis, The University of Central Florida.

Mirmiran, A. (2000) Private communication.

Mirmiran, A. and Shahawy, M. (1996) 'A new concrete-filled hollow FRP composite column', *Composites: Part B*, Vol. 27B, pp. 263–268.

Mirmiran, A. and Shahawy, M. (1997) 'Behavior of concrete columns confined by fiber composites', *Journal of Structural Engineering, ASCE*, Vol. 123, No. 5, pp. 583–590.

Mirmiran, A., Shahawy, M., Samaan, M. and El Echary, H. (1998a) 'Effect of column parameters on FRP-confined concrete', *Journal of Composites for Construction, ASCE*, Vol. 2, No. 4, pp. 175–185.

Mirmiran, A., Samaan, M., Cabrera, S. and Shahawy, M. (1998b) 'Design, manufacture and testing of a new hybrid column', *Construction and Building Materials*, Vol. 12, No. 1, pp. 39–49.

Mirmiran, A., Shahawy, M. and Samaan, M. (1999) 'Strength and ductility of hybrid FRP-concrete beam-columns', *Journal of Structural Engineering, ASCE*, Vol. 125, No. 10, pp. 1085–1093.

Mirmiran, A., Shahawy, M., El Khoury, C. and Naguib, W. (2000) 'Large beam-column tests on concrete-filled composite tubes', *ACI Structural Journal*, Vol. 97, No. 2, pp. 268–276.

Miyauchi, K., Inoue, S., Kuroda, T., and Kobayashi, A. (1999) 'Strengthening effects of concrete columns with carbon fiber sheet', *Transactions of The Japan Concrete Institute*, Vol. 21, pp. 143–150.

Nanni, A. and Bradford, N. M. (1995) 'FRP jacketed concrete under uniaxial compression', *Construction and Building Materials*, Vol. 9, No. 2, pp. 115–124.

Nanni, A. and Norris, M. S. (1995) 'FRP jacketed concrete under flexure and combined flexure-compression', *Construction and Building Materials*, Vol. 9, No. 5, pp. 273–281.

Neale, K. W. and Labossiere, P. (1997) 'State-of-the-art report on retrofitting and strengthening by continuous fibre in Canada', *Non-Metallic (FRP) Reinforcement for Concrete Structures, Proceedings of the Third International Symposium*, Sapporo, Japan, pp. 25–39.

Ohno, S., Miyauchi, Y., Kei, T. and Higashibata, Y. (1997) 'Bond properties of CFRP plate joint', *Non-Metallic (FRP) Reinforcement for Concrete Structures, Proceedings of the Third International Symposium*, Sapporo, Japan, pp. 241–248.

Park, R. and Paulay, T. (1975) *Reinforced Concrete Structures*, John Wiley & Sons, New York, USA.

Picher, F., Rochette, P. and Labossiere, P. (1996) 'Confinement of concrete cylinders with CFRP', *Proceedings of the First International Conference on Composites for Infrastructure*, Tucson, Arizona, USA, pp. 829–841.

Popovics, S. (1973) 'A numerical approach to the complete stress–strain curves for concrete', *Cement and Concrete Research*, Vol. 3, No. 5, pp. 583–599.

Priestley, M. J. N. and Seible, F. (1995) 'Design of seismic retrofit measures for concrete and masonry structures', *Construction and Building Materials*, Vol. 9, No. 6, pp. 365–377.

Priestley, M. J. N., Seible, F. and Calvi, G. M. (1996) *Seismic Design and Retrofit of Bridges*, John Wiley & Sons, New York, USA.

Purba, B. K. and Mufti, A. A. (1999) 'Investigation of the behavior of circular concrete columns reinforced with carbon fiber reinforced polymer (CFRP) jackets', *Canadian Journal of Civil Engineering*, Vol. 26, pp. 590–596.

Restrepol, J. I. and De Vino, B. (1996) 'Enhancement of the axial load-capacity of reinforced concrete columns by means of fiberglass-epoxy jackets', *Proceedings of the Second International Conference on Advanced Composite Materials in Bridges and Structures*, Montreal, Canada, pp. 547–553.

Richart, F. E., Brandtzaeg, A. and Brown, R. L. (1928) *A Study of the Failure of Concrete under Combined Compressive Stresses*, University of Illinois, Engineering Experimental Station, Illinois, USA.

Richart, F. E., Brandtzaeg, A. and Brown, R. L. (1929) *The Failure of Plain and Spirally Reinforced Concrete in Compression*, University of Illinois, Engineering Experimental Station, Illinois, USA.

Rochette, P. and Labossiere, P. (2000) 'Axial testing of rectangular column models confined with composites', *Journal of Composites for Construction, ASCE*, Vol. 4, No. 3, pp. 129–136.

Saadatmanesh, H., Ehsani, M. R. and Li, M. W. (1994) 'Strength and ductility of concrete columns externally reinforced with fiber composite straps', *ACI Structural Journal*, Vol. 91, No. 4, pp. 434–447.

Saadatmanesh, H., Ehsani, M. R. and Jin, L. (1996) 'Seismic strengthening of circular bridge pier models with fiber composites', *ACI Structural Journal*, Vol. 93, No. 6, pp. 639–647.

Saadatmanesh, H., Ehsani, M. R. and Jin, L. (1997) 'Seismic retrofitting of rectangular bridge columns with composite straps', *Earthquake Spectra*, Vol. 13, No. 2, pp. 281–304.

Saafi, M., Toutanji, H. A. and Li, Z. (1999) 'Behavior of concrete columns confined with fiber reinforced polymer tubes', *ACI Materials Journal*, Vol. 96, No. 4, pp. 500–509.

Samaan, M., Mirmiran, A. and Shahawy, M. (1998) 'Model of concrete confined by fiber composite', *Journal of Structural Engineering, ASCE*, Vol. 124, No. 9, pp. 1025–1031.
Scott, B. D., Park, R. and Priestley, M. J. N. (1982) 'Stress–strain behaviour of concrete columns confined by overlapping hoops at low and high strain rates', *ACI Journal, Proceedings*, Vol. 79, No. 1, pp. 13–27.
Seible, F., Priestley, M. J. N., Hegemier, G. A. and Innamorato, D. (1997) 'Seismic retrofit of RC columns with continuous carbon fiber jackets', *Journal of Composites for Construction, ASCE*, Vol. 1, No. 2, pp. 52–62.
Shahawy, M., Mirmiran, A. and Beitelman, T. (2000) 'Tests and modeling of carbon-wrapped concrete columns', *Composites: Part B*, Vol. 31, pp. 471–480.
Spoelstra, M. R. and Monti, G. (1999) 'FRP-confined concrete model', *Journal of Composites for Construction, ASCE*, Vol. 3, No. 3, pp. 143–150.
Tan, K. H. (1997) 'State-of-the-art report on retrofitting and strengthening by continuous fibers, Southeast Asian perspective–status, prospects and research needs', *Non-Metallic (FRP) Reinforcement for Concrete Structures, Proceedings of the Third International Symposium*, Sapporo, Japan, pp. 13–23.
Tegola, L. A. and Manni, D. (1999) 'Experimental investigation on concrete confined by fiber reinforced polymer and comparison with theoretical model', *Proceedings of Fourth International Symposium on Fiber Reinforced Polymer Reinforcement for Reinforced Concrete Structure*, Baltimore, Maryland, USA, edited by C. W. Dolan, S. H. Rizkalla and A. Nanni, pp. 243–254, American Concrete Institude, Farmington Hills, Michigan, USA.
Teng, J. G. and Lam, L. (2001) 'Compressive strength of FRP-confined elliptical columns', *to be published*.
Toutanji, H. A. (1999) 'Stress–strain characteristics of concrete columns externally confined with advanced fiber composite sheets', *ACI Materials Journal*, Vol. 96, No. 3, pp. 397–404.
Xiao, Y. (2001) Private communication.
Xiao, Y. and Ma, R. (1997) 'Seismic retrofit of RC circular columns using prefabricated composite jacketing', *Journal of Structural Engineering, ASCE*, Vol. 123, No. 10, pp. 1357–1364.
Xiao, Y. and Wu, H. (2000) 'Compressive behavior of concrete confined by carbon fiber composite jackets', *Journal of Materials in Civil Engineering, ASCE*, Vol. 12, No. 2, pp. 139–146.
Zhang, S., Ye, L. and Mai, Y. W. (2000) 'A study on polymer composites strengthening systems for concrete columns', *Applied Composite Materials*, Vol. 7, pp. 125–138.

7 Seismic Retrofit of Columns

7.1 Introduction

The collapse of and severe damage to many buildings and bridges in recent earthquakes (e.g. the 1987 Whittier Narrows, the 1989 Loma Prieta, the 1994 Northridge earthquakes in the USA, and the 1995 Kobe earthquake in Japan) have highlighted the need for the seismic retrofit of seismically insufficient structures. RC columns, being the key lateral and vertical load-resisting members in RC structures, are particularly vulnerable to failures in earthquakes (Elhassan and Hart 1995, Jaradat *et al.* 1998, Mutsuyoshi *et al.* 1999, Ma *et al.* 2000), so their retrofit is often the key to a successful seismic retrofit strategy. As a result, many retrofit projects of building and bridge columns using FRP composites have been carried out all over the world, supported by a great deal of research.

In the USA, most of the collapse or damage was experienced by columns built before the 1971 San Fernando earthquake under the old design and construction practices. Major changes in seismic design philosophies in the USA were implemented after this earthquake (Jaradat *et al.* 1998). In Japan, many of the collapsed and damaged structures were constructed prior to 1980 (Mutsuyoshi *et al.* 1999) under the old seismic design provisions. The Building Code in Japan was revised in 1981 to set more stringent requirements for seismic resistance, as a result of developments following the 1968 Tokachi-Oki and the 1978 Miyagi-ken Oki earthquakes (Kobatake 1998).

In general, structures designed using the new seismic design codes have been found to perform well in subsequent earthquakes (e.g. Jaradat *et al.* 1998). Those structures designed using the old codes, however, do not meet the more stringent seismic design requirements of the new codes. Seismic retrofit of these older structures, particularly RC columns, has thus been an important issue in the USA and Japan. A similar situation exists in many other countries.

When a column is subjected to seismic loading, its energy absorption capacity, rather than its load capacity, is the main concern. Traditionally, the former is enhanced by RC jacketing or steel jacketing (Ballinger *et al.* 1993, Ogata and Osada 2000). Both methods have a number of disadvantages as described in Section 6.1. In addition, steel and concrete jacketing results in a significant increase of column stiffness, which may lead to additional earthquake forces in the column.

In the mid 1980s, Katsumata *et al.* (1987, 1988) first proposed the use of FRP composite materials for the retrofit of existing RC columns to achieve improved seismic resistance. They tested 5 circular and 10 rectangular quarter-scaled columns with or without laterally wound carbon fibre strands to study their performance under combined axial and cyclic lateral loads. Later, Matsuda *et al.* (1990) identified two retrofit methods for RC bridge columns using FRP composites: (a) strength-oriented retrofit, and (b) ductility-oriented retrofit. In method (a), FRP plates are longitudinally bonded to increase the flexural strength of the column (i.e. longitudinal bonding of the FRP), whilst in method (b), FRPs with fibres or main fibres in the hoop direction are wrapped around the column (i.e. lateral bonding of the FRP) to enhance its ductility as proposed by Katsumata *et al.* (1987, 1988). Both methods can improve the energy absorption capacity of the column, as shown in Figure 7.1. Following their work, there have been numerous studies on and practical applications of seismic retrofit of existing RC columns using FRP composites (e.g. Priestley *et al.* 1992, Saadatmanesh and Ehsani 1992, 1994, Ballinger *et al.* 1993, Fyfe 1994, Priestley and Seible 1994, 1995, Elhassan and Hart 1995, Seible *et al.* 1995a, 1995b, 1997, Saadatmanesh *et al.* 1996, 1997, Fyfe and Arnold 1996, Xiao and Ma 1997, Haroun *et al.* 1998, Kobatake 1998, Steckel *et al.* 1998, Ma and Xiao 1999, Mullen *et al.* 1999, Mutsuyoshi *et al.* 1999, Xiao *et al.* 1999, Fukuyama and Sugano 2000, Ma *et al.* 2000, Ogata and Osada 2000, Youssef *et al.* 2000). Most of these studies were concerned with lateral bonding of the FRP, but longitudinal bonding of

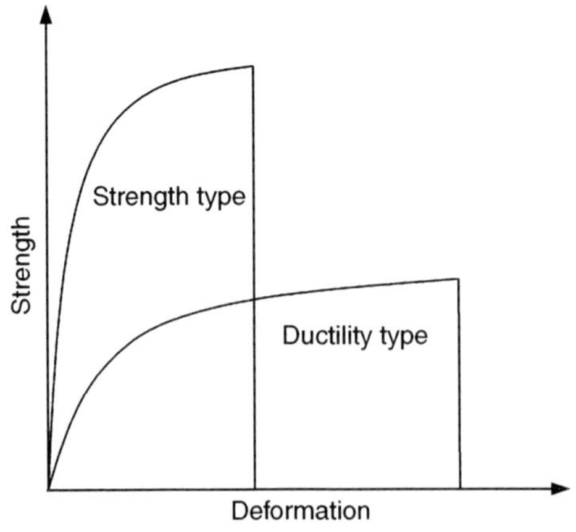

Figure 7.1 Strength-oriented and ductility-oriented seismic retrofit

the FRP has occasionally been considered when the amount of longitudinal steel reinforcement is insufficient (Kobatake 1998, Ogata and Osada 2000).

A successful example of FRP seismic retrofit of RC columns is the retrofit of a seven-storey hotel in Los Angeles (Elhassan and Hart 1995). The columns in this hotel suffered significant diagonal cracks during the magnitude 7.5 Landers earthquake of 28 June 1992, centred 175 km away from the building. The retrofit of these columns using GFRP jackets was completed a few weeks before the 1994 Northridge earthquake, and the building suffered no damage during this earthquake.

This chapter provides a review of existing knowledge on the seismic retrofit of RC columns using FRP composites. Typical failures of RC columns under seismic attacks are first described. Appropriate retrofit methods and the performance of retrofitted RC columns are next discussed. Finally, existing design recommendations are summarised.

7.2 Typical Failure Modes of RC Columns under Seismic Attack

7.2.1 General

During earthquakes, RC columns in bridges and buildings are subject to cyclic lateral loads, generally with coexisting axial loads. Failures of these columns can be due to insufficiency in shear strength, in flexural strength, or in ductility (Priestley and Seible 1995). Seible *et al.* (1997) identified the following three typical modes of failure of RC columns in earthquakes: (a) shear failure, (b) confinement failure in flexural plastic hinge regions, and (c) lap splice failure of longitudinal steel reinforcement.

7.2.2 Shear Failure

The most undesirable mode of failure is shear failure. In a typical shear failure, diagonal cracks in the concrete appear first, followed by the rupture or opening-up of transverse steel reinforcement, and then the buckling of longitudinal steel reinforcement, with the consequence being a brittle or explosive failure (Seible *et al.* 1995b, 1997), as shown in Figure 7.2. This mode of failure is primarily due to insufficient transverse steel reinforcement. In the USA, columns built before the 1971 San Fernando earthquake were usually designed with very little transverse steel reinforcement (Jaradat *et al.* 1998). Figure 7.2(a) shows the failure of a column in the 1971 San Fernando earthquake, which was laterally reinforced with 13 mm diameter hoops spaced at 305 mm. Column shear failure is also possible if the transverse reinforcement in the column is poorly detailed or inadequately anchored (Saadatmanesh *et al.* 1996). Examples of such detailing problems are steel hoops with insufficient lap splice lengths and those with insufficient anchorage of tails into the core concrete (Xiao *et al.* 1999). As shear

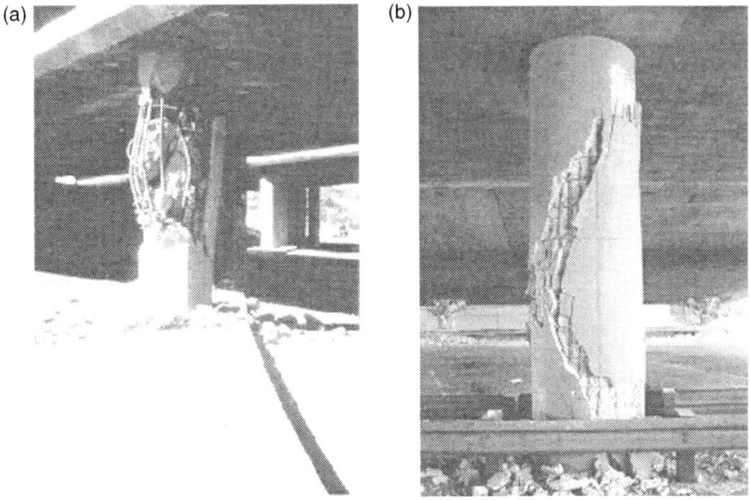

Figure 7.2 Shear failures of RC columns in earthquakes: (a) shear failure of an RC column in the 1971 San Fernando earthquake (Priestley *et al.* 1996); and (b) shear failure of an RC column in the 1994 Northridge earthquake (Priestley *et al.* 1996). Reproduced by permission of John Wiley & Sons, Inc.

failure of columns is catastrophic, it must be avoided. In seismic retrofit, existing columns with insufficient or poorly detailed transverse steel reinforcement should be shear strengthened so that shear failure is no longer the critical failure mode.

7.2.3 Flexural Plastic Hinge Failure

Flexural plastic hinge failure is also common in earthquakes. This mode of failure occurs at the column ends and is limited to a small region (Seible *et al.* 1997), characterised by the spalling of cover concrete, failure of transverse steel reinforcement and buckling of longitudinal steel reinforcement (Figure 7.3). This mode of failure is generally accompanied by large inelastic flexural deformation, and is thus a more ductile and desirable failure mode than the brittle shear failure mode.

In seismic retrofit of RC columns using FRP composites, the FRP provides lateral confinement to prevent the spalling of cover concrete and to restrain the longitudinal steel reinforcement from buckling so that more ductile responses can be developed and larger inelastic deformations can be sustained.

7.2.4 Lap Splice Failure

Lap splice failure is another type of seismic flexural failure. This mode occurs in columns in which the longitudinal steel reinforcement is lap spliced in the maximum moment regions near the column ends. In the USA, the old practices

Figure 7.3 Flexural plastic hinge failure of an RC column in the 1994 Northridge earthquake (Priestley *et al.* 1996). Reproduced by permission of John Wiley & Sons, Inc.

allowed the placement of starter bars for column longitudinal steel reinforcement during footing construction (Seible *et al.* 1995b), so that longitudinal steel reinforcing bars were lap spliced with the starter bars immediately above the base. These lap splices are thus located in a potential plastic hinge region where the moment demand is the highest. Under seismic loads, the lap splices may break down, leading to the loss of structural integrity for sustaining large inelastic deformations to achieve the required energy absorption capacity. Figure 7.4 shows the damage to a column caused by lap splice failure in the 1989 Loma Prieta earthquake (Priestley *et al.* 1996). In seismic retrofit, FRP composites can be used to provide a clamping force for the lap spliced longitudinal steel reinforcement so that lap splice failure is prevented and large inelastic deformations can be sustained.

7.2.5 Flexural Shear Failure of Columns with Longitudinal Reinforcement Cut-offs

Apart from the three failure modes discussed above, a mode of failure that is likely for columns with some of the longitudinal reinforcing bars cut off outside the potential plastic hinge regions at column ends has also been identified (Matsuda *et al.* 1990, Priestley *et al.* 1996, Kobatake 1998). In these columns, longitudinal reinforcing bars were terminated at several levels based on the design moment envelope, without accounting for the effects of tension shift due to diagonal shear cracking (Priestley *et al.* 1996). Flexural shear failure of such columns may occur at a cut-off section rather than at the ends of the column (Matsuda

Figure 7.4 Lap splice failure at column base in the 1989 Loma Prieta earthquake (Priestley *et al.* 1996). Reproduced by permission of John Wiley & Sons, Inc.

Figure 7.5 Flexural shear failure at pier mid-height in the 1995 Kobe earthquake due to the cut-off of longitudinal steel reinforcement (Priestley *et al.* 1996). Reproduced by permission of John Wiley & Sons, Inc.

et al. 1990, Kobatake 1998). Figure 7.5 shows the flexural shear failure of bridge piers at mid-height due to the cut-off of the longitudinal reinforcement in the 1995 Kobe earthquake (Priestley *et al.* 1996).

7.2.6 Footing Failure

Footing failure is less frequently encountered during earthquakes. The reasons for this mode of failure are discussed in detail in Xiao *et al.* (1996). In the USA, designs for bridge footings from the 1950s to the early 1970s were typically based on elastic analysis. The seismic force input to the footing corresponding to the formation of a column plastic hinge can be three to four times the elastic design level (Xiao *et al.* 1996). Consequently, footing failure may occur because of (a) inadequate pile capacities, (b) inadequate footing flexural strength, and (c) inadequate column/footing joint shear resistance (Xiao *et al.* 1996). Footing strengthening using FRP composites does not appear to have been explored so far, and is not further discussed in the following sections.

7.3 Methods of Retrofit

7.3.1 FRP Jackets with Horizontally Oriented Fibres

External FRP jackets can enhance both the shear capacity and the ductility of columns against seismic attacks. Such FRP jackets can be formed in a wet lay-up process by wrapping fibre sheets or winding fibre strands with resin around the column where the fibres/or the main fibres are generally horizontally oriented, and by prefabrication, as has been described in Section 6.2. Under shear forces, the tensile stresses in the FRP contribute to the overall shear resistance of the column, similar to its effect in shear strengthening of beams discussed in Chapter 4. Under flexure, the FRP provides confinement which enhances the strength and ultimate strain of the concrete, as discussed in Sections 6.4 and 6.5. The enhancement to the ultimate concrete strain is particularly important for seismic retrofit, as it allows a much greater ductility level to be achieved in inelastic deformations.

Details of FRP jacketing methods for columns are discussed in Section 6.2, so they are not repeated here. For shear strengthening, the FRP jacket is generally required to cover the entire column height, but for plastic hinge confinement and for lap splice clamping, the FRP jacket is generally only needed in the plastic hinge and nearby regions. Small gaps (around 20 mm) have been recommended between the ends of the FRP jacket and any adjacent transverse structural member (i.e. footing) by Seible *et al.* (1997) and Xiao *et al.* (1999) to prevent the jacket from direct axial loading.

If shape modification is adopted in the seismic retrofit of rectangular columns for more effective confinement, this shape modification needs to be implemented only for the plastic hinge and nearby regions, unlike for the strengthening of axially loaded columns for which this shape modification should extend over the full column height. The gaps between the ends of the column and the ends of the additional concrete should be appropriately larger than those for columns without shape modification to avoid direct loading of the additional concrete and the FRP jacket.

7.3.2 Longitudinal Bonding of FRP

Lateral FRP confinement to a column leads to an increase in its flexural capacity particularly if a significant axial load is present. If this increase is insufficient for the expected seismic loads, an additional increase may be achieved by bonding FRP sheets with the fibres or main fibres oriented in the longitudinal direction, leading to the combined use of lateral and longitudinal FRP. A good example where longitudinal bonding of FRP is necessary is when a high column is found to be flexurally deficient at cut-off sections of longitudinal steel reinforcement (Matsuda *et al.* 1990, Kobatake 1998, Ogata and Osada 2000). Experimental results showed that longitudinally bonded FRP over the cut-off region combined with laterally bonded FRP can effectively shift flexural failure back to potential plastic hinge regions under maximum moments (Matsuda *et al.* 1990, Ogata and Osada 2000).

7.4 Flexural Ductility of Retrofitted Columns

7.4.1 Ductility Factors

The purpose of seismic retrofit of RC columns is to achieve a sufficient level of deformation ductility to dissipate seismic energy before one of the failure modes reviewed above becomes critical. The displacement ductility factor μ_Δ (Priestley *et al.* 1996, Seible *et al.* 1997), defined below, has been commonly used to quantify the ductility of the seismic performance of a structure or a member:

$$\mu_\Delta = \frac{\Delta_u}{\Delta_y} \tag{7.1}$$

where Δ_y and Δ_u are the displacements at first yield and the ultimate limit state respectively. According to Park (1989), the ultimate displacement of a column can be taken as the value at which the load has reduced by 20 % from the peak load (Figure 7.6a) or the value at first fracture or buckling of a reinforcing element (Figure 7.6b), whichever occurs first.

Alternatively, the curvature ductility factor μ_Φ may be used (Priestley *et al.* 1996), which is given by

$$\mu_\Phi = \frac{\Phi_u}{\Phi_y} \tag{7.2}$$

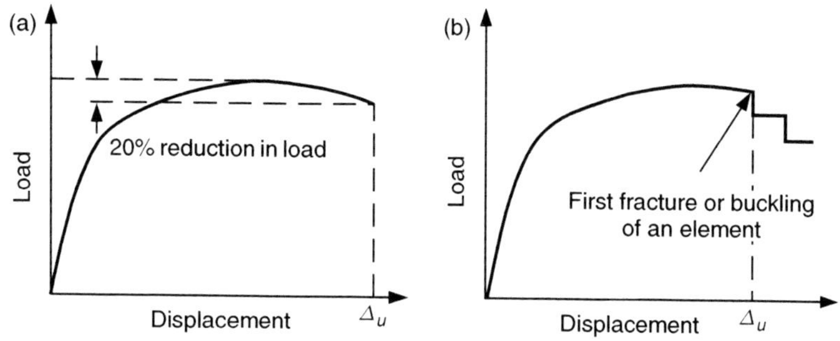

Figure 7.6 Definitions of ultimate displacement based on (a) a drop after peak load and (b) fracture or buckling

where Φ_y and Φ_u are the curvatures at first yield and at the ultimate limit state, respectively, which can be obtained from a sectional curvature analysis as described in Section 6.9. Furthermore, the plastic displacement Δ_p and the plastic curvature Φ_p are defined as follows (Priestley *et al.* 1996) and illustrated in Figure 7.7:

$$\Delta_p = \Delta_u - \Delta_y \tag{7.3}$$

and

$$\Phi_p = \Phi_u - \Phi_y \tag{7.4}$$

The plastic curvature Φ_p is assumed to be constant within a plastic hinge region. The relationship between displacement ductility and curvature ductility factors is given by (Seible *et al.* 1997)

$$\mu_\Delta = 1 + 3\left(\mu_\Phi - 1\right)\frac{L_p}{L}\left(1 - 0.5\frac{L_p}{L}\right) \tag{7.5}$$

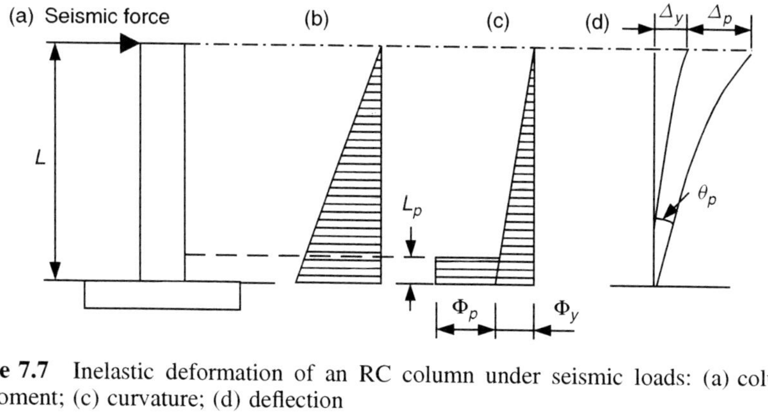

Figure 7.7 Inelastic deformation of an RC column under seismic loads: (a) column; (b) moment; (c) curvature; (d) deflection

where L is the distance from the critical section of the plastic hinge to the point of contraflexure (i.e. the shear span) and L_p is the length of the plastic hinge region, which is given by (Priestley *et al.* 1996)

$$L_p = 0.08L + 0.022 f_{yl} d_{bl} \geqslant 0.044 f_{yl} d_{bl} \tag{7.6}$$

where f_{yl} and d_{bl} are the yield strength and diameter of the longitudinal steel reinforcing bars. The plastic rotation θ_p is defined as the product of the plastic hinge length and the plastic curvature, that is

$$\theta_p = L_p \Phi_p \tag{7.7}$$

7.4.2 Ductility of Retrofitted RC Columns

A large number of studies have shown, through the hysteresis loops of tests of RC columns under cyclic loads (e.g. Priestley and Seible 1995, Saadatmanesh *et al.* 1996, 1997, Seible *et al.* 1997, Xiao and Ma 1997, Haroun *et al.* 1998, Ma *et al.* 2000), that lateral FRP confinement is effective in enhancing the ductility of RC columns under seismic attacks. For RC columns with limited transverse steel reinforcement, it is difficult to develop a flexural displacement ductility factor greater than $\mu_\Delta = 2.5$ to 3.5 because of the limited ultimate strain of unconfined cover concrete (Priestley and Seible 1995). However, with FRP confinement, much greater ductility (e.g. $\mu_\Delta > 6$) can be expected. Some of the existing test results are shown in Tables 7.1 and 7.2.

Table 7.1 presents a summary of the double bending tests conducted by Xiao *et al.* (1999). These authors tested three circular columns 610 mm in diameter and having a height-to-diameter ratio of 3. The cylinder compressive strength of concrete was 35.9 MPa. The columns were longitudinally reinforced with 20 deformed No. 19 bars (nominal diameter = 19.1 mm, yield strength = 303 MPa) and laterally reinforced with round No. 6 (nominal diameter = 6.4 mm, yield strength = 340 MPa) hoops spaced at 150 mm. Two of the three columns were retrofitted with prefabricated GFRP jackets and the remaining one was tested as an as-built column. The hysteretic responses of the as-built and retrofitted columns from Xiao (2001) are shown in Figure 7.8.

Table 7.2 summarises the single bending tests conducted by Saadatmanesh *et al.* (1996, 1997). These authors tested five circular columns and five rectangular columns. The circular columns had a diameter of 305 mm and a total height of 2.41 m (including a 406 mm thick base). The rectangular columns had the same height, with the cross-sectional dimensions being 368 mm by 241 mm. All the columns were reinforced and detailed to model approximately typical pre-1971 designs of highway bridge columns in the USA (Saadatmanesh *et al.* 1996, 1997). Each of the two five-column series included two as-built columns, one with and one without lap spliced longitudinal reinforcement, and three retrofitted columns confined by either active or passive GFRP jackets in the plastic hinge region.

Table 7.1 Summary of tests conducted by Xiao *et al.* (1999)

Model column	Retrofit details	Axial load (kN)	Ideal flexural strength (kN)	Max measured lateral load (kN)	Ductility[a] μ_Δ	Observations at end of test
CS-A	As built	654	600	560	+3, −2	Shear cracking and rupture of steel hoops
CS-ISJ-RT	Individual shell jacketing with four layers for the ends and three layers for middle portion	654	600	640	$\geqslant$12 in both directions	Horizontal cracks at large ductility levels
CS-CSJ-RT	Continuous shell jacketing with four layers for the ends and three layers for middle portion	654	600	600	$\geqslant$12 in both directions	Delamination of jacket at bottom of column

[a] '+' for push direction and '−' for pull direction.

Table 7.2 Summary of tests conducted by Saadatmanesh *et al.* (1996, 1997)

Column	Concrete cylinder strength (MPa)	Longitudinal steel details	Retrofit details	Calculated lateral strength according to ACI 318-89 (1989) (kN)	Max measured lateral load (kN)	Ductility[a] μ_Δ	Failure mode
C-1	36.5	Lap spliced bars	None	50.7	58.3	±1.5	Lap splice failure
C-2	38.3	Lap spliced bars	Passive	50.7	81.4	+6, −7	Not tested to failure
C-3	38.5	Lap spliced bars	Active	50.7	89.4	±6	Not reported
C-4	36.6	Continuous bars	None	50.7	71.6	+4, −3	Concrete failure and longitudinal bar buckling
C-5	36.5	Continuous bars	Passive	50.7	87.2	±6	Not reported
R-1	34.9	Lap spliced bars	None	102.7	96.5	±1.5	Lap splice failure
R-2	35.7	Lap spliced bars	Direct wrapping	102.7	138.8	+6, −5	Not reported
R-3	33.3	Continuous bars	None	132.5	161.5	±2	Flexural shear cracks within and beyond plastic hinge region and longitudinal bars separated from core concrete
R-4	35.6	Continuous bars	Direct wrapping	132.5	214	±6	Not tested to failure
R-5	36.2	Continuous bars	Modified to oval section followed by wrapping	132.5	226.8	±6	Not tested to failure

[a] '+' for push direction and '−' for pull direction.

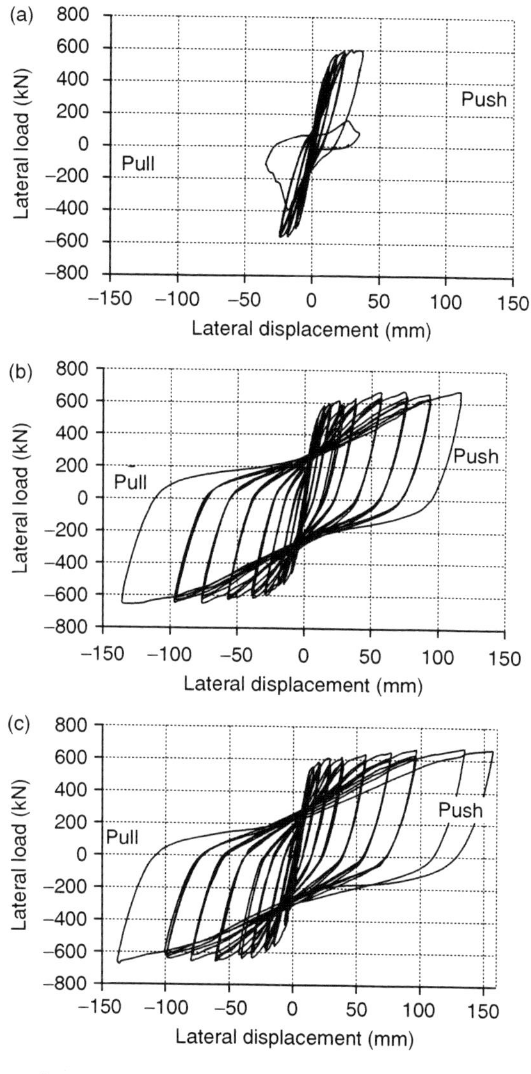

Figure 7.8 Hysteretic responses of the as-built (a) and FRP-retrofitted RC columns (b, CS-ISJ-RT; c, CS-CSJ-RT) tested by Xiao *et al.* (1999). Reproduced by kind permission of the author

7.5 Ultimate Conditions of Retrofitted Columns

7.5.1 General

Most existing tests were concerned with the overall performance of RC columns after FRP retrofit, in which the specimens were retrofitted to a degree that none of the typical failure modes as discussed in Section 7.2 were observed before the test was stopped at a sufficiently high ductility level or at the occurrence of a nonfatal secondary failure such as local delamination of FRP (e.g. Fyfe 1994, Saadatmanesh *et al.* 1996, 1997, Seible *et al.* 1997, Xiao *et al.* 1999). While these tests are useful in demonstrating the effectiveness of FRP seismic retrofit, they do not provide sufficient information for a precise definition of the ultimate condition of the column in any of the failure modes, except to show that a particular retrofit design is sufficient to prevent failure in these modes at the desired ductility level. Only a limited number of tests were concerned with the ultimate condition of retrofitted columns failing in a particular mode (e.g. Mutsuyoshi *et al.* 1999).

7.5.2 Shear Failure

For shear strengthening, Priestley *et al.* (1996) believed that, unlike steel, the full tensile capacity of the FRP could not be used as otherwise the relatively large strain reached in the FRP would lead to a reduction in the shear resistance contributed by concrete aggregate interlock under cyclic loads. There is, however, another reason that this full tensile capacity cannot be used, as explained in Chapter 4: the FRP is nonuniformly stressed in a shear failure and, owing to its lack of ductility, is unable to reach its full tensile strength over the entire shear crack.

Priestley and Seible (1995) suggested that in calculating the shear resistance contributed by FRP, the ultimate tensile strength of the FRP should be limited to a tensile stress corresponding to a strain of 0.004, to avoid degradation in concrete aggregate interlock. Mutsuyoshi *et al.* (1999) tested four rectangular columns with strong longitudinal steel reinforcement to study shear failure in RC columns under combined axial and cyclic lateral loads. The experimental results showed that all three columns wrapped with CFRP failed in shear by FRP rupture following the yielding of transverse steel reinforcement, although they suggested from finite-element analysis that shear failure by concrete crushing could also be possible for columns with a large amount of FRP.

7.5.3 Flexural Plastic Hinge Failure

For plastic hinge zone retrofit of columns without lap spliced longitudinal steel reinforcement, FRP confinement can enhance the ultimate strain of concrete. The ultimate condition of the FRP jacket is FRP rupture. Mutsuyoshi *et al.* (1999) tested four rectangular columns with strong shear reinforcement, without lap spliced longitudinal reinforcement under combined axial and cyclic lateral loads. Two of the three columns wrapped with CFRP showed rupture of the FRP at failure, whilst the third column, which had the highest amount of CFRP, showed crushing of concrete at the bottom without FRP rupture.

7.5.4 Lap Splice Failure

For plastic hinge zone retrofit of columns with lap spliced longitudinal steel reinforcement, FRP confinement is also required to provide a clamping stress to avoid lap splice failure. For these columns, the dilation strain of concrete needs to be limited to a value of 0.001 (Priestley and Seible 1995, Priestley *et al.* 1996), as otherwise relative movement and debonding of the longitudinal bars are likely to occur. Thus, the ultimate condition is characterised by the FRP jacket reaching a stress corresponding to a hoop tensile strain of 0.001.

Xiao and Ma (1997) developed an analytical procedure to predict the behaviour of columns with lap splice deterioration. They suggested that debonding of the lap splice be allowed in the retrofitted column as long as the yield of longitudinal bars can be developed. If this suggestion is adopted, a more liberal limit to the hoop strain in the FRP jacket is expected to be acceptable.

7.5.5 Flexural Failure of Columns with Longitudinal FRP

If the FRP is longitudinally bonded for flexural strengthening, it must be borne in mind that this can only be effective if the ends are sufficiently anchored. Therefore, this method is not suitable for enhancing the flexural strength at columns ends owing to the lack of an anchorage length for the FRP, unless the FRP is properly anchored in the adjacent structural elements. If the FRP is bonded to enhance the flexural strength outside the plastic hinge regions at column ends, as may be necessary in columns with longitudinal steel bar cut-offs, it is then possible to anchor the FRP sufficiently away from the critical sections. Debonding is still possible as in beams with the FRP bonded to the tension face. As longitudinally bonded FRP is usually used together with laterally wrapped FRP, the latter is likely to help resist debonding of the former, although the fact that the longitudinal FRP will be subject to tension–compression cycles means that there may be an adverse interaction between debonding and buckling of the FRP jacket (Matsuda *et al.* 1990). The failure mechanism of columns with longitudinally bonded FRP is likely to be complicated, particularly if laterally wrapped FRP is also present, and much further research is required to establish a clear understanding.

7.6 Design Recommendations

7.6.1 General

Although seismic retrofit may be achieved either by FRP jackets with the main fibres in the hoop direction (lateral FRP) or by the bonding of FRP composite sheets/plates with the main fibres in the longitudinal direction (longitudinal FRP), the former appears to have been far more popular. The latter is only required when there is insufficient flexural strength, whilst the former can enhance shear strength and ductility without significantly increasing the flexural strength unless the axial load is large. A large increase in the flexural strength, if unwanted, is undesirable, as it may lead to overloading of adjacent structural elements (Priestley *et al.* 1996).

For seismic retrofit by longitudinal bonding of the FRP, only limited tests have been reported in the open literature. Specific design equations do not appear to have been presented, but these can be easily established following the same principles as adopted in Section 3.4 if debonding is not considered. Little is currently known about debonding in these columns.

By contrast, a comprehensive set of design recommendations has been developed for seismic retrofit of columns using FRP jackets with the main fibres in the hoop direction by the research group at the University of California at San Diego, USA. The description here is largely based on the study presented in Seible *et al.* (1997). As a result, the recommendations do not fall within the framework of the British code for the structural use of concrete (BS 8110 1997) adopted in other chapters of the book.

7.6.2 Shear Strength

A shear strengthening design approach has been described by Priestley and Seible (1995), Priestley *et al.* (1996) and Seible *et al.* (1997), based on an earlier study by Priestley *et al.* (1994a). In this approach, the seismic shear strength of an unstrengthened column consists of three independent components, V_c, V_N and V_s, which represent the contributions by concrete, the axial load N, and the truss mechanism of transverse steel reinforcement, respectively. The concrete component V_c for both circular and rectangular columns is given by (Priestley *et al.* 1994a)

$$V_c = k\sqrt{f_c'}A_e \tag{7.8}$$

where k can be found from Figure 7.9 which is taken from Priestley *et al.* (1996), f_c' is the cylinder compressive strength of concrete in MPa, and A_e is the effective shear area and equal to $0.8A_g$ for both rectangular and circular columns with A_g being the gross sectional area of the column. It can be seen from Figure 7.9 that k depends on the ductility demand of the column. For portions of columns outside plastic hinge regions, k takes an initial value of 0.25 in design, but within plastic hinge regions, k varies with the required ductility level.

The axial load component V_N is given by

$$V_N = N \tan\alpha = \frac{d - x}{2L}N \tag{7.9}$$

where α is defined in Figure 7.10, d is the diameter of circular columns or the section depth in the lateral load direction for rectangular columns, x is the neutral

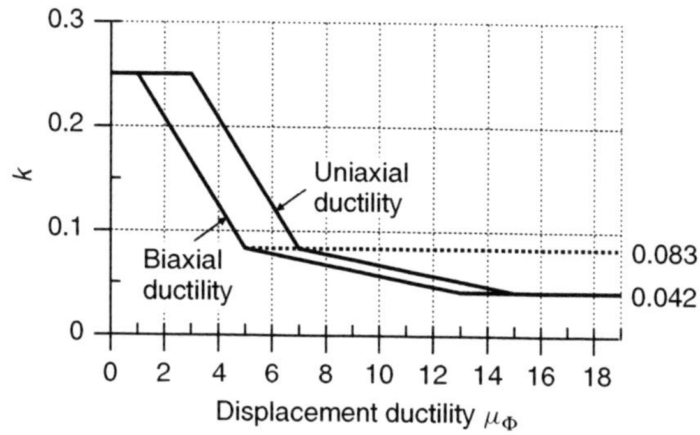

Figure 7.9 Dependence of k for design on curvature ductility demand of columns (Priestley *et al.* 1996). Reproduced by permission of John Wiley & Sons, Inc.

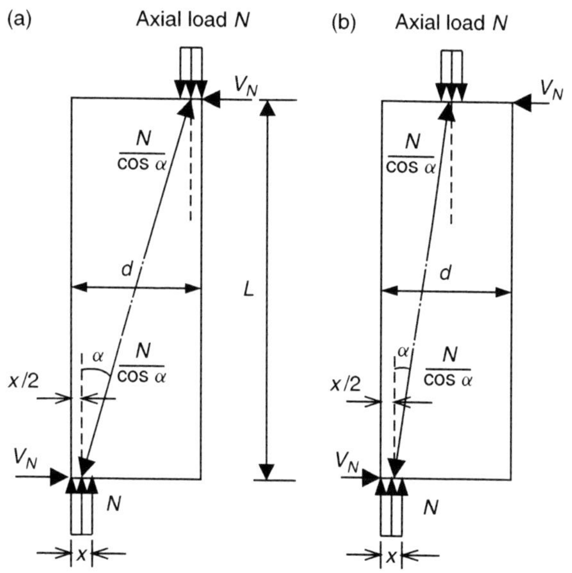

Figure 7.10 Contribution of axial force to column shear strength: (a) double bending; (b) single bending

axis depth, N is the axial load, and L is the shear span. For a cantilever column (single bending) L is the column height, and for a column in double bending, L is half the column height.

The truss mechanism component of transverse steel reinforcement, V_s, for circular columns is given by

$$V_s = \frac{\pi}{2}\frac{A_{sh} f_{yh} D'}{s}\cot\theta \tag{7.10}$$

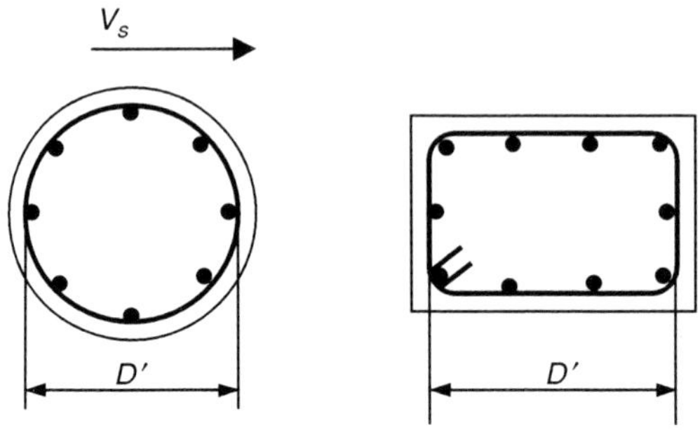

Figure 7.11 Definition of D' in truss mechanism

and that for rectangular columns is given by

$$V_s = \frac{A_{sh} f_{yh} D'}{s} \cot\theta \tag{7.11}$$

where D' is the distance between the centres of the peripheral hoop or spiral as defined in Figure 7.11, d_h, A_{sh}, f_{yh} and s are respectively the bar diameter, cross-sectional area, yield strength and spacing of transverse steel reinforcement in the form of a hoop or spiral, and θ is the inclination of the critical shear crack to the column axis and has been recommended to be 35° (Priestley and Seible 1995, Priestley *et al.* 1996), although Seible *et al.* (1997) used 45° which leads to more conservative results. A more accurate formula for the truss mechanism contribution in circular columns has been derived by Xiao *et al.* (1999), but its prediction is only slightly different from that of Equation 7.10.

Assuming that the FRP jacket for shear strengthening is allowed to reach a tensile stress limit $f_{frp,e}$ corresponding to a tensile strain limit which is small enough to avoid degradation in concrete aggregate interlock, the shear contribution of the FRP jacket is given by (Priestley and Seible 1995, Priestley *et al.* 1996)

$$V_{frp} = \frac{\pi}{2} f_{frp,e} t_{frp}\, d \cot\theta \tag{7.12}$$

for circular columns and by

$$V_{frp} = 2 f_{frp,e} t_{frp}\, d \cot\theta \tag{7.13}$$

for rectangular columns. Therefore, the thickness of the FRP jacket, t_{frp}, required is given by

$$t_{frp} = \frac{V_o/\phi_V - (V_c + V_N + V_s)}{(\pi/2) f_{frp,e}\, d \cot\theta} \tag{7.14}$$

for circular columns and by

$$t_{frp} = \frac{V_o/\phi_V - (V_c + V_N + V_s)}{2 f_{frp,e}\, d \cot\theta} \tag{7.15}$$

for rectangular columns, where V_o is the column shear demand based on full flexural overstrength in the potential plastic hinges, and ϕ_V is the shear strength reduction factor taken as 0.85. V_o is given by (Seible *et al.* 1997)

$$V_o = 1.5 \times \frac{M_i}{L} \tag{7.16}$$

where M_i is the ideal flexural strength of the column section based on measured material strengths. When an accurate analysis is available, V_o can be directly based on a close estimate from such an analysis of the maximum possible flexural strength of a column plastic hinge, which considers all possible overstrength factors of materials such as strain hardening of steel and confinement to concrete (Priestley *et al.* 1994b).

For rectangular columns strengthened by an elliptical FRP jacket following shape modification, the shear contribution of the jacket, V_{frp}, can be evaluated by (Priestley *et al.* 1996)

$$V_{frp} = 2 f_{frp,e} t_{frp} \left[2a - 2b\left(1 - \frac{\pi}{4}\right)\right] \cot\theta \quad (A\text{–}A' \text{ direction}) \tag{7.17}$$

and

$$V_{frp} = 2 f_{frp,e} t_{frp} \left[2b - 2a\left(1 - \frac{\pi}{4}\right)\right] \cot\theta \quad (B\text{–}B' \text{ direction}) \tag{7.18}$$

where a is the length of the semi-major axis and b is the length of the semi-minor axis of the elliptical section. The various properties of an ellipse are discussed in Section 6.3.

Priestley *et al.* (1996) proposed that in calculating the shear resistance contributed by the FRP, the stress in the FRP, $f_{frp,e}$, be limited to $0.004E_{frp}$ for a strain limit of 0.004 with E_{frp} being the elastic modulus of the FRP, to avoid degradation in concrete aggregate interlock. This strain limit may appear to be quite small, but a careful examination using the shear strength model developed by Chen and Teng (2001) as given in Chapter 4 reveals that the actual strain limit imposed is about 1 % owing to the nonuniform strain distribution in the FRP over the shear crack. A 1 % strain limit in Chen and Teng's model reduces to an average value of 0.004 for design use owing to a distribution factor of 0.5 for complete wrapping and a further empirical reduction factor of 0.8 (Equation 4.23). This 1 % strain limit appears to be a reasonable proposal, compared with the corresponding limit of 1.5 % proposed in Section 4.6 for static loading. Chen and Teng's model thus provides a useful interpretation of the strain limit proposed by Priestley *et al.* (1996).

Mutsuyoshi *et al.* (1999) proposed a different approach for determining the shear strength of FRP-strengthened RC columns. In their approach, the shear resistance contributed by the axial load is combined with the contribution by concrete, as is done in the approach of ASCE/ACI 426 (ASCE–ACI 1973).

7.6.3 Flexural Plastic Hinge Confinement

The design equations for plastic hinge confinement of RC columns have also been described by Seible *et al.* (1997). For a given displacement ductility μ_Δ, the required curvature ductility μ_Φ can be found from Equation 7.5. The required ultimate strain of confined concrete in the extreme compression fibre is given by

$$\varepsilon_{cc} = \Phi_u x = \mu_\Phi \Phi_y x \tag{7.19}$$

where ε_{cc} is the ultimate compressive strain of FRP-confined concrete. The yield curvature Φ_y can be found by a moment–curvature analysis such as that described in Section 6.9 or by a simplified approach (Priestley *et al.* 1996). The required thickness of the FRP for plastic hinge confinement for circular columns can then be computed from Equation 6.17, which yields (Seible *et al.* 1997)

$$t_{frp} = 0.09 \frac{d(\varepsilon_{cc} - 0.004) f'_{cc}}{\phi_F f_{frp} \varepsilon_{frp,rup}} \tag{7.20}$$

where f'_{cc} is the compressive strength of confined concrete and $f'_{cc} = 1.5 f'_c$ has been suggested for most retrofit design (Priestley *et al.* 1996), ϕ_F is the flexural strength reduction factor taken as 0.9, f_{frp} is the tensile strength of the FRP in the hoop direction, and $\varepsilon_{frp,rup}$ is the ultimate strain of the FRP.

It should be noted that Equation 6.17 is based on the stress–strain model proposed by Mander *et al.* (1988) for steel-confined concrete. The accuracy of this model has been discussed in Section 6.5, and more accurate estimates for the ultimate strain of FRP-confined concrete can be made using Lam and Teng's (2001) model presented in Section 6.5.

For an elliptical column, d should be replaced by the average of the major and minor axis lengths of the section (Priestley *et al.* 1996, Seible *et al.* 1997). For a rectangular column, the confinement effect was considered to be about 50 % smaller (Priestley and Seible 1995). Seible *et al.* (1997) thus suggested that for a rectangular column with an aspect ratio of depth/width $\leqslant$1.5, a rectangular FRP jacket with twice the theoretical thickness derived for a circular column with an equivalent diameter (the average side length) should be used.

In addition, Seible *et al.* (1997) suggested that for columns with $M/(Vd) > 4$, the amount of FRP for plastic hinge confinement should be checked to ensure that the buckling of longitudinal steel bars does not occur using the equations described in Priestley *et al.* (1996). Those equations were developed for RC columns confined by transverse steel reinforcement in the form of hoops or spirals, so their applicability to FRP confinement does not appear to have been established. For FRP-confined columns, provided the FRP jacket remains intact to prevent the spalling of concrete, buckling of longitudinal steel bars is believed to be unlikely.

7.6.4 Lap Splice Clamping

Seible *et al.* (1997) proposed that to prevent the occurrence of lap splice debonding or relative slippage of longitudinal steel bars, the required lateral confining pressure f_l (clamping pressure) should be not less than

$$f_l = \frac{A_{bl} f_{yl}}{[p/(2n) + 2(d_{bl} + c)]L_s} \tag{7.21}$$

where p is the length of the internal cracking line surrounded by the n longitudinal bars, A_{bl} is the area of a single longitudinal bar, f_{yl} and d_{bl} are the yield strength and diameter of longitudinal bars, c is the thickness of concrete cover to the longitudinal steel bars, and L_s is the length of lap splice. The term $p/(2n) + 2(d_{bl} + c)$ represents the perimeter of cracking around a longitudinal bar when debonding of lap spliced bars occurs (Figure 7.12).

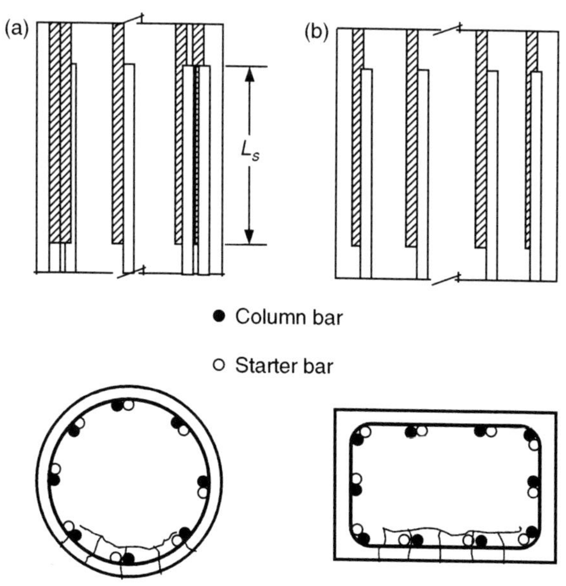

Figure 7.12 Cracking around longitudinal bars in (a) a circular and (b) a rectangular column at lap splice failure

On the other hand, Seible *et al.* (1997) suggested that the dilation strain of a column be limited to 0.001, because at dilation strain levels between 0.001 and 0.002 lap splice debonding starts according to their experimental results. This implies that the strain of the FRP in the hoop direction should not exceed 0.001. Correspondingly, the confining pressure provided by the FRP jacket for a circular column is

$$f_l = \frac{2 \times 0.001 E_{frp} t_{frp}}{d} \tag{7.22}$$

Thus, the thickness of the FRP jacket required for a circular column is then given by

$$t_{frp} = \frac{500\, d(f_l - f_h)}{E_{frp}} \tag{7.23}$$

for circular columns, where f_l is given by Equation 7.21, and f_h is the confining pressure provided by the transverse steel reinforcement. For rectangular columns, Seible *et al.* (1997) suggested modification of the section into an ellipse and use of the average of the major and minor axis lengths of the ellipse to replace d in Equation 7.23.

Xiao and Ma (1997) developed an analytical model for predicting the behaviour of lap spliced columns in which slips of the lap spliced longitudinal bars are allowed. By using this model and adopting a trial and error procedure, the required jacket thickness for a specific ductility demand to be achieved in a retrofit design can be determined. A less conservative estimate for the jacket thickness results from this method.

7.6.5 Detailing Considerations

When designing an FRP jacket for a given column, the three different failure modes (shear failure, plastic hinge confinement failure and lap splice failure) will require different jacket thicknesses, and for shear strengthening different jacket thicknesses may be used for the plastic hinge regions and for other portions of the column for cost-saving purposes. For any given location in the column, the total thickness of the jacket should be the sum of those determined for the three failure modes respectively.

Figure 7.13 illustrates a scheme recommended by Seible *et al.* (1997), where L_{c1} is the length of the primary confinement region for the plastic hinge, L_{c2} is the secondary confinement region adjacent to the plastic hinge, and L_V^i and L_V^o are the shear strengthening regions inside and outside the plastic hinge respectively. The secondary confinement region L_{c2} is necessary to avoid the formation of flexural plastic hinges above the primary plastic hinge region L_{c1} when confinement allows for significant overstrength development in the primary plastic hinge.

The lap splice length L_s is determined from the lap length of the starter bars and column bars while the lengths of the confinement regions are related to the

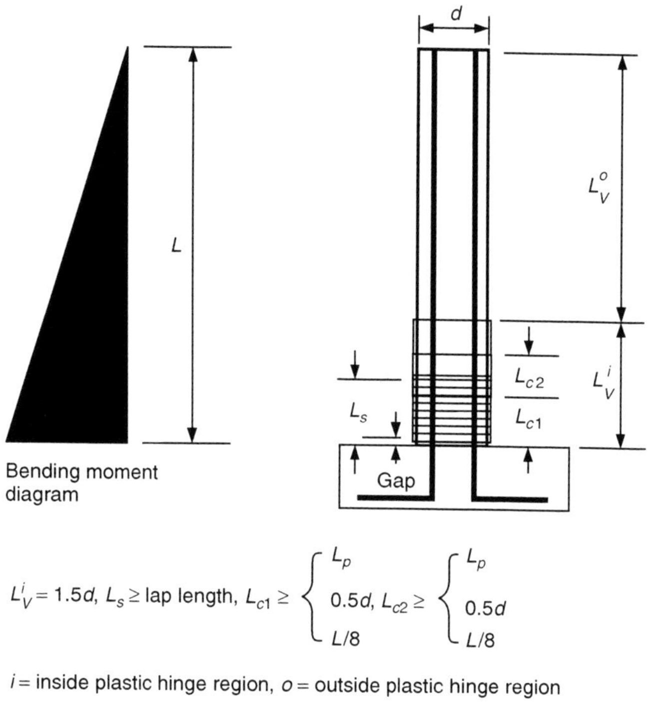

Figure 7.13 FRP jacket regions for column seismic retrofit. From Seible *et al.* (1997). Reproduced by permission of American Society of Civil Engineers

column geometry (Seible *et al.* 1997). Half the FRP thickness for L_{c1} has been used for L_{c2} in the design examples of Seible *et al.* (1997). Detailed design procedures and examples can be found in Seible *et al.* (1997), as well as other references (e.g. Priestley *et al.* 1996, Xiao *et al.* 1999, Ma *et al.* 2000).

7.7 Conclusions

This chapter has summarised existing knowledge on the seismic retrofit of RC columns in bridges and buildings. The purpose of seismic retrofit is to prevent these columns from failing in one of the following modes before a sufficient amount of inelastic deformation is developed: (a) shear failure, (b) flexural plastic hinge failure, (c) lap splice failure, and (d) flexural shear failure outside plastic hinge regions. The most popular method of seismic retrofit has been the provision of an FRP jacket, which works against the first three failure modes by contributing to the shear resistance as additional transverse reinforcement, confining the column so that concrete spalling and longitudinal steel bar buckling are suppressed to achieve large inelastic flexural deformations, and providing a clamping pressure to enhance the resistance of lap splices to debonding or slippage. A small increase in the flexural capacity may result, but this increase is much smaller than may be caused by steel or concrete jacketing and should not be a concern. For this method of retrofit, a comprehensive set of design recommendations is available and has been presented. These design recommendations enable the selection of suitable FRP jacket thicknesses. A method for quantifying the level of deformation ductility has also been presented. It should be pointed out that the design equations contain much empiricism, and there has been only limited research investigating the performance of columns failing in each of the distinct failure modes. Much further research is therefore required.

For seismic retrofit by bonding the FRP with longitudinal fibres with or without a coexistent FRP jacket with transverse fibres, only limited tests have been reported in the open literature. Specific design equations do not appear to have been presented, but the design equations for beams presented in Section 3.4 can be easily adapted for these columns if debonding is not considered. Little is currently known about debonding in such columns.

References

ACI 318-89 (1989) *Building Code Requirements for Reinforced Concrete (318-89) and Commentary (318R-89)*, American Concrete Institute (ACI), Farmington Hills, Michigan, USA.

ASCE–ACI Joint Task Committee 426 (1973) 'Shear strength of reinforced concrete members', *Journal of Structural Engineering, ASCE*, Vol. 86, No. 1, pp. 45–59.

Ballinger, C., Maeda, T. and Hoshijima, T. (1993) 'Strengthening of reinforced concrete chimneys, columns and beams with carbon fiber reinforced plastics', *Proceedings of the International Symposium on Fiber-Reinforced-Plastic Reinforcements for Concrete Structures*, edited by A. Nanni and C. W. Dolan, American Concrete Institute, Farmington Hills, Michigan, USA, pp. 243–248.

BS 8110 (1997) *Structural Use of Concrete, Part 1. Code of Practice for Design and Construction*, British Standards Institution, London, UK.

Chen, J. F. and Teng, J. G. (2001) 'Shear capacity of FRP-strengthened RC beams: FRP rupture', *to be published*.

Elhassan, R. M. and Hart, G. C. (1995) 'Analysis and seismic strengthening of concrete structures', *The Structural Design of Tall Buildings*, Vol. 4, pp. 71–90.

Fukuyama, H. and Sugano, S. (2000) 'Japanese seismic rehabilitation of concrete buildings after Hyogoken-Nanbu Earthquake', *Cement and Concrete Composites*, Vol. 22, pp. 59–79.

Fyfe, E. R. (1994) 'New concept for wrapping columns with a high strength fibre/epoxy system', *Proceedings of the Materials Engineering Conference 804, ASCE*, pp. 1156–1162.

Fyfe, E. R. and Arnold, S. (1996) 'The concept of the composite system and column seismic retrofit at 1-5 & HWY using the TYFO S Fibrwrap™ system', *Proceedings of the 41st International SAMPE Symposium and Exhibition*, pp. 1304–1310.

Haroun, M. A., Feng, M. Q., Bhatia, H. and Sultan, M. (1998) 'Cyclic qualification testing of jacketed bridge column in flexural and shear', *Proceedings of the International Model Analysis Conference –IMAC*, Vol. 2, pp. 1770–1776.

Jaradat, O. A., Mclean, D. I. and Marsh, M. L. (1998) 'Performance of existing bridge columns under cyclic loading – Part 1: Experimental results and observed behavior', *ACI Structural Journal*, Vol. 95, No. 6, pp. 695–704.

Katsumata, H., Kobatake, Y. and Takeda, T. (1987) 'A study on the strengthening with carbon fibre for earthquake-resistant capacity of existing reinforced concrete columns', *Proceedings of the Seminar on Repair and Retrofit of Structures, Workshop on Repair and Retrofit of Existing Structures*, US–Japan Panel on Wind and Seismic Effects, UJNR, Tsukuba, Japan.

Katsumata, H., Kobatake, Y. and Takeda, T. (1988) 'A study on strengthening with carbon fibre for earthquake-resistant capacity of existing reinforced concrete columns', *Proceedings of the Ninth World Conference on Earthquake Engineering*, Tokyo–Kyoto, Japan, Vol. VII, pp. 517–522.

Kobatake, Y. (1998) 'A seismic retrofitting method for existing reinforced concrete structures using CFRP', *Advanced Composite Materials*, Vol. 7, No. 1, pp. 1–22.

Lam, L. and Teng, J. G. (2001) 'Stress–strain models for FRP-confined concrete', *to be published.*

Ma, R. and Xiao, Y. (1999) 'Seismic retrofit and repair of circular bridge columns with advanced composite materials', *Earthquake Spectra*, Vol. 15, No. 4, pp. 747–764.

Ma, R., Xiao, Y. and Li, K. N. (2000) 'Full-scale testing of a parking structure column retrofitted with carbon fibre reinforced composites', *Construction and Building Materials*, Vol. 14, pp. 63–71.

Mander, J. B., Priestley, M. J. N. and Park, R. (1988) 'Theoretical stress–strain model for confined concrete', *Journal of structural Engineering, ASCE*, Vol. 114, No. 8, pp. 1804–1826.

Matsuda, T., Sato, T., Fujiwara, H. and Higashida, N. (1990) 'Effect of carbon fibre reinforcement as a strengthening measure for reinforced concrete bridge piers', *Proceedings of 1st US–Japan Workshop on Seismic Retrofit of Bridges*, Public Works Institute, Ministry of Construction, Tsukuba, Japan, pp. 356–374.

Mullen, C. L., Rice, J. R. and Hackett, R. M. (1999) 'Filament wound retrofitting of reinforced concrete bridge structures', *Journal of Reinforced Plastics and Composites*, Vol. 18, No. 12, pp. 1113–1121.

Mutsuyoshi, H., Ishibashi, T., Okano, M. and Katsuki, F. (1999) 'New design method for seismic retrofit of bridge columns with continuous fibre sheet – performance–based design', *Proceedings of the Fourth International Symposium on Fibre Reinforced Polymer Reinforcement for Reinforced Concrete Structures*, Baltimore, Maryland, USA, edited by C. W. Dolan, S. H. Rizkalla and A. Nanni, pp. 229–241, American Concrete Institute, Farmington Hills, Michigan, USA.

Ogata, T. and Osada, K. (2000) 'Seismic retrofitting of expressway bridges in Japan', *Cement and Concrete Composites*, Vol. 22, pp. 17–27.

Park, R. (1989) 'Evaluation of ductility of structures and structural assemblages from laboratory testing', *Bulletin of the New Zealand National Society for Earthquake Engineering*, Vol. 22, No. 3, pp. 155–166.

Priestley, M. J. N. and Seible, F. (1994) 'Seismic assessment of existing bridges', *Proceedings of Seismic International Workshop*, Queenstown, New Zealand, pp. 447–471.

Priestley, M. J. N. and Seible, F. (1995) 'Design of seismic retrofit measures for concrete and masonry structures', *Construction and Building Materials*, Vol. 9, No. 6, pp. 365–377.

Priestley, M. J. N., Seible, F. and Fyfe, E. (1992) 'Column seismic retrofit using fibreglass/epoxy jackets', *Proceedings of the First International Conference on Advanced Composite Materials in Bridges and Structures*, Sherbrooke, Canada, edited by K. W. Neale and P. Labossiere, Canadian Society for Civil Engineering, Montreal, Canada.

Priestley, M. J. N., Verma, R. and Xiao, Y. (1994a) 'Seismic shear strength of reinforced concrete columns', *Journal of Structural Engineering, ASCE*, Vol. 120, No. 8, pp. 2310–2329.

Priestley, M. J. N., Seible, F., Xiao, Y. and Verma, R. (1994b) 'Steel jacket retrofitting of reinforced concrete bridge columns for enhanced shear strength – part 1: theoretical considerations and test design', *ACI Structural Journal*, Vol. 91, No. 4, pp. 394–405.

Priestley, M. J. N., Seible, F. and Calvi, G. M. (1996) *Seismic Design and Retrofit of Bridges*, John Wiley & Sons, New York, USA.

Saadatmanesh, H. and Ehsani, M. R. (1992) 'Behavior of externally confined concrete columns', *Proceedings of Materials Engineering Congress, ASCE*, pp. 677–690.

Saadatmanesh, H. and Ehsani, M. R. (1994) 'Seismic retrofit of existing reinforced concrete columns by glass-fibre composites', *Proceedings of the Materials Engineering Conference 804, ASCE*, pp. 758–763.

Saadatmanesh, H., Ehsani, M. R. and Jin, L. (1996) 'Seismic strengthening of circular bridge pier models with fibre composites', *ACI Structural Journal*, Vol. 93, No. 6, pp. 639–647.

Saadatmanesh, H., Ehsani, M. R. and Jin, L. (1997) 'Seismic retrofitting of rectangular bridge columns with composite straps', *Earthquake Spectra*, Vol. 13, No. 2, pp. 281–304.

Seible, F., Abdallah, M. G. and Nuismer, R. (1995a) 'Advanced composite carbon shell systems for bridge columns under seismic loads', *Proceedings of the National Seismic Conference on Bridges and Highways*, Federal Highway Administration and California Department of Transportation, San Diego, USA.

Seible, F., Hegemier, G. A., Priestley, M. J. N. and Innamorato, D. (1995b) 'Developments in bridge column jacketing using advanced composites', *Proceedings of the National Seismic Conference on Bridges and Highways*, Federal Highway Administration and California Department of Transportation, San Diego, USA.

Seible, F., Priestley, M. J. N., Hegemier, G. A. and Innamorato, D. (1997) 'Seismic retrofit of RC columns with continuous carbon fiber jackets', *Journal of Composites for Construction, ASCE*, Vol. 1, No. 2, pp. 52–62.

Steckel, G. L., Hawkins, G. F. and Bauer, J. L. Jr. (1998) 'Qualification of composites for seismic retrofit of bridge columns', *Proceedings of the 43rd International SAMPE Symposium and Exhibition, Part 2*, pp. 1803–1817.

Xiao, Y. (2001) Private communication.

Xiao, Y. and Ma, R. (1997) 'Seismic retrofit of RC circular columns using prefabricated composite jacketing', *Journal of Structural Engineering, ASCE*, Vol. 123, No. 10, pp. 1357–1364.

Xiao, Y., Priestley, M. J. N. and Seible, F. (1996) 'Seismic assessment and retrofit of bridge column footings', *ACI Structural Journal*, Vol. 93, No. 1, pp. 79–94.

Xiao, Y., Wu, H. and Martin, G. R. (1999) 'Prefabricated composite jacketing of RC columns for enhanced shear strength', *Journal of Structural Engineering, ASCE*, Vol. 125, No. 3, pp. 255–264.

Youssef, M. N., Haroun, M., Feng, M. and Mosallam, A. (2000) 'Experimental study on RC bridge columns retrofitted using fibre composite materials', *Proceedings of the 45th International SAMPE Symposium and Exhibition (II)*, pp. 1803–1812.

Index

Page numbers in italic, e.g. *55*, signify references to figures.
Page numbers in bold, e.g. **6**, denote references to tables.

adhesives 44
adhesive-to-concrete (AC) interface
 54–5, *55*
Ahmed and Van Gemert's model 85–6
aramid-fibre-reinforced polymer (AFRP)
 3
 comparison of properties **6**
 mechanical properties 3–5, **3**
 safety factors 6–7, **7**
axially loaded columns, strengthening
 147–8, 206–8
 compressive strength of FRP-confined
  concrete 158–9
  experimental data 161–2
  experimental data trends 162–4,
   163, **164**
  Lam and Teng's model 167–8,
   167, *168*
  model performance 164–7, *165*,
   166, **167**, **167**, *168*
  strength models 159–61
  strength models **160**
 design recommendations
  general expression for axial
   strength 200, **200**
  ultimate strength 200–1
 failure modes and typical behaviour
  circular concrete columns
   153–4, *153*, *154*, *155*
  elliptical concrete columns
   156–7, *157*
  rectangular concrete columns
   155–6, *155*, *156*
 strengthening methods
  comparison of methods 150, **150**
  constructional aspects 150–1
  filament winding 149, **150**
  general methods 148
  prefabricated shell jacketing
   149–50, **150**
  shape modification 151–3, *152*
  wrapping 148, *149*, **150**
 stress–strain behaviour of
  FRP-confined concrete 168–9
  experimental data 175–6
  experimental data trends
   176–81, *177*, **178**, *179*, *181*,
   182
  FRP hoop rupture strain 180–1,
   182
  intercept of stress axis by second
   linear portion 180, *181*
  Lam and Teng's model 185–90,
   186, **187**, *188–9*
  model review 169–75, *171*,
   172–4, *175*
  performance of stress–strain
   models 182–4, **183**, *184–5*
  stress–strain curves 182–4,
   184–5
  ultimate axial strain 176–80,
   177, **178**, *179*, 182, **183**

beams, flexural strengthening 31–2,
 46–7, 80–1
 bonding with soffit plate *32*
 concrete stress–strain relationships
  47–50, *47*
 crack-induced interfacial debonding
  strength models 73–4

beams, flexural strengthening (*continued*)
flexural crack-induced debonding 74–7
flexural shear crack-induced debonding 77
test-to-predicted debonding strength ratios **75**
design equations 47–51
design recommendations
application of procedure 80
critical sections and plate-end anchorage 78, *78*
partial safety factors 80
strength check 79
failure modes and typical behaviour
classification of failure modes 36, *37*
crack-induced interfacial debonding 42–4, *43*
debonding precautions 44–5
flexural failure (FRP rupture) 36–8, *37*, *38*
other debonding modes 44
plate-end debonding failure *37*, 39–41, *39*, *40*, *42*
shear failure *37*, 39
U strip end anchorage 45–6, *45*
interfacial stresses 53, 56–7
approximate analytical solutions 53–4, *53*
finite element (FE) results 54–6, *54*
higher order analytical solutions 56
methods of strengthening 35–6
end anchorage for unstressed soffit plates 33–4, *33*
general methods 32–3
load–deflection curves *38*
prestressed soffit plates 34–5, *35*
unstressed soffit plates 33
plate end debonding strength models 57–8, 70–1
assumptions for geometric and material properties 61–3
concrete cover separation models 67–8
concrete tooth models 59–60, *59*, **64–6**, 86–90
experimental data 61–3
graphical comparisons 69–70, *70*, *72*
interfacial stress-based models 60, **64–6**, 90–5
limitations 60–1
Oehler's model 67, 69, *69*, 71–2
performance of models 63–71, **64–6**
review of models 58–61
selection criteria for Smith and Teng's database 61
shear-capacity-based models 58–9, **64–6**, 83–6
Smith and Teng's model 71, 72–3, *72*, 79, 81
presence of preloading 51–2, *52*
beams, shear strengthening 103–4, 130
Chen and Teng's model
comparison with experimental observations 127, *128*, *129*, **129**
general expressions 115–17, *116*, *117*
shear failure by FRP debonding 119–23, *120*, *122*
shear failure by FRP rupture 117–19, *120*
comparison of models with experimental observations 127, *128*, *129*, **129**
design recommendations 123
general equations 123–4
indeterminate beams 127
shear failure controlled by FRP debonding 124–5
shear failure controlled by FRP rupture 124
spacing requirement for FRP strips 126–7
existing models 112–15
failure modes 109
failures near mechanical anchors 112
shear failure due to FRP debonding 110, *110*, *111*
shear failure with FRP rupture 109–10, *109*
shear failure without FRP rupture 110
strengthening methods
advantages and disadvantages **108**
bonding configurations **105**
selection of schemes 107–9
strengthening schemes 104–7, *106*, *107*

- bond strength 11–13
  - bond behaviour 13
  - crack-induced interfacial debonding 11–12, *12*, 42–3, *43*
    - flexural 43
    - flexural shear 43–4
  - failure mode 13–14
  - models 14, 26–7
    - Chen and Teng's model 20–6, **20**, *22*, *23*, *24*, **24**, *25*
    - design proposals 18–19
    - design recommendations 26
    - empirical models 14–15
    - fracture-mechanics-based models 15–18
    - performance of models 19–20, **20**
    - shear–slip models 16–18, *17*, 21–2
  - shear tests *12*
  - test methods 13

- cantilever slabs 138–40, *138*, *140*
  - failure modes and typical behaviour 140–2, *140*, *141*
  - laboratory tests *139*, *141*
- carbon farm sheets *2*
- carbon fibre sheets (CFS), bond strength determinations 14
- carbon-fibre-reinforced polymer (CFRP) 1, *2*, 3
  - comparison of properties 5, **6**
  - mechanical properties 3–5, **3**, *4*
  - safety factors 6–7, **7**
- Chen and Teng's bond strength model 20–6, **20**, *22*, *23*, *24*, **24**, *25*
  - flexural crack-induced debonding **75**, 76–7, *77*, 79, 81
- Chen and Teng's shear strength model
  - comparison with experimental observations 127, *128*, *129*, **129**
  - general expressions 115–17, *116*, *117*
  - shear failure controlled by FRP debonding 119–23, *120*, *122*
  - shear failure controlled by FRP rupture 117–19, *120*
- circular concrete columns
  - failure modes and typical behaviour 153–4, *153*, *154*, *155*
- concrete
  - British Standards 6
  - compressive strength when FRP-confined 158–9
    - experimental data 161–2
    - experimental data trends 162–4, *163*, **164**
    - Lam and Teng's model 167–8, **167**, *168*
    - model performance 164–7, *165*, *166*, **167**, **167**, *168*
    - strength models 159–61, **160**
  - concrete tooth models 86
    - Raoof and Hassanen's model 89–90
    - Raoof and Zhang's model 86–8
    - Wang and Ling's model 88–9
  - stress–strain relationships 47–50, *47*, *155*, *156*, *158*, 168–9
    - Lam and Teng's model 185–90, *186*, **187**, *188–9*
    - review of models 169–75, *171*, **172–4**, *175*
    - stress–strain curves 182–4, *184–5*
    - ultimate axial strain 182, **183**
  - surface preparation 44
  - tooth models 59–60, *59*, **64–6**
- concrete-splitting tensile strength 19
- crack-induced interfacial debonding 11–12, *12*, 42–3, *43*
  - flexural 43
  - flexural shear 43–4

- earthquake resistance 213
- eccentrically loaded columns, strengthening 147–8, 206–8
  - compressive strength of FRP-confined concrete 158–9
    - experimental data 161–2
    - experimental data trends 162–4, *163*, **164**
    - Lam and Teng's model 167–8, **167**, *168*
    - model performance 164–7, *165*, *166*, **167**, **167**, *168*
    - strength models 159–61, **160**
  - design recommendations 201
    - existing studies 201–2, **202**
    - moment–curvature analysis 204, *206*
    - typical interaction curves 205–6, *205*
    - ultimate strength 202–4, *203*

eccentrically loaded columns,
strengthening (*continued*)
failure modes and typical behaviour
circular concrete columns 153–4, *153*, *154*, *155*
elliptical concrete columns 156–7, *157*
rectangular concrete columns 155–6, *155*, *156*
strengthening methods
comparison of methods 150, **150**
constructional methods 150–1
filament winding 149, **150**
general methods 148
prefabricated shell jacketing 149–50, **150**
shape modification 151–3, *152*
wrapping 148, *149*, **150**
stress–strain behaviour of FRP-confined concrete 168–9
experimental data 175–6
experimental data trends 176–81, *177*, **178**, *179*, *181*, *182*
FRP hoop rupture strain 180–1, *182*
intercept of stress axis by second linear portion 180, *181*
Lam and Teng's model 185–90, *186*, **187**, *188–9*
model review 169–75, *171*, **172–4**, *175*
performance of stress–strain models 182–4, **183**, *184–5*
stress–strain curves 182–4, *184–5*
ultimate axial strain 176–80, *177*, **178**, *179*, 182, **183**
ellipse, definition *157*
elliptical concrete columns,
FRP-confinedfailure modes and typical behaviour 156–7, *157*
modelling stress–strain relationships 199–200, *199*
epoxy resins 5, 11

failure of concrete constructs
seismic failure 215
flexural failure of columns with longitudinal FRP 227
flexural plastic hinge failure 216, *217*, 226
flexural shear failure 217–19, *218*
footing failure 219
lap splice failure 216–17, *218*, 227
shear failure 215–16, *216*, 226
ultimate conditions of retrofitted columns 227
failure of FRP-to-concrete joints 13–14
cantilever slabs 140–2, *140*, *141*
circular concrete columns 153–4, *153*, *154*, *155*
classification of modes 36, *37*
crack-induced interfacial debonding 11–12, 14, 42–3, *43*
flexural 43
flexural shear 43–4
debonding precautions 44–5
elliptical concrete columns 156–7, *157*
flexural failure (FRP rupture) 36–8, *37*, *38*
one-way slabs 140–2, *140*, *141*
other debonding modes 44
plate-end debonding failure *37*, 39–40, *39*
comparative test results 68–9
concrete cover separation 40–1, *40*, *42*, 67–8
plate-end interfacial debonding 41
rectangular concrete columns 155–6, *155*, *156*
shear failure *37*, 39
two-way slabs 142–3
U strip end anchorage 45–6, *45*
Fibre Anchor® 142
fibre-reinforced polymer (FRP) 1–2
fibre-reinforced polymer (FRP) composites 7–8
comparison of properties 5, **6**
formation methods 2–3
mechanical properties 3–5, **3**, *4*
strain–strain relationships *52*
safety factors 6–7, **7**
flexural failure (FRP rupture) 36–8, *37*, *38*
FRP-to-concrete joints, bond strength 11–13
bond behaviour 13
crack-induced interfacial debonding 11–12, *12*, 42–4, *43*
failure mode 13–14

models 14, 26–7
Chen and Teng's model 20–6, **20**, *22*, *23*, *24*, **24**, *25*
design proposals 18–19
design recommendations 26
empirical models 14–15
fracture-mechanics-based models 15–18
performance of models 19–20, **20**
shear–slip models 16–18, *17*, 21–2
shear tests *12*
test methods 13

glass-fibre-reinforced polymer (GFRP) *2*, 3
comparison of properties 5, **6**
mechanical properties 3–5, **3**, *4*
safety factors 6–7, **7**

interfacial stress-based models 90
Saadatmanesh and Malek's model 93–5
Tumialan *et al.*'s models 95
Varastehpour and Hamelin's model 92–3
Ziraba *et al.*'s models 90
concrete cover separation 91–2
plate-end interfacial debonding 90–1

Jansze's model 85

Lam and Teng's model 167–8, **167**, *168*
best-fit stress–strain model 185–7, *186*, **187**, *188–9*
elliptical concrete columns 199–200, *199*
rectangular concrete columns 195–8, *196*, *197*, *198*
stress–strain model for design use 188–90
linear elastic fracture mechanics (LEFM) 16

middle adhesive (MA) layer 54–5, *55*
modelling
finite element (FE) results 54–6, *54*
FRP-to-concrete bond strength 14, 26–7
Chen and Teng's model 20–6, **20**, *22*, *23*, *24*, **24**, *25*
design proposals 18–19
design recommendations 26
empirical models 14–15
fracture-mechanics-based models 15–18
performance of models 19–20, **20**, *23*, *24*, **24**, *25*
shear–slip models 16–18, *17*, 21–2
interfacial stresses 53–6, *53*
strength of FRP-confined concrete 159–61, **160**
Lam and Tang's model 167–8, **167**, *168*
model performance 164–7, *165*, *166*, **167**, *168*
stress–strain behaviour of FRP-confined concrete 168–9, 169–75, *171*, **172–4**, *175*
diameter equivalence 193
elliptical columns 199–200, *199*
general equations 190–2, *191*
Lam and Teng's model 185–90, *186*, **187**, *188–9*, 195–8, *196*, *197*, *198*
rectangular columns 190–3
shape factor 192, *197*
stress–strain curves 182–4, *184–5*
ultimate axial strain 182, **183**
Mohr–Coulomb materials 18–19, 90, 92
moment–curvature analysis 204, *206*

nonlinear fracture mechanics (NLFM) 15–16

Oehler's model 67, 69, *69*, 71–2, 84–5
one-way slabs 136–8, *136*, *137*
design recommendations 143
failure modes and typical behaviour 140–2

plate-end debonding failure *37*, 39–40, *39*
concrete cover separation 40–1, *40*, *42*, 67–8
plate-end interfacial debonding 41
plate-to-adhesive (PA) interface 54–5, *55*
pultrusion 2, 3, 33, 34

Raoof and Hassanen's model 89–90
Raoof and Zhang's model 86–8
rectangular concrete columns, FRP-confined
  failure modes and typical behaviour 155–6, *155*, *156*
  modelling stress–strain relationships 190
    diameter of equivalent circular column 193
    experimental data 193–5, *194*, *195*
    general equations 190–2
    general equations *191*
    Lam and Teng's model 195–8, *196*, *197*, *198*
    shape factor 192, *197*
reinforced concrete (RC), FRP strengthening 7–8
resin infusion 3, 33
resins 5, 11

Saadatmanesh and Malek's model 93–5
safety factors
  fibre-reinforced polymer (FRP) composites 6–7, **7**
  flexural strengthening of beams 80
seismic attack
  design recommendations 227–8
    detailing considerations 234–5, *235*
    flexural plastic hinge confinement 232–3
    lap splice clamping 233–4, *233*
    shear strength 228–32, *229*, *230*
  failure modes 215
    flexural failure of columns with longitudinal FRP 227
    flexural plastic hinge failure 216, *217*, 226
    flexural shear failure 217–19, *218*
    footing failure 219
    lap splice failure 216–17, *218*, 227
    shear failure 215–16, *216*, 226
  flexural ductility of retrofitted columns 222, **223**, **224**
    ductility factors 220–2
    hysteresis loops 222, *225*
    inelastic deformation *221*
    ultimate displacement *221*
  retrofit methods
    FRP jackets with horizontally oriented fibres 219–20
    longitudinal bonding of FRP 220
  retrofit of RC columns 213–15, *214*, 235–6
ultimate conditions of retrofitted columns 225–2
    flexural plastic hinge failure 226
    lap splice failure 227
    shear failure 226
shear failure *37*, 39
shear tests *12*
shear-capacity-based models 83–4
  Ahmed and Van Gemert's model 85–6
  Jansze's model 85
  Oehler's model 84–5
slabs, flexural strengthening 135, 145
  design recommendations
    one-way slabs 143
    two-way slabs 143–5
  failure modes and typical behaviour
    cantilever slabs and one-way slabs 140–2, *140*, *141*
    two-way slabs 142–3
  strengthening methods
    cantilever slabs 138–40, *138*, *139*, *140*
    general methods 135–6
    simply supported one-way or two-way slabs 136–8, *136*, *137*
  yield lines *137*
Smith and Teng's model 71, 72–3, *72*, 79, 81
  advantages 73
  analytical solution 81–3, **83**
  data selection criteria 61
steel, stress–strain behaviour *4*

test methods, bond strength 13
Tumialan *et al.*'s models 95
two-way slabs 136–8, *136*, *137*
  design recommendations
    anchorage of FRP strips/sheets 145
    debonding strengths 144
    flexural strength by moment coefficient method 143–4
    flexural strength by yield line analysis 144

failure modes and typical behaviour 142–3

U strip end anchorage 33–4, *33*
failure modes 45–6, *45*

Wang and Ling's model 88–9
wet lay-up method 3, 33, *63*
width-to-strength ratio for bonds *22*
woven glass fabric *2*
wrapping methods 148, *149*, **150**

yield lines *137*, 144

Ziraba *et al.*'s models 90
concrete cover separation 91–2
plate-end interfacial debonding 90–1